Failure Is Not an Option

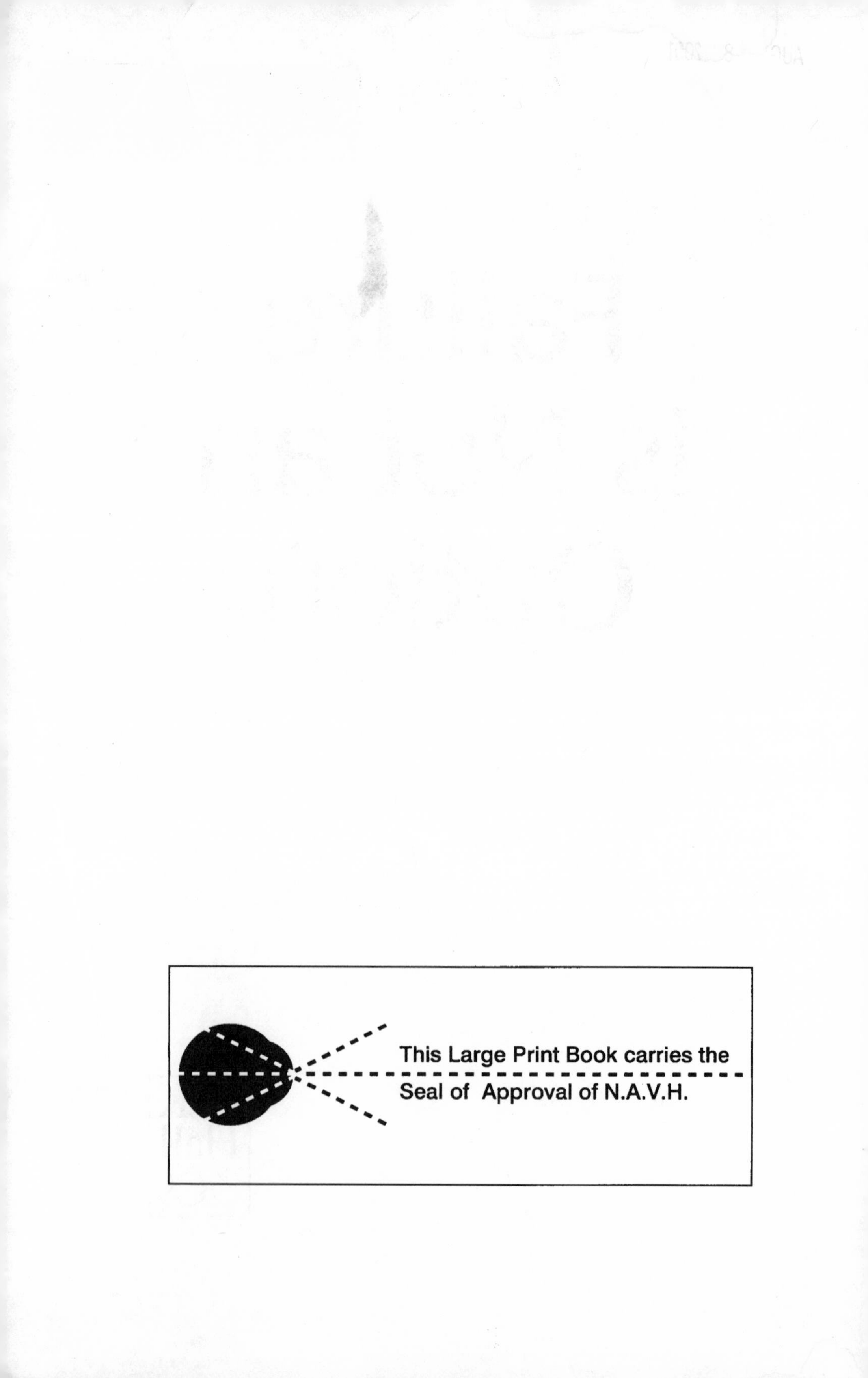
This Large Print Book carries the
Seal of Approval of N.A.V.H.

Failure Is Not an Option

MISSION CONTROL FROM MERCURY TO APOLLO 13 AND BEYOND

GENE KRANZ

G.K. Hall & Co. • Thorndike, Maine

Published in 2000 by arrangement with Simon & Schuster, Inc.

All photographs and art courtesy of NASA unless otherwise noted

G.K. Hall Large Print Nonfiction Series.

The text of this Large Print edition is unabridged.
Other aspects of the book may vary from the original edition.

Set in 16 pt. Plantin by Warren S. Doersam.

Printed in the United States on permanent paper.

Library of Congress Cataloging-in-Publication Data

Kranz, Gene.
Failure is not an option : mission control from Mercury to Apollo 13 and beyond / Gene Kranz.
p. cm.
ISBN 0-7838-9136-9 (lg. print : hc : alk. paper)
ISBN 0-7838-9137-7 (lg. print : sc : alk. paper)
1. Manned space flight — Systems engineering — United States — History. 2. Ground support systems (Astronautics) — History. 3. United States National Aeronautics and Space Administration. 4. Large type books. I. Title.
TL873.K73 2000b
629.45′3′0973—dc21 00-040700

With love to my wife, Marta,
and our children,
Carmen, Lucy, Joan,
Mark, Brigid, and Jean

CONTENTS

1 The Four-Inch Flight 9
2 "Liftoff; the Clock Is Running" . . . 51
3 "God Speed, John Glenn" 85
4 The Brotherhood 122
5 The Making of a Rocket Man . . . 172
6 Gemini — The Twins 202
7 White Flight 224
8 The Spirit of 76. 260
9 The Angry Alligator 297
10 A Fire on the Pad. 324
11 Out of the Ashes 351
12 The X Mission 375
13 The Christmas Story 394
14 1969 — The Year of Apollo 415
15 SimSup Wins the Final Round . . . 430
16 "We Copy You Down, Eagle" . . . 457
17 "What the Hell Was That?" 497
18 The Age of Aquarius 514
19 Coming Home 546
20 Shepard's Return 570
21 What Do You Do After the Moon? 589
22 The Last Liftoff. 621

Epilogue 636
Where They Are 643
Acknowledgments 647
Appendix: Foundations of
Mission Control 655
Glossary of Terms 656

1

THE FOUR-INCH FLIGHT

"Houston, we have a problem."

At some time in the hours that followed that terse announcement from Apollo 13, many of us in NASA's Mission Control Center wondered if we were going to lose the crew. Each of us had indelible memories of that awful day three years before when three other astronauts sat in an Apollo spacecraft firmly anchored to the ground. Running a systems test. Routine. In terms of the distances involved in spaceflight, we could almost reach out and touch them.

Moments after the first intimation that something had gone terribly wrong, technicians were up in the gantry, desperately trying to open the hatch. It took only seconds for an electrical glitch to ignite the oxygen-rich atmosphere of the cabin, creating a fire that was virtually a contained explosion. In those few seconds, the men inside the capsule knew what was happening — and they must have realized, at the last moment, that there was no escape. We simply could not reach them in time.

Now, three equally brave men were far beyond

us in distance, far out in the vast absolute zero world of space, the most deadly and unforgiving environment ever experienced by man. We could measure the distances in miles. But with so many miles, the number was an abstraction, albeit one we had become used to dealing with in matter-of-fact fashion.

We could reach them only with our voices, and they could speak to us only through the tenuous link of radio signals from precisely directional radio antennas. This time they were truly beyond our reach. Time and distance. So close were we in the Apollo fire that claimed the first three Americans to be killed in a spacecraft.

Now we were so far, so very far, away.

Once again, technology had failed us. We had not anticipated what happened back then, on Earth. We had not anticipated what had happened this time. In fact, it would be hours before we really understood what had happened. There was one big difference in this case. We could buy time. What we could not accomplish through technology, or procedures and operating manuals, we might be able to manage by drawing on a priceless fund of experience, accumulated over almost a decade of sending men into places far beyond the envelope of Earth's protective, nurturing atmosphere. All we had to work with was time and experience. The term we used was "workaround" — options, other ways of doing things, solutions to problems that weren't to be found in manuals and schematics. These three astronauts were beyond our physical reach. But not beyond the reach of human imagination, inventiveness, and a creed that we all lived by: "Failure is not an option."

That was not true in the beginning of the space program. There had been many early failures back then — because we hadn't learned enough about the perilous business in which we were engaged.

Would it happen again — the loss of three men? We had failed our crew in Apollo 1. This time we had a few hours to do something. But did we have the wisdom? And could we somehow build not just on our own years of experience but the courage and resourcefulness of three astronauts far, far from home?

Sociologists and engineers call it "the human factor." It's what we must depend on when all the glittering technology seems, suddenly, useless.

For me, and others sitting safely in Mission Control in Houston, we could depend only on a learning curve that started at a place that wasn't more than a complex of sand, marsh, and new, raw concrete and asphalt. It wasn't even Kennedy Space Center then. But it was our first classroom and laboratory. And all we had learned since those first, uncertain years would be what we had to work with to figure out what had happened — and what to do about it.

November, 1960

As a former Air Force fighter pilot, I am not usually a nervous passenger, constantly staring out the window to make sure a wing hasn't fallen off or monitoring the noise of the engines. But for once, on that fateful day, November 2, 1960, I couldn't wait to get on the ground.

East Coast Airlines had only one flight a day from Langley Air Force Base in Virginia to South Florida, using creaky, old twin-engine Martins and Convairs. How long the flight took on one of those old prop aircraft on any given day depended on the size of the bugs that hit the windshield and slowed it down.

This time my eagerness had nothing to do with the condition of the aircraft. This was my first trip to Cape Canaveral, Florida, the launching site for the infant American space program. During the brief flight on the shaky Convair, I was absorbed in thoughts about the new battle in which I had elected to play a part. As an American, I hated to see our nation second in anything — and I had no doubt we were second in space. I had seen an example of what Soviet technology could do as I watched MiG aircraft making contrails high in the sky over the demilitarized zone in Korea, higher than our F-86 fighters could climb. Now the Russians had utterly surprised us by launching the space race. This was a race we had to win and I wanted to be part of it. In a matter of weeks, I had given up my exhilarating work in aircraft testing to take a job with the National Aeronautics and Space Administration (NASA), officially coming on board on October 17. Two weeks later I was on my way to the Cape, and my family — my wife, Marta, and our two young daughters — was camping out at a motel near Space Task Group headquarters at Langley. My instructions were pretty simple: get to Mercury Control and report for work.

Well, I thought, here I am, looking around for launch towers and gantries — but all I could see looked like a regular old Air Force base. It turned out that my knowledge of the local geography was just a little bit hazy. We had landed at Patrick AFB and I literally did not know whether we were north or south of my destination. After the plane rolled to a stop and a couple of guys from base operations rolled a metal stairway out to the aircraft's door, a shiny new Chevrolet convertible wheeled to a halt just beyond the wingtip. An Air Force enlisted man popped out, saluted, and held open the car's door for a curly-haired guy in civilian clothes, a fellow passenger who deplaned ahead of me. That was unusual — a nonmilitary vehicle cruising around the ramp of a military base. As I stepped onto the tarmac, I looked around for the man my boss had said would meet me. I didn't see anyone who seemed to be looking for me, so I started searching for a taxi or any form of transportation. I felt like a foreigner in a strange land.

The plane's baggage was being offloaded next to the operations building when the tall, thin, curly-haired guy now driving the Chevy yelled out, "C'mon, I bet you're going to the Cape." I suppose my military-style crew cut and ramrod-straight posture gave me away. As I nodded, he said, "Climb aboard."

After clearing the plane, he peeled into a 180-degree turn and raced along the ramp for

100 yards, my neck snapping back as he floored the Chevy. I had never driven this fast on a military base in my life. I was thinking I had hitched a ride with a madman, or at least someone who apparently had no concern about being pulled over by the Air Police for speeding and breaking every regulation in the book. This feeling was reinforced as we took a few hard rights and lefts, then roared toward the gate, momentarily braking as an Air Force military policeman snapped a salute and waved us through. I took a closer look at the stranger behind the wheel. He was hatless, wearing a Ban-Lon shirt. There was no gold braid on him. I wasn't accustomed to seeing a guy in a Ban-Lon shirt rate salutes.

Hitting the highway, he made a wide turn and a hard left, burning rubber. In no time, he had the needle quivering between eighty and ninety miles an hour. After a joyful cry of "Eeeee hah," he turned and offered his hand, saying, "Hi, I'm Gordo Cooper." I had just met my first Mercury astronaut. As I soon learned, if you saw someone wearing a short-sleeved Ban-Lon sport shirt and aviator sunglasses, you were looking at an astronaut. We humble ground-pounders wore ties and white shirts, and yes, those nerdy pencil-holding pocket protectors.

I thought of that handshake often in the many years that followed. Mercury worked because of the raw courage of a handful of men like Cooper, who sat in heavy metal eggcups jammed on the top of rockets, and trusted those of us on the ground. That trust tied the entire team into a common effort.

I took it as a good omen that Cooper, taking pity on a befuddled stranger, offered me a lift to the base. He was one of the seven former test pilots selected for the first class of astronauts. They had been introduced, unveiled like sculptures, in April of 1959. Instantly the media compared them to Christopher Columbus and Charles Lindbergh. Today, I wonder how many of them the average American could name. They were John Glenn, Alan Shepard, Virgil I. (Gus) Grissom, Wally Schirra, Donald K. (Deke) Slayton, Scott Carpenter, and Cooper. They were similar in size and build, partly because the design of the capsule ruled out anyone over five-foot-eleven.

All of them were white, all from small towns, all middle-class, and all Protestant. This was not the result of deliberate discrimination, but because at the time that was the kind of man who became a military test pilot. At this period it was hard for Americans from any minority to get into flight training. But the military, like the rest of the country, grew up and lived up to its fundamental commitment to equality, thanks in large measure to the civil rights movement that, like the space program in the same era, demanded conviction and courage.

That day when I arrived in Florida I stumbled into the future. I didn't have enough time even to learn the recently coined space jargon before the Mercury flight director, Chris Kraft, gave me the task of writing the operating procedures for Mercury flight controllers. Without knowing much about anything, I was telling people how to do everything, writing the rules for the control team that would support the Mercury-Redstone launch. Not only had I never laid eyes on the Mercury Control Center, I had never even seen, close up, any rocket big enough to carry a human payload.

I did not really research the program before I joined. I knew that it was called the "man in space project." Lyndon Johnson, then the Senate majority leader, was given the job by President Dwight Eisenhower of determining how we should respond to the Soviets' launch of Sputnik on October 4, 1957. The impact of the first orbiting satellite, visible to the naked eye as it passed through the night sky over America, was profound. Sputnik was a shock to national pride — Russian science had put the first object in outer space, giving Americans both an inferiority complex and a heightened sense of vulnerability in what was then the most intense phase of the Cold War. Out of this was born the "missile gap" between ourselves and the Soviet Union.

Years later we would discover that this "gap" was an intelligence myth. But the Soviet Union was indeed ahead in a space race that this tiny, rather primitive satellite had effectively initiated. Our adversary had developed rockets with greater thrust and throw weight — for the military this meant ballistic missiles that could "throw" a heavier warhead a greater distance than anything in our arsenal.

The reverberations of that little sphere emitting its "beep-beep" radio signal as it sailed unrestricted through space were far reaching.

Among other things, it would spark a massive federal education funding program, significantly called "The National Defense Education Act," to stimulate better teaching of math and science as well as foreign languages to more students throughout the country. A sleeping giant suddenly woke up.

One of the other immediate results of Sputnik was the National Space Act of 1958 and the creation of the National Aeronautics and Space Administration. To me, our leap into space was the logical next step beyond the X-15 rocket-powered aircraft. The problem was that our first "leaps" would be some fairly short hops. All of these factors had influenced my decision to join this embryonic program. It had been cautiously funded, was working from a somewhat thin base — and was also a crash effort for everyone involved in it. I don't think that at the time I realized just how caught up I was in the excitement and challenge of this race. Nor could I have anticipated just how thrilling and dangerous, frustrating and inspiring the first lap in it would be. All of those involved were obsessed by a driving dream, working with an intensity that fused NASA employees and contractors, launch and flight operations into one powerful organism.

Cooper dropped me off at Mercury Control and I was greeted by the familiar face of the only person in the program I knew down at the Cape, Paul Johnson, a troubleshooter working for Western Electric, one of the subcontractors to

Bendix in building the control center and the tracking network. Western Electric's responsibilities included radars, telemetry (radio signals to and from the rocket and spacecraft that told us how things were working — and what wasn't) control consoles, and communications. These were the core systems. Western Electric quickly parlayed this into a responsibility for integrating operations, training, maintenance, and network communications. Paul was amazingly young for his responsibilities. He had an intuitive feel for this unprecedented development and deployment of technology, writing the specifications and testing procedures and doing everything that needed to be done to check out the largest "test range" in history, one that went around the globe.

"Kraft said you were coming down," he greeted me, smiling, "and I thought I'd give you a hand." During the next two days, Johnson gave me a master's degree in the art of mission control. He had been at the Cape for the preceding week and had been writing the manual on the team structure and operations. He handed it over to me to finish defining the standard operating procedures for Mercury Control, such as how to check out the console displays and communications, set the format for Teletype communications, and how specifically to request data from the technicians (politely but urgently).

I soon began to think of Paul as my guardian angel. From the moment I came on the job it seemed that whatever I was doing, wherever I turned, he was there. He always appeared when the pressure was on and I was happy to see him.

As I felt my way through a program inventing itself, Marta was moving from the motel into a new house in Hampton, Virginia. It was our fifth move in the four years since our marriage, setting up successive households in South Carolina, Texas, Missouri, New Mexico, and now Virginia. Carmen, our two-and-a-half-year-old daughter, had been born in Texas, and Lucy, fourteen months, in New Mexico. Like most service families they were ready for anything, anytime.

The Space Task Group's launch team was permanently stationed at the Cape to support the test and checkout of the rocket and capsule. The flight team of which I was now a member, the astronauts, engineers, and program office operated from Langley Air Force Base and traveled to the Cape for each mission. I had been on the job in Virginia only two weeks, hardly long enough to figure out the pecking order, when Kraft walked up to my desk and said, "Everyone else is tied up. You're all I've got. We're coming up on our first Redstone launch. I'd like you to go down to the Cape, get with the test conductors and write a countdown. Then write some mission rules. When you finish give me a call and we'll come down and start training."

The shock on my face must have registered as Kraft continued: "I'll tell Paul Johnson to meet you at Mercury Control to give you a hand." When Kraft talked, his eyes never left mine. I was given this assignment mainly because I was available. In this period of intensive development, jobs were open all over the place; NASA was forming organizations for mission planning, recovery operations, astronaut training, launch operations, and Mercury Control. Every new hire with the requisite technical and scientific credentials was put into a job slot the minute he came on board.

Kraft was one of the original thirty-six members of the Space Task Group, most of whom stepped forward to do a job that had never been done. He recognized that someone had to be in charge of the ground effort and he volunteered to lead that effort. A graduate of Virginia Polytechnic Institute, Chris had worked at Langley in the aircraft stability and control laboratory. My senior by nine years, he did not immediately impress me as a leader, the way some of my early mentors had. Kraft led a step at a time, and each Mercury mission added a new dimension to his presence and style.

My days as an observer were over, my chance to get up to speed ended. This was the first indication that my job slot would be in Mercury Control. Some people in Mercury Control had technical experience working on the tracking stations or at the Cape on the Vanguard, Explorer, and Pioneer missions. Others, like me, came from aircraft flight testing or were engineers from the pilotless aircraft research program at Langley. From my work, most recently at Holloman AFB in New Mexico, I knew about flying, systems, procedures, and checklists. I could figure out what a countdown should contain. Mission rules were different. There had never before been such a mission in U.S. history — I would just have to give it a shot. Since there were no books written on the actual methodology of space flight, we had to write them as we went along.

There was a relatively small group working down at Mercury Control, forty to fifty people. Some of them had grown up launching the early U.S. rockets derived from the German V-2 of the Second World War. Now, in a few months, we would attempt to send the first American into space. It was a scary thought, but not for anyone who had been around test pilots.

I had flown supersonic F-100s, which needed at least a mile to get off the runway on a good day. When you took off at 230 miles per hour, if the engine crapped out or you lost the afterburner, it could quickly become a bad day. But when you punched through the sound barrier it was a jolt of pure adrenaline. The SuperSabre looked like it was more than capable of carrying out its air superiority mission. But you had better be ready when you strapped yourself in. No matter how skilled you were in handling it, you were never sure when the elements or the aircraft, in a perverse way, would decide to test you. Every time I climbed aboard I could feel the thrill of tension and anticipation.

At Holloman AFB, where I had worked as a flight test engineer, we had been putting people into scary situations for years. It was not unusual for a guy to climb to an altitude of 100,000 feet in a balloon and then bail out in a parachute, falling 90,000 feet before his parachute opened. This was the environment of risk and these were the kinds of people who had been picked as the Mercury Seven astronauts.

Looking back, I can see now how minimal, even primitive, our facilities were at the time, both in the control center and in the blockhouse — a massively reinforced structure placed as close as prudently possible to the launch pad where the guys who were responsible for the actual functioning of the rocket manned their posts. We tended to talk about "the Germans in the blockhouse" largely because Wernher von Braun and his cohorts, who had worked on the rocket programs, came to the United States after Germany's defeat in World War II. They were originally stationed near El Paso, Texas, and tested captured V-2 rockets for the military at the White Sands, New Mexico, test range. Later they were moved to permanent facilities at Huntsville, Alabama, and worked for the Army Redstone Arsenal. Most of the Germans became American citizens, adopting Huntsville as their home. In 1960 rocket development at the Redstone Arsenal was transferred to the newly formed Marshall Space Flight Center (MSFC), and von Braun, along with nearly 100 other German scientists and technicians, began work on a powerful series of rockets called Saturn I.

At this point in the space program, our communications network was actually run out of NASA's Goddard Space Flight Center in Greenbelt, Maryland. It had been named after Dr. Robert Goddard, the American pioneer in rocketry, who had developed rocket engine and guidance technology in the 1930s equal, if not superior in some respects, to what von Braun and his colleagues were working on as late as 1945. Goddard, one of my boyhood heroes, had had the backing of Charles Lindbergh, which enabled him to test his rockets in New Mexico, not far from the site where von Braun and his Germans would fire the first captured V-2 rockets in the late 1940s and test those that evolved from V-2 technology in the years that followed.

The German scientists and technicians would come back to the Cape occasionally for selected launches (particularly high-profile manned missions), but they had their hands full at Marshall developing a new generation of rockets. By the time NASA launch operations were forming up, American engineers were well acquainted with rockets, building on the experience of the Germans, as were the contractors producing the Redstone and Atlas missiles. While the new generation of American scientists and engineers was now doing the job, the first boosters in the manned spaceflight effort were barely adequate, as events would demonstrate. In many ways this technology was as "out on a limb" as Charles Lindbergh's Ryan monoplane. He didn't have any manuals either, and his facilities were primitive. Roosevelt Field in 1927 and Canaveral in 1960 had a few things in common. The massive Cape facility that would grow up in the next decade and soon become the Kennedy Space Center (which would include the largest enclosed space in the world, the vertical assembly building) was beyond our wildest dreams at the time.

In 1960 the Cape looked like an oil field, with towering structures, dirt, and asphalt roads newly carved out of the palmetto scrub. The alligators were reluctantly surrendering to the onslaught of newly arrived civilization. If you didn't have a good sense of direction you were in trouble. There were few directional signs and once you got off the road visibility narrowed. At night you could easily imagine the gators and snakes taking their revenge on any intruder foolish enough to be wandering around on foot, lost in the boondocks.

The man in space program was simple in concept, difficult in execution. Every mission was a first, a new chapter in the book. Many, if not most, of the components in both rockets and capsules had to be invented and handmade as we went along, adapting what we could from existing aviation and rocket engine technology. Before putting a man on top of a rocket, we would first fly one or two tests with a "mechanical man," a box full of electronics weighing about as much as an astronaut to simulate the conditions that would be present when an astronaut was on board the capsule. The capsule would send back some prerecorded messages to test our communications. Then we graduated to spider monkeys — and then to chimpanzees, working our way up the evolutionary ladder, so to speak. The missions were initially to be twenty-minute lob shots, using the Army's

Redstone rocket; then we would go into orbit with the Air Force's first-generation Atlas intercontinental missiles. The military boosters were barely ready for operational use. Here the missile gap was indeed real — except the gap was between what the hardware was supposed to do and what it had shown it could do.

The day after I arrived Paul Johnson and I went to the launch pad. I was shocked when I first saw the Redstone rocket. It was stark, awkward, and crude, a large black-and-white stovepipe atop a simple cradle. It had none of the obvious coiled power and distinctive personality of an airplane; it was not graceful in form, not something you could come to love and rely on. The Mercury capsule squatted atop the rocket, black in color and seemingly constructed of corrugated sheet metal. With its tall red escape tower it looked more like a buoy in a harbor than a rocket ship from a science fiction novel. Given the oil field–like setting in the wilderness and the crude appearance of the rocket, I felt more like a drilling rig roughneck than a rocket scientist when I made my way into the bar of the Holiday Inn that evening.

Putting any reservations aside, I plunged into working with Paul on defining the joint tests of control and communications systems, as well as the Go NoGo points for telemetry display, command, and communications in Mercury Control. My next step was to synchronize the Mercury Control Center (MCC) countdown with the capsule and booster countdowns. Paul

Johnson returned to Langley to deal with a set of problems in the tracking network while I completed the work down at the Cape. To this day I feel enormous gratitude to Paul for giving me a running start. This was one of the critical moments in my life when someone stepped in and pointed me in the right direction.

I had left behind a world where airplanes were flying at roughly five miles a minute. In this new, virtually uncharted world we would be moving at five miles per second. During a mission countdown, or even a flight test, so many things would be happening so fast that you did not have any time for second thoughts or arguments. You wanted the debate behind you. So before the mission, you held meetings to decide what to do if anything went wrong.

You wrote down on paper the outcome of these meetings and this became what you needed for a launch, your personal list of Go NoGo's. There was no room in the process for emotion, no space for fear or doubt, no time to stop and think things over. A launch is an existential moment, much like combat. With no time to think about anything, you had to be prepared to respond to any contingency — and those contingencies had to be as fully anticipated as possible before you pushed the button. You also had to be thoroughly knowledgeable about the responsibilities of launch control and range safety. During a launch the only mission alternative to save the capsule was an abort, and we had to pick the points to act before the range safety officer (RSO) stepped in to blow up the rocket and the capsule after launch if things went to hell.

By the end of the first week we had just finished the initial paperwork for the countdown procedures and mission rules, but had yet to run a simulated countdown. I was finally breathing easier. Johnson had taken me from Kraft's few words on what he wanted done to a point where I finally knew what he was talking about. It was only some six days after my "rocket ride" with Gordo Cooper, but my job was starting to seem real to me. The few days of hands-on familiarity with MCC systems now tied into the concept of the MCC team. The MCC was coming together as a working reality. My newfound knowledge, while only paper thin, was as good as anyone else's. It was now time to put it to the test. I called Kraft and said the countdown and the operational rules were ready.

Shortly before the rest of the team that would be involved in the first launch arrived from Langley, I made a thoughtful walk-through inspection of the relatively small — in comparison to later control centers — space that contained the operating elements of Mercury Control.

When a fighter pilot arrives at a base the first thing he does is go down to the flight line and look at the new airplane he is going to fly. You walk around it, feel the skin, climb up on the wing, and look in the cockpit, knowing that soon this airplane is going to be yours. It is a time when you feel a bit cocky, knowing that you are one of the few who will be privileged to live in this highly charged new world of high-speed flight.

I felt the same way on my solitary walk around the Mercury control room; I felt like I was meeting an airplane. I was, at long last, feeling at home. The telemetry, communications, and display areas were like the facilities at Holloman, but there was no counterpart for the control room itself. The room was square, about sixty feet on each side, dominated by a world map in the front. The map contained a series of circles, bull's-eyes centered on the worldwide network of tracking stations. Below each were boxes containing many different and, for the uninitiated, unintelligible symbols. A toylike spacecraft model, suspended by wires, moved across the map to trace the orbit. On each side of the map were boards, where sixteen critical measurements were plotted by sliding beads, like those on an abacus, up and down wires as the capsule circled the world. In less than four years much of this technology would be obsolete — only the concept of Mission Control would remain. The meters and console displays would eventually be replaced by television displays driven by computers, which provided the controllers virtually instantaneous access to every bit (or byte) of the spacecraft's data. Digital systems would enable ground control of the space systems. This would make it possible for controllers on the ground to work in partnership with a spacecraft's crew to achieve the objectives of any flight. But this was yet to come; now we had to control the missions with fragile communications, a first-generation solid-state computer, slide rules, and guts. We were in the Lindbergh stage of spaceflight.

Given my aircraft test flight background, the control room felt vaguely familiar, with the exception of the three rows of consoles on elevated platforms. Each console was configured differently. Consoles on the top row were flat pedestals with communications boxes on top. When I first arrived at the Cape, Paul Johnson had taken me on a tour of the control room and pointed out the procedures console. I sat at the console, staring at the flat gray face and writing desk. The only instruments were a clock and an intercom panel with a rotary (!) phone at the top. This was the state-of-the-art work station that Paul and his colleagues from Western Electric had designed from scratch. It was on the left, in the middle row, and closest to the Teletype room. As I sat down at my console, two people came over and introduced themselves.

Andy Anderson, tall and skinny with long, sandy hair, was the boss of the communications center. His hotshot Teletype operator, a short redhead with a brush cut, was simply "Eshelman." No one called him anything else. During a launch, I reeled off a running account of key data on the sequence of events to Eshelman, who typed them out and transmitted them by landline and radio links to remote tracking stations in Bermuda, Africa, Australia, and distant islands and ships in the Atlantic and Pacific. Eshelman had the skill and grace of a concert pianist as he stood, intently bent over the Teletype keyboard, interacting in real time with the Bermuda Teletype operator, just as if they were having a conversation. The tools we used in Mercury were primitive, but the dedication of highly trained people offset the limitations of the equipment available to us in these early days and kept the very real risks under control. But at a price; this was high-sweat, high-risk activity, demanding a degree of coordination between the ground and the capsule exceeding what I had experienced even in the testing of experimental aircraft.

During the next two years, Anderson, Eshelman, and I controlled virtually all the Teletype message traffic originating from Mercury Control at the Cape. This was the heart of the ground control system, tied to that tenuously linked chain of tracking stations and manned remote sites by a variety of communications systems. Low-speed Teletype provided the backbone, and the controllers became adept at moving messages rapidly between the tracking sites as the spacecraft passed overhead.

The tracking network voice system used a massive manual switchboard up at Goddard; its operator plugged cables into a bewildering assortment of jacks as he performed a frenetic ballet. He carried a thick bundle of cables wrapped around his arm, darting from one part of the big switchboard to another, making connections manually so we could talk to tracking sites and working around bad circuits to provide alternative connections. This remarkable guy, known as "Goddard voice," was another guardian angel.

Since we never knew whether every link had heard the voice exchanges, as a cross-check I transcribed every major communication into a Teletype message. We didn't have computers in Mercury Control. So the radar information from the launch, orbit, and reentry was transmitted by tracking sites around the world to the computers at Goddard for processing, and then sent down to drive the plot boards in Mercury Control. Advanced as they were at the time, and filling whole large rooms, those computers had a speed and processing capacity easily exceeded by desktop PCs today. So our margins for error were made even thinner by the limitations of these resources.

While waiting for Kraft's full team to arrive from Langley I explored everything from the launch pad to Hangar S, where they checked out the spacecraft prior to launch. I was welcomed everywhere by engineers and technicians who were as new to their jobs as I was. All of them were eager to discuss their work, trade ideas, and figure out how each of us fit into the total picture. I felt that I was not alone, that virtually everyone was writing their game plan as they went along. I felt an undercurrent of organization that was emerging from a leadership structure still solidifying.

By the time Kraft and the rest arrived at the Cape, I was no longer feeling like a rookie. I had spent every available moment in Mercury Control, prowling through the room and listening to the check-out, observing how the technicians handled communications with "Goddard voice," the tracking stations, and the blockhouse.

Project Mercury was literally having trouble getting off the ground. In August of 1960, after the first Mercury-Atlas exploded in flight, the major journal in the aerospace business, *Missiles and Rockets*, stated: "NASA's Mercury manned satellite program appears to be plummeting the United States toward a new humiliating disaster in the East-West space race. The program is more than one year behind the original schedule and is expected to slip to two. It no longer offers any realistic hope of beating Russia in launching the first man into orbit, much less to serve as an early stepping-stone for reaching the Moon."

The testing of the Mercury capsule escape system was carried out at the Wallops Island Station just below the Maryland-Virginia border. This was a Langley test facility for all sorts of "sounding," or high-altitude research rockets. The tests of the escape system were about 50 percent successful. While we were getting ready at the Cape, one of the Mercury tests at Wallops failed spectacularly on November 8. Sixteen seconds after launch the escape and jettison rockets fired prematurely, thus leaving the capsule attached to the booster rocket, which reached a ten-mile apex and then came screaming back to Earth, destroying the capsule at impact.

The Mercury program used two booster rockets — the Redstone and the Atlas. Both were derivatives of military systems but with vastly different capabilities. The Redstone was an Army battlefield rocket. It would be used to start the capsule systems qualification test flight and, if that was successful, for two ballistic manned missions. The ballistic missions were to be about twenty minutes in duration, reaching a maximum altitude of about 130 miles and providing a short weightless period before reentry. The Atlas was an Air Force intercontinental missile and was to be used for both ballistic and orbital Mercury missions. The first three missions were ballistic, to continue the booster and capsule qualification, test the tracking network, and provide experience for the MCC team. The orbital testing would continue the qualification testing using the mechanical man and a chimpanzee before the manned orbital flights.

When Mercury-Atlas 1 exploded in flight, we fell about one year behind in the schedule, so a lot was riding on the first Mercury-Redstone flight, MR-1. Kraft's team arrived on November 13 for the MR-1 launch, now only eight days away. Once again my guardian angel, Johnson, arrived to save my bacon. He took a place to the right of the console and punched up the buttons of the intercom during our simulation dress rehearsal.

Immediately, a half dozen different conversations flooded through my headset. It reminded me of the cool, almost casual but terse and clear voice chatter that came up on the tactical frequency when things heated up during the time I was in Korea directing air strikes on ground targets. As I listened, I picked up the voices of the test conductors. Johnson broke out some thick documents and advised Kraft of the page and sequence of the countdown. It was fortunate that this was just a test. It gave Johnson a chance to brief me on the countdown process, get to know the people talking on the loops and Mercury Control's role in the test.

At intervals, Johnson encouraged me as the MCC procedures controller to make suggestions to Kraft or one of the other controllers. Throughout the test, he glanced around the room and made mental notes about what people at the various console positions should be doing or doing in a different way. Periodically, I would print out a message and call the ground communications controller. Eshelman would rush into the room to pick up the message, put it on the Teletype line, and then rush back with a confirming copy of the transmitted text. After several hours, I picked up the routine of the count and felt comfortable, as long as all was going well.

This first mission in which I would play a role was a ballistic test of a Redstone booster rocket and a Mercury capsule. The Redstone's engine was scheduled to burn for two and one half minutes. After the booster engine cut off, the escape tower separated from the capsule by firing the tower ring attachment bolts and igniting the tower escape rocket. Then after the booster thrust had decayed, explosive bolts would fire, followed by the firing of three small posigrade solid rockets, to separate the capsule from the booster. ("Posigrade" is the term used for adding velocity in the direction of flight, in this instance the small rockets used to separate the Mercury capsule from the booster. Retrograde rockets fire opposite to the direction of flight.) The Redstone boosted the spacecraft to an altitude of 130 miles before it started to arc downward.

At 20,000 feet, the capsule's drogue parachutes would deploy, stabilize its motion, and slow it down sufficiently to allow the main parachutes to deploy safely. Then the landing sequence would begin. The entire mission was planned to last only sixteen minutes. After the countdown simulation test we began training for the brief actual flight of the Redstone. For three days we rehearsed, calling out events and issuing backup commands to the automatic sequences.

During the simulation run-through our instructors sat watching us from their vantage point at the top row of the consoles and played magnetic tapes into the telemetry and radar systems, which in turn drove the controller's meters and plot board displays. If all else failed, we would be handed a written question, like a pop quiz in school. You had to stand up in front of the entire Mission Control Center team and say, "Flight! A new problem has shown up and this is what I am going to do about it." You took it seriously. God help you if you couldn't come up with an answer — instantly.

The launch complex from which we would fire the Redstone consisted of the launch pad, service tower, and a blockhouse for launch site command and control. Servicing the pad was a network of power and communications cables, and pipes carrying fuel and other fluids. The blockhouses were igloolike structures that sat about 230 meters from the pad and looked somewhat like squat World War II pillboxes. Atlas blockhouses were a bit different — twelve-sided concrete bunkers with walls three meters thick and domed tops, embedded in thirteen meters of sand. Earlier blockhouses used a viewing slit with a thick quartz/glass window. Later on periscopes were added; they afforded a view of such things as the fuel tanks and cable trenches. The blockhouses had to be close enough for direct viewing and far enough from the booster to survive an explosion. (Five percent of the early boosters exploded shortly after liftoff.)

The blockhouse team was a mixture of German rocket scientists, former Army technicians, and booster contractors. The Germans were most often found talking in their native language, huddled over their displays and praying for things to go right. In addition to von Braun's colleagues, the blockhouse capsule engineers were drawn from a broad talent pool, mostly from aircraft (soon to be called aerospace) contractors. Kraft's team was made up of engineers who came from all sorts of backgrounds, put together like a pickup softball team. The capsule engineers came from the NASA Space Task Group, network controllers from the Air Force, facility technicians from Bendix, and the Mercury Control Center CapComs from the initial group of seven astronauts. Three-man operations teams were deployed to the thirteen

Project Mercury manned tracking stations to provide global tracking, data, and voice communications data coverage. The leader of the three-man team was the CapCom, and he was responsible for site mission readiness, real-time mission support, and status reporting to the Mercury Control flight director. During the manned missions he provided the communications with the astronaut in the capsule, hence the term "CapCom."

The teams at the thirteen manned tracking stations were provided by Kraft's operations organization. With high-risk time-critical decisions, the astronaut corps believed that only astronauts should talk to the astronaut in the capsule. In the Mercury Control Center and at the blockhouse, CapComs were selected from the Mercury astronauts and they were often sent to tracking stations designated as mission critical.

During this period, the American space program drew on some military resources as well as those of NASA. By contrast, the Russian program was part of their military. Soviet hardware, software, and personnel were military, albeit with some modification — spacecraft instead of warheads sitting on top of boosters that, like our Atlas, originated in military programs developed for strategic warfare. Over time the Russian effort would become somewhat more civilian in nature, but from inception, NASA's operations would be separated by a kind of firewall from military operations and personnel.

In the first decades of the race into space the

Russians enjoyed the advantages of running a program powered by the virtually unrestricted resources and funding of a military that, in a command economy, came first in economic priority. We were on a somewhat more modest footing in the early days. That would change dramatically thanks to President John F. Kennedy and Vice President Johnson pushing for the funding and resources that would enable an explicitly civilian space program to succeed.

In Mercury Control the only controllers reporting administratively to Chris Kraft were the trajectory operators, Tecwyn (Tec) Roberts, and Carl Huss, whom I had yet to meet; Howard Kyle, who doubled for Kraft at the flight director console; Paul Johnson; and myself. The capsule engineers assigned to work with the MCC team understandably focused on hardware rather than flight operations and had their hands full checking out the spacecraft in Hangar S, and they looked at training in the Mercury Control Center as a waste of their time. The core of the engineering staff was based at Space Task Group headquarters at Langley Research Center. When McDonnell Aircraft sent a capsule to the Cape the engineers would come down and check it out in Hangar S.

Despite a certain amount of confusion about who should be doing what, since we were inventing it all as we went along, we were moving quickly to the crunch point. To my dismay, two days before launch at the readiness review, Johnson took me aside and said he would not be there. He had to go to the Canary Islands to work out a tracking site problem. He wished me good luck.

November 21, 1960, Mercury-Redstone 1

Only one month and four days after I was hired, I was at the procedures console. Thanks to Johnson's unflagging coaching and the training we had done, I had no problems and felt comfortable with the mechanics. But I had a long way to go before I would have that sense of "being ahead of the airplane" or "ahead of the power curve" as pilots put it — having the experience to anticipate what *could* happen rather than just reacting to what *was* happening at the moment.

As the countdown proceeded, I noticed a change in the intensity of the atmosphere in the control room. I had felt that before when I signed off my aircraft — accepted it as ready for flight — at Holloman. Although the job was different, the emotional content was the same. Controllers were going through the same gut churning as we had had prior to a B-52 test flight. During a hold in the countdown to fix a leak in the Redstone's hydrogen peroxide system that fueled the control thrusters, Kraft turned to me and said, "How about getting me a couple of cartons of milk from the roach coach?" (Mercury Control, like all the other Cape facilities, was out in a vast palmetto swamp, about half a mile off the main road. A lunch wagon, known as the roach coach, pulled by a pickup truck, made its rounds to the test stands, camera trackers, and other stations. The loudspeakers would announce its arrival when it stopped at Mercury Control three times each day. The menu was pretty limited — but it did offer milk.) For the first time I realized that behind Kraft's calm exterior he had the same sensations I felt — of squirrels running around in my stomach — as we approached launch.

The launch countdown progressed without any major incident. As liftoff approached, I leaned back and peered at the video image on Kraft's console coming from a camera focused on the Redstone standing on the launch pad. Precisely at zero on the clock, there was a great cloud of smoke. Kraft looked startled, and then he leaned forward intently. The TV cameraman momentarily lost track as he panned the camera upward, and, for a few seconds, there was nothing on the screen but a smoky sky. From my position, it looked much like the rockets I had launched from aircraft. I was surprised at how quickly the Redstone had accelerated and moved out of sight.

Then after a few seconds the camera panned down. Although smoke still obscured the launch pad, the vague outline of the Redstone was still there. Kraft's face was incredulous. He stammered for a few seconds, then called out, "Booster, what the hell happened?"

Our booster engineer in the control room came from the Marshall Redstone team. After liftoff he was responsible for reporting booster status — engine and guidance — to Kraft's team. If something was wrong he was supposed to give us a heads-up so the trajectory people in Mercury Control could better assess what they were seeing on their radar plot boards. Now he reverted to his native German as he tried to figure out what happened and, more importantly, what the blockhouse team should do about it. All hell broke loose. The Redstone had lifted a few inches off the launch pad and then the engine shut down. By some miracle, the rocket had landed back on the launcher cradle.

When the escape tower cut loose, unguided and spewing flame, it corkscrewed to an altitude of about 4,000 feet. As it plummeted back to Earth, loudspeakers around the facility blared out warnings to the astronauts, engineers, and VIPs in the viewing area to take cover. The escape tower ultimately landed some 1,200 feet from the launch pad.

The launch team in the blockhouse was as stunned as Mercury Control. Booster continued talking in German to his counterparts in the blockhouse, oblivious to Kraft's repeated calls. Television cameras showed the events on the pad as the main and reserve landing parachutes popped out of their stowage compartment at the nose of the capsule, ejected upward and partially opened up. Initially, they hung limply, then they slowly inflated and caught the sea breeze. At the same time strips of tinfoil were ejected from the capsule and spilled over the sides of the booster and capsule. (The strips, or chaff, were used to help radar track the capsule as it was slowed down by the parachute and headed for splashdown, or water landing, in the Atlantic Ocean.)

Every controller held his breath, afraid the parachute would topple the rocket and cause an explosion.

The intercom that had been quiet was now busily filled with directions, observations, and opinions. Everything that happened, although it had taken only seconds, passed before me in slow motion. Then it finally clicked. *We had launched the escape tower.*

The Redstone rocket, surrounded by smoke, was armed and fueled but still sitting on the launch pad. Kraft told everyone to calm down, but Booster was still on the hot line, interrogating the blockhouse in German. We all could see the anger glowing in Kraft's eyes as he walked over and yanked Booster's headset cord loose from the console, saying, "Speak to *me*, dammit!"

Chaos continued for several minutes until Booster, in halting English, told us that the Redstone engines had fired and the rocket had lifted off the pad enough to drop the umbilical, the bundle of cables that connected launch control with the booster and that normally fell away right after liftoff. Then the engine inexplicably shut down. The Mercury capsule, sensing the booster's engine shutdown, acted as if it were in orbit and sent the command to jettison the escape tower. The capsule, not sensing any further acceleration, acted as if it were in the recovery phase and so deployed the parachutes. We now had a live rocket on the pad, a fully pressurized Redstone; and, since the umbilicals had disconnected, we had no control of it. The booster's destruct system was armed and there was no way to secure the system. No one had any idea what to do next.

Kraft walked over to me, eyes blazing. Pointing at Booster he snapped, "The damn Germans still haven't learned who they work for. Everyone in this control room must work for me." We sat stunned, helpless in Mercury Control. We had no technical data on the spacecraft or the launch system beyond a simple manual, the equivalent of an owner's manual for a new car. All of us were still thinking in aircraft, not rocket, terms — and we were definitely behind the power curve. We had no data to work with because we weren't smart enough to know what we really needed. We were dealing with a *new* control room, a *new* network, *new* procedures, and entirely *new* jobs, doing something that we had never done before, something almost alien to our nature.

A tentative proposal came from the blockhouse to reconnect the umbilical. The chances of people getting killed doing this were discussed and we decided that it could not be done safely. The next, and equally desperate, suggestion was to get a cherry picker (a kind of crane or boom with a man-holding bucket on the end of it, like those used by telephone and utility line repair crews) and cut the nylon parachute risers. This would at least eliminate the threat of wind filling the parachute and toppling the Redstone, but this idea was also discarded because of the risk to personnel. All the while we were apprehensively watching a partially inflated parachute and praying that the sea breeze did not pick up, fill the parachute, and topple the whole damn rocket over.

After impatiently listening to a pretty far-out proposal to depressurize the rocket by using a rifle to shoot holes in the fuel and oxidizer tanks, Kraft sputtered and growled, "Dammit, that's no way to do it! They sound like a bunch that just started spring training!" Even to a rookie like me, shooting a hole in the tanks did not seem to be a sound plan.

Kraft listened intently as each of the crazy schemes came across the loops, everybody desperately searching for a way out. Then one of the test conductors came up with a plan that made sense. "The winds are forecast to remain calm, so if we wait until tomorrow morning, the batteries will deplete, the relays and valves will go to the normally open condition. As the oxidizer warms up, the tank vents will open, removing the flight pressure. With the booster depressurized and batteries depleted, it will then be safe to approach the rocket."

Kraft nodded and growled at his controllers, "That is the first rule of flight control. If you don't know what to do, *don't do anything!*" We secured Mercury Control and the blockhouse gang got saddled with the unenviable job of nervously watching over the Redstone throughout the night. "Doing nothing" worked: by early the next morning, the batteries were depleted, the destruct system disarmed, and the pressure relieved. The capsule and the Redstone rocket had survived with only minor damages to the tail fins.

This fiasco was the most embarrassing episode yet for the young engineers of the Mercury program. The history books call this mission, sardonically but accurately, "The Four-Inch Flight." While the badly shaken booster engineers frantically worked on finding out just what had happened, I promised myself that when I returned to Langley I would use the same technique that had worked for me flying airplanes or flight-testing at Holloman. I would get to know as many technicians, designers, testers, and planners as possible and find out what data they had that would be useful to me. I would then compile a book that contained essential, carefully organized, and easily accessible information so in future emergencies we would have what we needed to know right at our fingertips. The engineers at Langley were tremendously cooperative and even gave me a drafting board where I could study blueprints. The learning curve of my first mission had been steep. But I had gained something precious. I now knew how much I didn't know.

We went back to Langley and regrouped. The launch team debriefed, fixed the launch umbilical circuit problem that had caused the premature engine shutdown, and a month later we sent the rocket aloft. Project Mercury closed out 1960 with its first successful Redstone launch.

In January of 1961, our second mission gave the chimpanzee Ham a hell of a ride. The Redstone failed to shut down when commanded and went to fuel depletion, landing almost 120 miles downrange from the recovery forces. We wrote our reports and classified the mission a success; after all, Ham survived and the rocket had not blown up.

In these early months, we were plain lucky that America understood there was no achievement without risk, and there were no guarantees in this new business called spaceflight.

To hurl a man into space and bring him back alive, we needed to wire the world. This meant stringing communications across three continents and oceans, building tracking stations, installing the most powerful computers we could lay our hands on, and learning the business of real-time spaceflight with our teams.

We established the thirteen manned network stations, which provided optimal coverage only during the initial three orbits. Cable connections from the United States stretched to switching centers in London, Hawaii, and Australia. The Mercury voice and Teletype communications were controlled by Jim McDowell from the central switching center at Goddard. After logging thousands of hours at his end of the lines, McDowell had an instinctive feel for each of his communications links and was able to predict and anticipate problems to an uncanny degree, bringing alternate circuits on line moments before the prime circuits failed. It was common to lose communications because of construction workers severing cables, or cranes knocking down power lines, sunspot activity, or even fog.

After only a few weeks of training, in March 1961 the controllers went to the most remote outposts, installations connected to Mercury Control by a communications system best described as brittle. Text messages were prepared by Teletype operators at machines that punched holes into a narrow paper tape. When the message was completed, the tape was fed into a machine and transmitted to the tracking stations. This took at least twenty times longer to transmit a data packet than a present-day $100 fax machine would take to transmit the same amount of information. The constant chattering of the Teletype machines provided the audible backdrop for virtually all of the work at a site.

The CapCom was the remote site team boss and handled all air-to-ground communications. His systems monitor assessed the capsule status with a bank of twenty-one meters and a couple of eight-pen recorders, like those used on lie detector, or polygraph, machines. The team flight surgeon had even fewer displays, thirteen meters, a scope to monitor the astronaut's electrocardiogram, and an eight-pen recorder. Each three-man remote site team, with their brief contacts of eight minutes or less with the capsule during each orbit, provided the global coverage for early spaceflight. They were our eyes and ears as the spacecraft passed overhead. Their charge was simple: stay out of trouble, keep the mission on track, and provide any needed assistance to the crew. Easier said than done.

The Mercury remote site CapComs were all

fresh college graduates; this was their first job. They were paired with systems monitors, also young, who worked for Philco, a high-technology (in those days) electronics company. The systems monitors had no more than two years' experience working at the early global satellite tracking stations. Only the very young seemed to have the guts to volunteer for these assignments, living on their own in distant and remote places.

The tracking stations were often fairly primitive corrugated steel buildings like the hootches I lived in while in Korea, housing the electronic equipment and consoles. The sites were easily identified by their myriad of antennas. Primary site communications were provided by sixty-word-per-minute Teletype and a radio voice link from relay stations at London, Honolulu, and Sydney. The system was a daily crapshoot, susceptible to a variety of problems. When communications failed, the remote site teams were on their own, improvising and taking any action necessary during the period the capsule was in view to restore contact.

The key sites were located at the points for the major Go NoGo decisions, and the locations of the deorbit maneuver. These included Bermuda; Australia; Hawaii; Guaymas, Mexico; and the California coast. These sites were usually designated as critical, and the team was augmented with an astronaut CapCom. The more remote facilities were the Canary Islands, Nigeria, Zanzibar, Canton Island, and ships sailing the Atlantic, Pacific, and Indian oceans.

World War II cargo vessels had been converted into floating sites to track satellites. They were the length of a football field, manned by a makeshift crew recruited from the hiring halls at the local ports. Since the ships carried no cargo, a foot of concrete was poured on the top deck to make them ride lower, and the superstructure was filled with antennas and electronics.

Chris Kraft developed the concept of Mercury Control and taught the first generation of controllers. Like everyone else, he was drinking from a fire hose and needed every bit of help he could get. I was the operations and procedures officer. The job description consisted of keeping anything from falling through the cracks before or during the mission. I wrote the countdowns, prepared all message traffic, made sure the communications were working, briefed the tracking stations on the mission, and gave Kraft any assistance he needed. In effect, I was the flight director's wingman.

I became the scribe of Mercury Control, originating and approving every outgoing Teletype message and most voice communications. Within weeks after I had come on the job in 1960, my relationship with Kraft was solid enough for me to take on responsibility to clear virtually all of the messages without having to bother him. On the first Mercury deployment this got me into big trouble with the U.S. State Department and President Kennedy's Peace Corps. I sent out a message to one of our controllers requesting information on the health conditions at one of our sites in Nigeria. The controller replied that "a hospital of doubtful cleanliness is nearby" and noted that the local people were "extremely poor, local government performance rather feeble. There are no night-clubs or bars. Temperatures are as high as 115 with frequent dust storms." The bad news? "When the rainy season begins it will get worse."

The Nigerian government intercepted the message and threatened to remove the Peace Corps unless the U.S. government apologized. The flap filtered down through the NASA chain of command until it got to me. The message from Kraft was clear: "You screwed up. Next time you're gone." The U.S. apology kept the lid on the issue. The teams stayed on site and I got my first lesson in international diplomacy.

On the last day of March 1961, five months after my arrival, the tracking network was declared operational. We had twenty-one sites, thirteen of which were manned. The total cost of the network, built in one year, was $60 million.

2

"LIFTOFF; THE CLOCK IS RUNNING"

The Soviet Union was our rival in space. While we were blowing up rockets, they were impacting the Moon with a probe. They even photographed the far side. Each Russian breakthrough came as a shock. (Our "intelligence" on the Russian space program was pretty hot stuff — notebooks with newspaper and trade journal clips pasted in them. The military apparently didn't feel that the civilians in NASA had a need to know whatever it was they knew.)

Most Americans followed the selection and training and further adventures of the seven original astronauts. That was about all they really knew about our infant manned space program. The astronauts were instant celebrities, not so much selected as anointed. The public, as well as the Mission Control team, was caught up in the beauty pageant aspect of the first manned launch: Which astronaut would be first? Who was the best?

April 1961

In April, as we were deploying for a pair of missions, the Russians beat us again. Yuri Gagarin became the first human in space, and in orbit to boot, and we neophytes in the Space Task Group viewed the Russian success with both frustration and admiration. We packed up our bags, kissed the wife and kids goodbye, and, a few hours later, were once again at Mercury Control. Marta was now expecting our third child. We were launching increasingly complex missions from the Cape every month. Over half of the year we were TDY, on temporary duty, at the Cape. Unlike the later years in Houston, our wives did not know each other and often lived pretty far apart, so it was a lonely time for them. Compounding the problem was the dispersal of so many of our people to far-flung remote sites. Working in Mercury Control, I was fortunate: I could easily stay in touch by phone — and I could share with Marta the excitement and pride that we felt as the program went forward.

Following two successful Redstone launches, we moved on to the unmanned Atlas mission, which was designed to test the spacecraft and the global network. The mission that would follow was the one we had been waiting for. It was planned to launch the first American into space.

After we arrived at the Cape, we found that the military, which actually ran the Cape and nearby Patrick AFB, as well as the recovery forces, had pulled the plug on our resources and reallocated them to deal with one of the worst crises of the

Cold War. A force of 1,300 Cuban exiles, who had been trained and armed by the CIA and given decidedly insufficient American tactical support, had landed at the Bay of Pigs in Fidel Castro's Cuba in the predawn hours of April 17. Initially planned under the Eisenhower administration, this ill-advised invasion had probably been doomed from the outset, but its fate was sealed when President Kennedy, only a few months into his term and ambivalent about the entire operation, withheld American air support. Castro's small air force decimated the exiles bent on his overthrow. All this was happening a few hundred miles to the south of the Cape. We sat in our hotel rooms, anxiously waiting to recover the resources we needed for the next two missions, our eyes glued to the television sets.

The combination of Gagarin's flight and the U.S. humiliation at the Bay of Pigs provided a sobering background to our deployment. The press focused on America's pitiful space record, while touting Russia's successes. It was reminiscent of what had happened a year earlier, when *Newsweek* lowered the boom on the Mercury program: "To lose to the Russians all we needed to do was start late, downgrade Russian feats, fragment authority, pinch pennies, think small, and shirk decisions." I don't recall anyone disagreeing with that assessment. The message was understood in Washington, and it was taken to heart at the Cape.

I find it difficult today to convey the intense frustration and near despair as we picked ourselves up after each setback, determined to break the jinx on the program. Now we were going for two back-to-back missions — launching an unmanned Atlas downrange and then carrying out our first manned Redstone mission. We tried not to think about the gaps in knowledge, experience, and technology in our program — they were big enough to drive a truck through — and we could never forget that while we were screwing around with baby steps in suborbital missions the Russians had put a man in orbit. So we would continue with our preparations at the Cape, tired of being one step behind. It seemed like no matter what we did, the Russians were always one step ahead.

In those dark days our only thought was, "This time it has to work."

Testing went smoothly once we regained the test range and network resources. The Mercury network, operational less than a month when we deployed, was working beautifully. NASA, playing catch-up with the Russians, changed the Mercury-Atlas launch (MA-3) from a ballistic mission to an orbital one. We split the Canary Islands team and sent a small group to Nigeria and Zanzibar. We planned two launches in the next ten days: an unmanned Atlas orbital mission and America's first manned mission on a Redstone.

The simulation team was composed of another small group of controllers. Their task was to create what we now call virtual reality — to replicate, in chillingly convincing detail, every element of the mission, from countdown to completion. The simulation supervisor (Sim-

Sup) had five people playing the roles of thirty. They would supply a data stream — telemetry, command, radar tracking, voice reports — and our controllers would have to respond. SimSup's team would provide the voice calls and responses of the three test conductors, range safety officer, recovery team, and everyone else involved in a launch. One guy might play a dozen different people responding to controllers' calls. The SimSup's objective was to test the judgment of each individual and the competence of the total mission team. How quickly would they recognize and solve problems? How well did the mission rules and the procedures used in the various facilities and the network function in real time? Were we ready?

SimSup would prepare and send out magnetic tapes to each of the Mercury facilities. For instance, a single orbit takes ninety minutes. The tapes would be played in sequence, starting at the Mercury Control Center at the Cape. At four minutes after simulated launch, Bermuda would start playing their tape — so for about six minutes MCC and Bermuda could compare data — then MCC would lose data and a few minutes after that the Bermuda tape would end. There would be an eight-minute gap before the tape at the Canary Islands site would start running. Each of the tapes had to contain timing and data replicating what was expected to happen during the actual flight. The simulation team would introduce various malfunctions on the tapes sent out to each site and the controllers there would have to deal with them.

While all this was going on, an astronaut sitting in the capsule simulator had to deal with the same malfunctions thrown at the controllers at the various sites. If either the astronaut or the site controllers wanted to take an action that conflicted with the data on the tape, then the whole simulation would start to fall apart.

Few of the early simulations achieved their objectives. Everything was new and untested — new equipment, new procedures. Attempts to conduct "seconds critical" training often failed. Our final MA-3 training run was no exception. Our novices stumbled from the start, when the wrong tapes were selected. After several restarts, tempers flared as controllers at the separate sites began improvising in an attempt to complete the test. Later in the afternoon, a disgusted Kraft called off the exercise and told me to sort out what had happened.

Three days before the Atlas orbital test launch, I prepared the Teletype daily advisory, critiquing the previous day's run and offering an apology. The message to all the tracking sites was blunt:

> TODAY'S OPERATION WAS A COMEDOWN AND INDICATES WE STILL HAVE PROBLEMS. MISSION CONTROL'S PERFORMANCE WAS SUBSTANDARD. WE APOLOGIZE FOR THE WAY WE CLOSED DOWN THE VOICE NETWORK. OUR DISCOURAGEMENT WITH THE DRILL LED US TO WALK AWAY UNCOMPLETED.

LAUNCH FORECAST — NO DELAYS ANTICIPATED. BOOSTER, CAPSULE AND RECOVERY GO. FLIGHT CONTROL IS GO ASSUMING ADEQUATE TRAINING.

It was customary to add the latest news headlines, which included these:

U.S. WILL ACT AGAINST CUBA TO GUARD ITS SECURITY.
KENNEDY'S ANTI-CASTRO INVASION STAMPED OUT.

REPORTS CLAIM CASTRO SUFFERING FROM MENTAL COLLAPSE.

MARILYN MONROE SAYS NO TO REMARRIAGE TO JOE DIMAGGIO.

ALLISON AND MANTLE HIT THREE HOME RUNS.

AT&T 126-$^{1}/_{4}$ DOWN $^{3}/_{8}$ BECAUSE OF PROJECT MERCURY SLIP.

We could only hope now that our training would prove sufficient. It was time to launch.

April 25, 1961, Mercury-Atlas 3

The Atlas rocket, shedding ice formed by condensation on the sides of the liquid oxygen tanks, lurched skyward on three shafts of liquid fire, steam billowing from the flame bucket that channeled the fiery exhaust away from the blockhouse, support tower, cables, and pad. Inside the control room, I could not hear the engines roar, but the sense of my first Atlas launch seeped from my fingertips as I scribbled the liftoff time in the Teletype message and handed it to a waiting runner.

Without pausing, I picked up the running Teletype dialogue over the order wire, advising Bermuda of the launch time and status. Seconds after the launch, a note of anxiety crept into the Welsh accent of Tec Roberts, the flight dynamics officer (FIDO) responsible for launch and orbital trajectory control, as he reported, "Flight, negative roll-and-pitch program."

A collective shudder went through everyone in the control room as the controllers absorbed the chilling significance of Roberts's terse report. The roll-and-pitch program normally changed the initial vertical trajectory of the launch into a more horizontal one that would take the Atlas out over the Atlantic. This Atlas was still inexplicably flying straight up, threatening the Cape and the surrounding communities. The worst-case scenario would be for it to pitch back toward land or explode. The higher it flew before it exploded, the wider the "footprint" of debris scattered all over the Cape and surrounding area would be.

The RSO (range safety officer) monitoring the launch confirmed the lack of a roll-and-pitch program, then continued to give the Atlas an opportunity to recover and start its track across the Atlantic. The RSO lifted the cover on the command button and watched as the Atlas raced to a fatal convergence with the limits on his plot board.

At forty-three seconds after liftoff, Roberts reported, "The range safety officer has transmitted the destruct command." Seconds later, Kraft's TV glowed a vivid black and white with the explosion. (We did not have color monitors.) We waited, not speaking, counting the seconds, listening for the telltale, muffled *krump* that would signal the mission was over. Carl Huss, the retro controller (RETRO), responsible for reentry trajectory planning and operations, reported, "Radar tracking multiple targets." Roberts's response echoed all our feelings: "Chris, I'm sorry."

We sat by the consoles, not talking for several seconds. Then, one by one, the controllers closed their countdown books and started to pack their documents.

My message to the remote site teams was succinct: "MA-3 WAS TERMINATED BY RANGE SAFETY AT 43 SECONDS INTO FLIGHT. STAND BY FOR DEBRIEFING."

The destruction of the Mercury-Atlas 3 left us dazed and disheartened. We had gone to launch feeling that with three successful unmanned suborbital missions the jinx seemed broken, the odds were turning in our favor. Then the Atlas had to be blown up.

There was no cheer as the Mercury team retired to the bar at the Holiday Inn in Cocoa Beach and waited for Walt Williams, the operations director, and the launch team to finish the press conference. We knew they would take a beating. We had no time to lick our wounds or feel sorry for ourselves and no one had much to say. Numb with shock, frustration, and anger, we were uncertain about the impact of this spectacular failure on the entire program. We did know that it would mean more time at the Cape, more time away from our families. But did we really know what the hell we were doing? The Atlas was the key to orbital flight and we had racked up two Atlas failures in three Atlas missions.

When we debriefed and vented our frustration over drinks at the bar, it was all in the family. During the missions, Kraft set the tone when we lost our composure. He knew that he could not show any uncertainty to his team. In Mercury Control there was no room for displays of emotion. Kraft allowed himself an occasional broad smile and lit up a cigar when his gang did well. When we failed, maybe he would mutter a curse word or two. His message from the earliest Mercury days: we were the visible point men for the program and we had to maintain a calm, professional, and confident image. But, as long as no outsiders, especially press, were around, we could let it hang out when we left the Cape and hit the bar.

We didn't even have the luxury of a day off. We debriefed, wrote our reports, and, on April 26, 1961, returned to Mercury Control. With only seven days to prepare for our first manned flight, our self-confidence was badly shaken and we all shared the same unspoken fear: in only a few days, would we be crowning our first space hero — or picking up pieces of him all along the Eastern seaboard? We attacked the final days of the training schedule with a renewed urgency. Everyone connected with the program was now consumed with one goal: the successful flight of America's first man in space. We simply could not accept, or even contemplate, another failure.

It was the top layer of NASA management which was directly in the line of fire coming from Congress and the executive branch. Our new administrator, James Webb, who was everybody's boss, didn't exactly deflect criticism when he said, "This present program is not designed to match what the Russians may do." Reporting to him for manned space flight were Wernher von Braun at Marshall Space Flight Center, Harry Goett at Goddard Space Flight Center, and Robert Gilruth at the Langley Research Center. (Mercury preflight operations at the Cape were run by a division headed by G. Merritt Preston and reported to Gilruth.)

Gilruth, director of Project Mercury and boss of the Space Task Group, was the pioneer who focused the Mercury effort. He was regarded with great affection by those who worked under him. Starting at Langley in 1937, he had a sterling record of achievement in flight-testing. By the late 1950s his people were the most knowledgeable on high-speed flight research and he was the obvious choice to form and lead the Space Task Group.

Gilruth's Washington counterpart was George

Low, an Austrian whose family fled to America when Germany invaded and occupied his native land. After serving in the Army in World War II, he worked for Convair and then was chief of special projects at the National Advisory Committee for Aeronautics — NACA — the predecessor of NASA. In the summer of 1958 Low was tasked to organize a new space agency and became the chief of manned space flight during the agency's early years. He was absolutely determined to put a man on the Moon, believing only such a bold goal would sustain the manned space effort. Personally, I would settle for rockets that worked.

In the first two years, we spent almost half our time on temporary duty at the Cape. When we left our homes and families behind, we never knew whether to pack for a few hours or a month. We lived out of our suitcases, in nondescript motels, coping with loneliness, too little per diem, and too many failed countdowns. Invariably we would go home to find our wives feeling neglected and frustrated. You have no doubt heard the saying "Behind every great man is a woman" — and behind her is the plumber, the electrician, the Maytag repairman, and one or more sick kids. And the car needs to go into the shop.

To make matters worse, most of the wives had a postcard image of sun-washed beaches and a kind of intergalactic glamour. We felt this pressure and knew we could not resolve it. The beach was only a rumor to most of us, and glamour was a tub of beer on ice. The Cape Canaveral area was hardly the glossy Florida of

Miami or Fort Lauderdale, but it was a community full of solid, decent working people, who encouraged us by their support for our fledgling space program.

We didn't have nine-to-five jobs. Pad tests, simulations, and practice countdowns are driven by launch pad activity. Launch preparation is more an art than a science, with the high priests (test conductors) of the blockhouse, in charge of the countdown, dictating the overall schedule. Our job at the Mission Control Center was to be ready whenever the launch team needed us. The MCC support for the capsule and booster launch pad tests usually started with the test conductors' call to console stations about 2 A.M., so the controllers usually hit the sack about sundown and arrived at Mercury Control about 1 A.M. to prepare for the test.

The flight rules that were a mystery in the beginning now anchored me at the center of the mission policy and decision process. They were an upscale equivalent of the Go NoGo criteria we used in aircraft flight tests. I had inherited the rule-recording job when Kraft sent me to the Cape for the first Mercury-Redstone test. This task opened the door for me to every technical aspect of Mercury operations. In the early months, I was the scribe sitting in on Kraft's meetings with crews, controllers, and management. I wrote out the record of the decisions of the meetings, putting them in the flight rule format, defining each decision as a series of conditions followed by the action procedure.

The Mercury rules summary was thirty pages

long. Many of the rules were procedural in nature and placed a heavy weight on the judgment of the controllers. Go or NoGo conditions were underlined in vivid green or red. The value of compiling and defining rules was not in the document itself as much as it was in hashing out Go NoGo stipulations in team meetings. The rules were set up on each page in this fashion:

ACTION	CONDITION/ MALFUNCTION	RULING	NOTES/ COMMENTS/ SOPS
FD	Capsule failure to separate.	15. MCC will send abort command. As last resort may request destruct command from RSO.	
FIDO	Booster package failure to separate. (No staging.)	16. Abort at velocity for impact in Area D due allowance for delta V due to escape rocket firing.	16. Automatic abort may occur prior to above listed time because of lack of guidance or fuel depletion.

[FD=Flight Director; FIDO=Flight Dynamics Officer; Area D= Landing zone located in the east Atlantic Ocean.]

Frequently, Kraft would leave a team meeting disagreeing with the results. The controllers and doctors were too conservative for his way of thinking. After reflecting for a while, Kraft would say, "I didn't like their input. When you write the rules, let's do it this way." Many of the critical flight policies were determined over dinner at Ramon's restaurant, late at night in one of the bull sessions, or during a lull after a tough training run.

No matter what the rules said, the ultimate authority was the operations director, Walt Williams. Born in New Orleans and an LSU graduate, Williams worked for Martin Aircraft and then at Langley, where he concentrated on control and stability systems during World War II. In 1956 he was head of engineering for the X-1 project (our first aircraft to break the sound barrier). He went on to be the director of virtually every supersonic test program for jet- and rocket-propelled aircraft and by 1958 was head of the X-15 test committee. He had been assigned as Gilruth's deputy for Project Mercury, and was the feared but highly respected boss of every aspect of Mercury operations.

The days remaining before the manned launch passed rapidly. Walt Williams gave the operation's Go at the mission review on launch minus one day, May 1, 1961, and the capsule servicing began as we conducted our final simulations. We did our data checks and returned to the motel. Normally, I sleep like a dead man, but I tossed and turned most of the night.

Since I had spent most of my time with the team at the Cape in unmanned rocket testing, most of what I knew about the astronauts came from the newspapers. When I did see them, they seemed to live in a world apart from the rest of us, communicating only with the top managers and spending most of their time in hands-on training, learning all they could about their spacecraft and rockets. I didn't blame them. The early space hardware had much in common with the imaginary technology in a Jules Verne science fiction novel. The capsule used steam-powered thrusters for control (the steam being generated by the reaction between hydrogen peroxide and a catalyst in a nickel chamber attached to the thrusters), a periscope for visual tracking, and an electric Earth globe for determining position. Half of the systems were primitive by today's standards for aircraft avionics and control systems, while the other half were untested first-of-its-kind technology.

Their test pilot backgrounds made the Mercury Seven highly independent — and fiercely competitive. Each one of them was determined to be the first man in space; each believed his performance during the months of training and testing would win him the coveted prize. As launch time neared, Gilruth took the extraordinary step of asking each astronaut to rank all the others in the order of who was best qualified for the mission. Some thought he did this just to confirm his personal selection. Alan Shepard, Gus Grissom, and John Glenn were chosen from the seven; one of that trio would be the first among equals. Shortly thereafter *Life* magazine got into the act, dubbing the first three the Gold Team, the remainder the Red Team.

The brass felt a need to reassure the public there was no split in the ranks. Wally Schirra, Deke Slayton, Gordon Cooper, and Scott Carpenter stepped forward at a press conference a few days later to confirm that there was plenty to do and they all were still on one team, not divided into "red" or "gold" factions. Their reassuring words and smiles could not cover up the fact that someone else had indeed been chosen to be first. NASA did its best to soothe the egos of those who did not make the first cut. The lyrics to the music went like this: Each astronaut had an important role to play. There were plenty of flights for all. And, sure as hell, there was enough work to go around.

We in MCC were not surprised by the choice of Glenn, Grissom, and Shepard to make the initial flights. The controllers had worked briefly with all three, but only our bosses and the astronauts knew who would fly this very first mission. Since we had seen more of Shepard, we were betting on Al. The media seemed to favor Glenn. In part it was his image as a God-fearing, clean-cut American patriot and good family man. But journalists may have reported this in hopes of provoking someone into saying who it would be.

Shorty Powers, a lieutenant colonel in the Air Force, was the man in charge of public relations and the "voice of Mercury Control." To Shorty, dealing with the press was just like a game of poker: never let on what is in your hand — and try to bluff them into showing you theirs.

Like Williams, Shorty never seemed to sleep

and his exhausting work schedule only compounded his irritability; it took very little provocation to make him lose his temper. When he did, it was like a launch, a great deal of energy and noise expended amid the fire and smoke. He was a bantam cock of a guy, about five-foot-four or -five, always dapper and a bit of a strutter. For the most part, Powers had an unenviable job, setting up a public relations barrier behind which the engineers and the astronauts could work in peace, while at the same time trying to feed reporters' insatiable demands for information. He really needed two sets of astronauts, one to do the mission, the other to perform for the media.

The astronauts and controllers stayed at the Holiday Inn at Cocoa Beach. It was not unusual to see Walt Williams charging into the bar late at night to get an answer to a problem from one of the controllers or engineers. He was likely to find the man he wanted there, because all of us sat around and chewed over rules and procedures endlessly and tuned in on Williams's questioning to get the latest word.

We knew that three days prior to the launch the selected astronaut would check out of the Holiday Inn and quietly try to move out to the crew quarters at the Cape. Reporters hung around the Holiday Inn fishing for information, striking up a conversation, occasionally trying to pass as a member of the launch team. The press, which had become somewhat negative and adversarial after the Gagarin flight and the Atlas 3 failure, was now caught up in the excitement of this new chapter. Realizing that a man's life and the future of the program were at stake, their coverage finally began to capture the excitement and suspense surrounding the imminent mission. The control team, taking our lead from Powers, neither confirmed nor denied any of the press speculation and ignored all the rumors that buzzed around the program like a swarm of busy bees.

"It's a Go"

The weather was stormy at midnight on May 2, 1961, when the control team arrived to support the initial check-out. The countdown progressed through the fueling of the Redstone while the launch team held the transfer van (used to move the suited astronaut from the hangar to the launch pad) at the hangar until the weather improved, or the launch was scrubbed

We had adopted the call sign "Freedom 7" during our final training runs, but I never knew who had named the capsule. Just one week after the Mercury-Atlas 3 failure we were once again in Mercury Control counting down to launch for our first manned mission. The headlines read: "AS HOP NEARS ASTRONAUT X IN SECLUSION." Soon the world would find out — and so would we. I think only Kraft, Williams, and the flight surgeon knew it was Shepard. It had been decided in Washington that the identity of the first man would remain a secret until he stepped forward to climb atop the rocket.

The Russians did not announce any launch in

advance; in fact they didn't release any news about their manned spacecraft effort until they were good and ready, and even then they gave only carefully selected details about a flight. They could do this quite easily in a closed society in which news was strictly controlled by the government. We did not have this luxury. From its earliest days, NASA had followed a policy of maximum, though prudent, disclosure. We had to do everything openly — and soon under intensive, live TV coverage. In their own good time, the Soviets had announced that Gagarin spent 108 minutes in orbit before returning safely to Earth in a parachute-cushioned landing.

We wanted to catch up and we believed that, at last, we were ready to do it — at least for the first step, a suborbital mission of limited duration. Dressed in his silvery space suit, Alan Shepard stood behind the door of Hangar S. Outside, the van that would deliver him to the launch site waited, along with a large group of reporters and photographers who were eager to tell the world which astronaut would step through the door.

Lightning and rain had been playing about the Cape all morning, and when the clouds had not cleared by 7:25 A.M. local time, the flight was canceled. No one could be responsible for the weather, but it struck us as another in an unending series of tough breaks. Who would stop the rain?

Shepard shimmied out of his suit and downed a shot of brandy. An alert reporter standing by the hangar door had seen him and broke the story: "FIRST U.S. ATTEMPT TO PUT MAN IN SPACE POSTPONED 48 HOURS. SHEPARD GIVEN FIRST CALL FOR HISTORIC VENTURE." The secret was out. Hard-charging Al Shepard was at the head of the line.

We drowned our disappointment in the usual way — with a mission scrub party. No matter what hour the test was scrubbed, we would return to the motel wide awake, after the lounges had closed, or before they opened. We had stashed beer and snacks at the Holiday Inn, which often donated food left over from the previous day's menu. We would eat and drink, and talk about what had gone wrong. It was a little like throwing a rueful party after a nonbirth; at least we hadn't experienced another disaster, and the baby was still there in the womb, ready to go. All we could do was pace the floor and wait.

In the heat of the following day, some of us headed to the ocean or the pool, or played full-contact volleyball, to sweat off the beer we had tossed down. The chalked volleyball sidelines didn't last long and there was the usual quibbling on out-of-bounds calls. The solution was to dig a trench and mark the sidelines with gravel embedded in concrete. Out-of-bounds calls were a lot easier when you came up with bloody forearms after a diving save. Bruises, sprains, jammed fingers, and nasty cuts were the order of the day.

Kraft, a standout baseball catcher and center fielder in college, was a fierce competitor. But on the volleyball court he was no longer the boss, just one of the team members. Carl Huss, the MCC RETRO, was a burly five-foot-eleven guy with black bushy eyebrows and hair. When playing volleyball he had a habit of rising on his toes and shifting from left to right with a rolling motion. His menacing visage, combined with this motion and his perpetual growl, earned him the nickname "Dancing Bear."

During one match, Huss spiked a shot straight into Kraft's face; the ball drove the prongs of his sunglasses deep into the flesh of his nose. Without flinching, Kraft pulled out the prongs, wiped the blood off his face, looked at Huss, and growled, "Nice shot. Try me again!"

After multiple injuries to his team members, Kraft set the rule: no volleyball after L – 3 — launch minus three days. Any controller violating the rule and unable to perform his console duties would be "disciplined." No one was willing to find out exactly what kind or degree of discipline Kraft meant.

The beverage of choice after these matches was Swan Lager, an Australian beer, and our supplier was Jack Dowling, the Australian government's envoy to NASA. Jack was the picture of a typical Aussie. He was a bit older than most of us, stocky, with wavy black hair, flecks of gray in his mustache, and eyebrows like caterpillars. All he needed to complete the image was the Crocodile Dundee bush hat. He had the accent, the one you never missed when you called "Goddard voice" and the switchboard operator patched the voice communications to the Australian tracking stations. You had to be careful about confusing their language with the one we were developing for space.

When we had needed a tracking station in the Southern Hemisphere, the people Down Under were quick to respond. They have always been our stout allies, and their very isolation inspired them to sign up for any new adventure. They sent their volunteers to train with us, and Dowling was one of those who learned to love the States so much he never left.

The personal relationships that developed in the early years at Cape Canaveral provided the foundation of a brotherhood that extended through Mercury, Gemini, and Apollo. This bonding got us through the difficult times. We worked together, played together, and lived together.

If I close my eyes I can recall images of that strip of coastal Florida — a line of motels, restaurants, and bars (some pretty funky) lining the highway that ran south from Daytona Beach through Titusville and Cocoa, where a causeway across the Indian and Banana rivers took you to the small town of Cocoa Beach. It was a two-lane blacktop that shimmered in the hot sun and paralleled the swamp that stood between the Cape and the mainland. Although Orlando was only a short drive inland, it might as well have been on Mars. Our world was confined to a small, tight circle centered on those strange new structures — gantries and launch pads and telemetry antennas — sprouting up on an overgrown sandbar. NASA's arrival in this once calm and sleepy area would change it forever — and make it perhaps a greater tourist attraction than the locals ever dreamed it could be (and perhaps more than many wanted it to be).

In this setting our bonding produced a spirit that responded to the challenge of John Kennedy's inaugural address: "Ask not what your country can do for you; ask what you can do for your country."

This was the spirit that a few days later would bring space and the astronauts to the front pages of newspapers and into the homes and hearts of America.

May 5, 1961, Mercury-Redstone 3

When the launch was scrubbed on May 2, it was reset for May 4 — and then scrubbed again because of weather. But then the weatherman gave us a solid Go for the next day. The weather was windy but clearing when Huss and I left the motel shortly after midnight. As we left, I drove around the east end of the motel to see if the searchlights at the launch pad were on. If we saw the lights, we would know that the launch complex was active and the countdown progressing. The lights drew me like a magnet; when I saw them I picked up speed, and to hell with the local cops. Our sleepiness quickly vanished during the twenty-minute drive.

Highway A1A took you through the heart of Cocoa Beach. With only a single stoplight, it was a small town in the process of trying to grow and live with its newfound fame. The brilliant, garish neon of the motels and restaurants and go-go bars seemed more like Las Vegas, but they were soon behind us. The traffic was heavy, as it usually was, cars pulling out of motels and dark streets to join you as you passed by. You drove into an inky darkness after passing Fat Boy's restaurant, which marked the city limits of Cocoa Beach. This traffic was different from the usual relaxed pace of tourists and locals. From this point the cars on the road were moving swiftly, their drivers knowing exactly where they were headed — a personal rendezvous with history.

The interior of the space capsule that Alan Shepard would soon climb into was so small that a human being could barely fit. The back of his couch was within inches of the heat shield. The instrument panel was less than two feet from his face and the parachutes only five feet forward. John Glenn had hung a sign on the panel: "No Handball Playing in This Area."

The Redstone rocket would lift Shepard's two-ton capsule on a fifteen-minute foray into space. He would reach an altitude of 100 miles, experience five minutes of weightlessness followed by a crushing 11-G entry, and land 260 miles downrange from the launch pad.

Closing in on the Cape, I could see that the lights looked much like the lights of the small cities amid swamplands when I trained and flew in Georgia. After a brief stop for a badge check, we continued toward Mercury Control, passing the security roadblocks that are always erected as launch time nears.

Huss was not a man given to making small

talk. Like a bear coming out of hibernation, he would wake up slowly, so we didn't talk to each other. I enjoyed the silence, which allowed me to think about what lay ahead as I made the long swooping turn to the north and the searchlights. The stars were occasionally visible through the muggy haze of the Cape. The outline of the launch tower was not yet visible through the palmetto as we approached Mercury Control. After parking the car, I started to psych myself up, just as I did when I was flying. In my mind, I could hear the marches of John Philip Sousa and the cadence quickened my step.

The support team finished its systems checks and I sat down to check out the communications at my console. There had been no countdown delays and the capsule test conductor (the team leader for the capsule check-out) came on and reported, "The count is on schedule." I then moved to Kraft's console to check communications from his voice panel to the pad team. On this day, I was covering his tail, watching for problems while Chris went about the business of being the flight director. A half hour later, he arrived, dressed nattily as usual, regardless of the time of day. He made his standard comment: "How's it going, young man?" I gave him a thumbs-up. He reached in the drawer for his headset, adjusted it, and then sat down.

At his console, Kraft projected the image of a general reviewing his forces prior to battle. The only time he showed any uncertainty was when the IBM engineers, Ira Sachs and Al Layton, periodically reported on the results of the network data flow testing. Tec Roberts and Carl Huss were monitoring the flow of data coming in from the tracking stations to Goddard Space Flight Center in Maryland, which in turn came down, via dedicated telephone lines, to the data display on the four plot boards in front of them. Sachs and Layton were concerned that there was a lot of ambient noise on the data lines and, like the rest of us, were not sure how a burst of static might affect the two computers at Goddard.

Sachs and Layton retested the lines and announced, "We had 270 failures out of the last 11,250 transmissions." The perplexed expression on Kraft's face indicated his annoyance. Softly, he asked Sachs, "Dammit, will you please tell me if that is okay?" That was the one thing they could not do. Getting a less than satisfactory answer, Kraft frowned, pretending to make sense of their report. Nobody knew how much bad data the computers could digest and still come up with acceptable answers. Computers just seemed to work, crash, or go off on tangents for no reason, with Huss and Roberts at their mercy.

Shepard had awakened at 1:00 A.M. and, after breakfast with John Glenn and Dr. Bill Douglas, began to undergo his physical exam, sensoring, and suiting. The weather was definitely better than the day before and a feeling of bullishness pervaded the control room. More controllers arrived and began making their checks.

Walt Williams, rumpled as usual, showed up about 2:00 A.M., after a quick, how-goes-it conversation with Shepard and Glenn. Chain-smoking, Walt talked briefly with Shorty Powers, then moved to his desk behind Kraft, his voice hoarse, grunting a question to Kraft. (Another sign of the times was that most people smoked in those days — and smoked a lot. Mercury Control would have been a nightmare for those who object to secondhand smoke because the air was regularly blue with tobacco haze.) Chris gave Walt a high sign, happy about the weather forecast.

The report that Shepard had entered the transfer van and was en route to the capsule gave me a chill. I passed the report to the controllers on the two tracking ships in the Atlantic about 280 miles downrange, north of Grand Bahama Island. When the van arrived at the pad, we saw a flurry around the base of the Redstone. It was surreal. The brilliant floodlights turned the night to day. The silver-suited Shepard paused, looked up, and then strode to the gantry for his sixty-five-foot elevator ride up to the capsule. I noted in the log that Shepard entered the capsule at 5:20 A.M. Again, I felt a shiver. This was history. I hoped that the other controllers were doing a better job of keeping their minds on their work than I was at that instant.

The countdown continued sporadically with five holds. During one of them I got Kraft his customary pint of milk to calm his ulcer. Williams spent more time outside checking on the weather, which was becoming increasingly overcast. With the launch gantry removed, the Redstone rocket stood starkly silent and alone on its platform.

It was now a little before seven o'clock in the morning. Looking up at the countdown clock I thought, If there ever was a time and place to get it all together it was now. It was time to kick in the afterburners and regain our confidence as Americans and as leaders.

Williams called a weather hold at launch minus fifteen minutes. Shortly thereafter, a problem developed in a Redstone power supply, and the decision was made to roll back the gantry and recycle the countdown to thirty-five minutes and hold. I got up to get a cup of coffee and stretch. Nerves were taut. It wasn't a subject anyone talked about openly, but we in MCC fully expected to lose one or two astronauts in Mercury. The prayer at that moment was, "Not now, Lord, please, not today."

Pilots don't growl at their crew chiefs, so it came as a surprise when Shepard, in *Freedom 7*, the tiny Mercury capsule atop the Redstone rocket, growled at his ground crew, "Why don't you guys fix your problems and light this candle?"

The countdown had been holding for weather when our computer at Goddard crashed, requiring a complete check run. This was our third attempt to launch MR-3 and the pressure from Washington was mounting.

Goddard estimated a delay of ten minutes for the computer check. Kraft, sensing the tension that had built up in the control room over the delay, told his keyed-up team to "take five and get a cup of coffee."

When I returned from my coffee break I lit

another cigarette, and as the test conductors completed the recycle and announced the hold, the air-ground loop to the capsule came alive. To my astonishment, I heard the then popular comedian Bill Dana's high-pitched parody of a reluctant astronaut:

> "My name . . . José Jimenez . . . Do you know what it really takes to be an astronaut?"
>
> "No, José. Tell me."
>
> "You should have courage and the right blood pressure and four legs."
>
> "Why four legs, José?"
>
> "Because they really wanted to send a dog, but they decided that would be too cruel."

As the José Jimenez routine continued, I punched the loops on my intercom to see if the recording was coming from Mercury Control. "Dammit," I thought, "who the hell is playing a nightclub act on the countdown loops?" I sure hoped it was not coming from Mercury Control. If it was, I knew I would catch hell from both Kraft and Williams. I expelled a sigh of relief when it became clear that the comedy routine was being piped in from the blockhouse. A wonderful discovery: our German colleagues had a sense of humor. But was now the time to display it? It would be a distraction for the launch and flight team — and if the mission had been scrubbed, the bosses would have been on the warpath. As it turned out, however, Gordo Cooper and Bill Douglas, the surgeon, had conspired to patch Dana's recording of José Jimenez into the capsule. They felt Shepard needed to relax a bit during the hold. This informality added a degree of unreality to the fact that we were only minutes away from launching the first American into space.

Not everybody was amused; I could see that Kraft was not happy. He did not like surprises that would distract his team. But by now the countdown was forgotten momentarily. The controllers were drinking coffee, joking and enjoying Bill Dana's comic monologue. Dana had been dubbed the Eighth Astronaut by Shepard and Schirra and was a favorite of everyone working on Mercury. Later, in the bar after the launch, I would decide that this bit of humor was exactly what we needed to relax a bit and get loose and ready for launch. But I am damn sure the Russians wouldn't have tolerated such shenanigans.

Shepard had been in the capsule for more than four hours when the count again resumed. It went smoothly and, after a brief hold at two minutes, continued toward liftoff. During the last seconds I saw Kraft's hand move to the liftoff switch on his console. I just hoped he didn't throw it early and start the mission clocks. At T-equals-zero, I glanced at Kraft's TV, saw the rocket ignition, and then heard Shepard say, "Ahhh, Roger, liftoff and the clock is started."

I logged the liftoff time (9:34 Eastern Standard Time) as 1434Z (Z for Zulu, or Greenwich mean, time, used to establish a standard time for all the tracking stations scattered throughout different time zones) in the Teletype message, turned in my chair, and took off to the Teletype center. Oops. I had not removed my headset. After I'd run about fifteen feet, the headset cord stretched to the maximum, snagged a chair, and sent it tumbling to the floor. Kraft, distracted, looked in my direction, frowned, then returned to the business of launching the first American into space.

Sheepishly, I picked up the chair and returned to the console as Shepard made his thirty-second status report. Shorty Powers announced to the world that "everything is A-OK," a phrase hated by the controllers and crews as "too Hollywood," but one that soon became a part of the American vocabulary. It seems quaint now, all these years later, virtually unused, almost forgotten.

The two previous Redstone missions taught me that a ballistic mission is over in a flash. An energy-charged 142-second rocket launch followed by a five-minute weightless period, retrofire, and then a reentry. The drive from the hotel to Mercury Control was longer than the fifteen-minute flight time. Shepard's mission was just like my first jet solo, a blur of noise and motion, an event long anticipated that was over far too soon.

Indeed things were A-OK. After over four hours in the capsule, Shepard was in peak form reporting launch events. At liftoff his heart rate had briefly increased to 120, peaking five minutes later at 140 beats per minute as Shepard called, "Booster cutoff." Now weightless and traveling one mile per second, Al was in the test pilot's nirvana. I was damn happy that the mission was going well and that Al's performance would answer the medical scientists' concerns about whether man could function in space. I had always felt that the flight surgeons were too plodding, too conservative for the rapidly evolving program.

After capsule separation from the booster the automatic system turned the capsule into a heat-shield-forward position. Approaching the 116-mile-high apogee (the highest altitude on the trajectory), Shepard took over manual capsule attitude control, maneuvering in the roll, yaw, and pitch axes and reporting that the capsule responded much as the simulators had. Using the periscope he reported seeing the western coast of Florida and the Gulf of Mexico.

Only moments later, Carl Huss broke into the communications loops, beginning the countdown to retro sequence. His words hung briefly, then were echoed by the MCC CapCom, "5 . . . 4 . . . 3 . . . 2 . . . 1 . . . Retro sequence!" Shepard confirmed he had maneuvered the capsule to attitude for rocket firing.

The weeks of frustration and training were finally paying off in a perfect mission. Now all we needed were the parachutes.

I listened, amazed at the professionalism that had developed in the Mercury team in the six months since I had joined. The pad team, Mercury Control, and the recovery forces were working in perfect synchronization, with an almost casual tone in their voices as if they had done this many times before. In less than fifteen minutes our first manned mission was over.

Shepard was safely aboard the aircraft carrier *Lake Champlain* eleven minutes after landing. While dictating his pilot's report on the carrier Shepard was called to the carrier's flag bridge to answer an unexpected telephone call. President Kennedy had watched the launch and landing closely via television and was now one of the first to congratulate America's new space hero. Kraft rapidly wrote out his mission summary report. It was less than two pages in length. It was now time to celebrate.

3

"GOD SPEED, JOHN GLENN"

We consumed large quantities of beer and barbecue at Fat Boy's in honor of Al, and reveled in the day's headlines: "U.S. SCORES SUCCESS IN FIRST TRY TO PUT MAN IN SPACE." "ALL AMERICANS ARE REJOICING, SAYS KENNEDY." Alan Shepard had become America's hero, but more so he was uniquely *our* hero.

The Space Task Group had little time to savor the wine and smell the roses before we had to do it all over again. In two months the second astronaut would go up. This gave Kraft's team an opportunity to apply the lessons we had learned during the turnaround to our next mission. The combination of our technical successes and President Kennedy's vow to land a man on the Moon in this decade spurred us on. When I read Kennedy's speech, delivered to a joint session of Congress on May 25, 1961, "A Special Message from the President on Urgent National Needs," I found it almost impossible to grasp that our nation had established the lunar target as the prize in the space race. I had always yearned to be involved in an undertaking that would challenge the imagination of man. Any doubt I ever had about moving into space vanished.

To those of us who had watched our rockets keel over, spin out of control, or blow up, the idea of putting a man on the Moon seemed almost too breathtakingly ambitious. Word about the speech spread like wildfire through our offices at Langley; all work virtually came to a halt and people began to offer various opinions. Most wondered if this was for real. It seemed, at that moment, like a pipe dream. I thought, "Well, let them get on with their great plans; I'm gonna get a man into orbit first." But it became real for us in the months that followed, particularly when Kennedy gave a speech sixteen months later at Rice University affirming his commitment to a lunar landing. I saw Kennedy when he came to visit Mercury Control at the Cape with Shepard and Glenn. His energy and charisma were electrifying; he made believers out of all of us, even the most skeptical. Our hopes had been renewed; maybe Kennedy really understood the towering odds we faced — and were willing to overcome.

When Kennedy first announced this ambitious goal we were back at Langley, preparing for the second manned Mercury mission, which would be flown by Gus Grissom, who had been Shepard's backup. For a change we had a breathing space of almost two months before we deployed to the Cape. Now that we had flown Shepard successfully, the pressure eased for a while. Then, on July 21, 1961, at 7:20 Eastern Standard Time, Grissom made the second suborbital flight on another Redstone, his capsule bearing the name *Liberty Bell 7*. His flight was virtually a carbon copy of Alan's. This launch doubled our manned time in space; we reached one half hour.

Grissom was a likable guy, seasoned, decisive, and taciturn. An Air Force cadet at eighteen, he had flown a hundred combat missions in Korea. When he first arrived in Korea, he found that until they had been shot at by a MiG, pilots were not allowed a seat on the bus to the flight line. Gus stood only once. On his first mission, he went looking for a MiG, found one, shot it out, and qualified for a seat on the bus.

Everything went well during the flight but shortly after Grissom's splashdown, the hatch inadvertently released. With the hatch open, the capsule started to take on water. In his silver space suit, Gus fought his way to the surface and swam a short distance away. In Mission Control, we sat helpless as a nearby helicopter grappled the loop at the top of the capsule to prevent it from sinking. By the time the helicopter took up the slack in the lift cable, the capsule had partially submerged. With the capsule under water and sinking fast, a tug-of-war began between the helicopter and the sinking spacecraft.

Almost unnoticed, Grissom was struggling as the downwash from the chopper's rotor blades whipped the sea to a froth. I was sweating bullets, as I recalled the times I stood near the runway, or sat in my aircraft watching and listening as a squadron pilot called "Mayday . . . Mayday" and attempted to nurse a damaged or flamed-out (no engine power) aircraft back to the runway. The feeling of utter helplessness gets you in the gut. I had seen a student pilot in jet training try to make the runway, finally ejecting too late when he realized he was short. I had that same sick feeling now watching Gus struggling in the water. Water was leaking through the suit inlet hose fitting, and as the seconds passed the space suit buoyancy was no longer supporting him. I kept murmuring aloud at the console, "Dammit, get Gus, forget the damn spacecraft." As a second helicopter went to Grissom's rescue, the drama of the capsule recovery seemed to play out forever on our television sets in the control room. The helicopter strained against the load, briefly lifting the capsule almost clear of the water, only to lose the battle as the engine began to overheat. (Through a remarkable feat in deep sea exploration technology, the capsule was located and recovered in extraordinarily good shape in July of 1999.)

The *Liberty Bell 7* mission was perfect until the very end and then it turned to worms. Rookies and veterans alike were once again reminded: there are no free rides in a flight test.

A rumor soon circulated that Gus had panicked when a small amount of seawater entered the capsule. I thought that was baloney. Everything I had seen of Gus and the astronauts indicated that they had the "right stuff." Grissom's hatch was the first flown with an explosive primer. Shepard's had had a mechanical design. Something must have screwed up in the new hatch.

I returned to Virginia from the Cape the next

day, and Marta greeted me with a sigh of relief, and a holiday of hugs and kisses. "Thank God, you're here," she said. "I didn't think you'd ever get that thing off the pad. How did you make it so fast? Did you write your report on the plane?"

As a rule, we spent most of a day writing the post-mission report and did not leave until it was finished. I made quick work of this one. I had to because I was racing the stork. The next day I drove Marta to the hospital and, after a brief countdown, we were blessed with our third daughter, Joan Frances. Marta's timing was impeccable. But she didn't leave much to chance. With the births of each of our children, she had labor induced to assure that I would be on hand.

That happy event was in stark contrast to my on-the-job progress. Anger has always served me well, and I had been smoldering with anger and frustration for months at the Cape. I was largely angry with myself. I knew what to do but did not have the hands-on knowledge to do it. I felt I was not carrying my share of the load. My first step was to learn every detail of every system in the Mercury spacecraft and boosters. Then I intended to help Kraft train members of the mission teams to make sure we had the competence and knowledge that I knew we needed to carry out our mission.

I believed Kraft had several problems looming on the horizon. His systems controllers were good engineers, but as I had learned at Holloman, engineers needed to become operators. There was a hell of a difference. An engineer can explain how a system should work (in theory) but an operator has to know what the engineer knows and then has to know how the systems tie together to get the mission accomplished. If the systems break down the operator must make rapid decisions on fixing or working around the problem to keep the mission moving.

We also had to change the way we were thinking. We had not crossed the bridge from an aircraft test mentality to one suited to space flight. Aircraft pilots have an option to get to a runway or else to eject. In space, we were always hours away from a landing site. If there was some kind of glitch, Mission Control had to be prepared to develop options to keep the spacecraft going until it could be returned from orbit.

The third problem that I identified would take time to solve. Mission Control was a new idea to the astronauts. They responded directly to Kraft because of his authority, but were generally cool to the controllers. They would double- and triple-check what the controllers said and did. We had to earn their respect and trust. To do that we had to be smarter than they were in each of our technical specialties, and we had to be utterly precise and timely in every action. This was a hell of a chore when we were writing the book as we went along.

After returning to Langley from the Cape following the Grissom mission I was assigned the responsibility for the remote site teams, with eight NASA civil servants and sixteen technical reps from the pioneer electronics company

Philco. The technical challenge was to develop the remote site team skills sufficiently so that any one of the site teams could take the actions required to keep the mission going when things went wrong. The Philco personnel were from across America, virtually self-taught with an incredible mastery of communications, electronics, and data systems. They had staffed the tracking stations in Alaska, California, and at Kwajalein in the Marshall Islands in the South Pacific for the early Air Force satellite missions. The civil service recruits were mostly fresh college graduates. The shotgun wedding between the young scholars and the Philco controllers was just what we needed to bring instant maturity and poise to the fledgling remote site teams.

The bible on the Mercury capsule was the pocket checklist, a five-by-seven-inch set of schematics sized to fit in the pocket of the astronaut's flight suit. I knew from my own flight test experience that this wasn't adequate. Based on everything I had learned at Holloman, you lived or died by the data at your fingertips. This slim volume was the sum total of our knowledge, and it reflected the naïveté of the engineers who produced it. I wanted to build a data set like the one I had had at Holloman. I got two engineers from McDonnell assigned to work directly with me at the Cape and at Langley. They would get the engineering data and test reports for me from McDonnell Aircraft's St. Louis plant and help me boil it all down to information that was useful and instantly accessible.

Kraft was skeptical but gave me the benefit of

the doubt. Two fine engineers, Ed Nieman and Dana Boatman, joined my remote site group. I would have my people train themselves in the same way I came up to speed as a flight test engineer. Each controller would research a capsule system to get a deep understanding of it, with the contractor's help. When you are sitting at the console, you want a set of handbooks with certified correct data, formatted so the information can be accessed and used in seconds. We needed to separate the "nice to know" from the "must know." We had to get the data that would enable us to work problems that were workable, and discard all the data that applied to problems that could not be solved by the crew or controller.

To manage this large mass of information, I assigned a single capsule system to each controller. He had to grind through it and digest it, put it in a readily usable format, and cross-check it with test reports and specifications. We built a comprehensive handbook out of this data. Once we had the words on paper, each controller taught the entire team about the system he had studied. This gave us a common frame of reference among the crew, the remote site controllers, and those in the control center. Once we got the data right then, in similar fashion, we wrote the troubleshooting procedures the crew would use, and from there we moved on to codify all the operational rules.

The potential for a superb real-time team was realized as the controllers fanned out to the remote tracking sites. Each mission was a final exam for the controllers; some passed the test and some didn't. Those who survived the early missions became leaders in the rapidly developing art of space flight operations.

September 1961, Mercury-Atlas 4

The next mission would use the capsule recovered from the flight that had been snuffed by the RSO on the third Mercury-Atlas mission. We had lost the booster but saved the capsule. When the destruct signal was sent, the escape tower fired and pulled the capsule away from the explosion. Then the tower separated from the capsule, the parachutes deployed, and the capsule landed safe and sound in the ocean.

Mercury-Atlas 4 was a simple orbital test of the Mercury spacecraft and control teams prior to a three-orbit mission with a chimpanzee. But there was a greater significance. This would be the first worldwide mission deployment since the global network had been completed five months earlier. For us, it was like planning the Normandy invasion. Thirteen teams of three controllers were sent, in those days mostly by slow prop-driven aircraft, to the distant corners of the Earth to function as self-contained mini-control centers. The teams had to be self-sufficient, so they carried everything needed for mission support, often including maintenance manuals and spare parts. The teams were mostly made up of men in their twenties; only one of the CapCom's team leaders had been overseas before.

The controllers assigned to the converted, aging merchant ships deployed first and had the toughest and loneliest assignments, wallowing haplessly in the ocean swells and occasionally hanging on for dear life during severe storms, fighting forty-five-foot waves and losing deck gear and communications antennas in the process.

The African sites were extremely risky. The young controllers found themselves in the middle of internal conflicts and were not warmly welcomed by many of the local citizenry. But the mission could not be done without them manning these remote and isolated sites. Throughout much of the mission, the remote tracking stations were the only sites in radio and telemetry communications with our crews. If communications were good with Mercury Control, the sites would report capsule and crew status to the MCC flight director and then receive instructions. If the communications were poor, as often was the case, they were out there on their own, improvising as they went along, dealing with often marginal technology and keeping their cool when things were falling apart.

A prime example was John Llewellyn, a stocky, square-jawed former Marine and an early member of the Space Task Group. He was chosen to lead the control team at the Zanzibar site, on the large island off the east coast of Africa in the Indian Ocean. John had fought through the First Marine Division's bitter retreat from the Chosin Reservoir in Korea in

December of 1950. After discharge he went to the College of William and Mary. A couple of years older than the rest of my controllers, John thrived in the crisis atmosphere of Zanzibar and drank many a toast with the Gordon Highlanders, the British unit stationed there during the period when, as British Prime Minister Harold Macmillan put it, "winds of change" were blowing through Africa.

This first deployment of a fully global network was a time of trial, the first test during a mission. Hundreds of people were learning to work together as a team for the first time and had had only a few weeks to put it all together. Things were particularly tense in Zanzibar, which was in turmoil. After a particularly bad day of rioting there the day before launch, I received a cryptic Teletype from the site: "MY TEAM WILL REMAIN AT SITE UNTIL TOMORROW'S TEST. WE HAVE THEM WHERE WE WANT THEM. SIGNED, LLEWELLYN." It took me a while to realize that John's team was surrounded. John was not going to take a chance that he could not get to the site the next day to support the mission. (Many of those who survived the hazards of remote site conditions would become the leaders in the Apollo program.)

After we accomplished a successful one-orbit mission, the remote site control teams returned to Langley to prepare for John Glenn's orbital flight, scheduled for December 19. The final tune-up with the network and control teams was a planned three-orbit mission with a chimpanzee named Enos.

Walt Williams was the toughest leader I have ever known. A bit of a brawler, with the build of a barroom bouncer, he manned the console behind Kraft's, surrounded by a blue haze from his chain-smoking. Walt had to be a tough customer — over the years he would have Presidents, members of Congress, the media, and prima donna astronauts to contend with. Because he had to be on top of everything, briefing him was a scary experience. God help you if you were unprepared. Here was a man who must have never slept. In the middle of a briefing he would start snoring, then grab a handful of his favorite Necco mints, chomp a few, and go back to sleep. Then, after the briefing, he would crisply summarize every key point in your presentation, isolate the important issues, and say, "Okay, let's do it."

On the crucial day when we were sending up Enos for the pre-Glenn shot, the launch countdown was ragged, with numerous holds for missed procedures and repairs to the data system. The telemetry was noisy, the glitches were visible on the console meters and recorders. I think Williams was moments away from calling it off when the Atlas finally lifted off. Then, after liftoff, Gus Grissom, at the CapCom console in Mercury Control, used his override switches to correct erroneous liftoff and booster engine shutdown data that had been fed into the Goddard computers by noisy data lines from the launch pad.

During the second orbit, several of the sites saw an increase in cabin temperatures and unexpected attitude control jet firings. The reports poured in by voice and Teletype as the hapless chimp sailed across Africa. The rate of fuel usage increased rapidly on the second orbit. The cabin temperature conditions were triggering an increase in Enos's body temperature. With the capsule moving five miles closer to the deorbit point each second, decision time was rapidly approaching. Kraft knew that if he got Enos home alive, even if the mission was one orbit short, he could declare it a success and clear the way for Glenn's launch. He didn't dare try to imagine the headlines if Enos was left stranded and died in orbit.

When Chris couldn't get a clear Go or NoGo from his systems monitor, he ordered the California site to go ahead with the retrofire. California then crisply reported, "Retro one fired . . . retro two fired . . . retro three. All fired and capsule attitudes were good." Enos was coming home. After confirming that the retros had fired, Kraft turned to me, smiled, and said, "That's a good show!" I knew then that Kraft would build a team, his team. The real-time role of the flight director and Mission Control had been demonstrated. For the first time, the control team had intervened in a deteriorating situation, made time-critical decisions, and saved the mission.

And now the team would have a new, permanent home: the decision had been made to relocate the controllers, the crews, and their families to what would become the permanent Manned

Spacecraft Center in Houston. Gilruth's staff and the logistics and engineering people would be the first to move in the new year. Then when they were in place, Kraft's operations organization would begin relocating. I could not believe we would relocate the operations teams at the peak point of manned missions. But the launch pressure eased slightly when the agency backed off the December 19, 1961, date for John Glenn's first American orbital flight. A postponement of up to five weeks would allow time for refurbishing and upgrading the launch pad and give the flight teams a needed rest. We had flown seven missions in 1961, weathered a few crises, put Shepard and Grissom into space, and brought them back alive. It was one hell of a first year for America's manned space program, but we still had a long way to go. We wanted to be the leading rather than the trailing edge in spaceflight.

January 1962

Trouble with hardware caused NASA to scrub Glenn's flight, reset for January 27. We returned to Langley and landed in a blinding snowstorm. As we got off the aircraft, we were told to return to the Cape; the Atlas problem had been found. Air travel was out of the question, so we were driven to Richmond, Virginia, where we boarded a train and rode through the night to Orlando.

During their few hours at Langley, several of the controllers had picked up their wives and children. After getting the order to return, they had the wives and children frantically pack for the trip to the Cape. Our families met us at Langley to get on the buses that would take us to the train in Richmond. The wives and kids were assigned chair cars. The rest of us had no places reserved to sit or sleep, so we sat through the night in the dining car, playing poker or rehashing how we would resume our support to the launch countdown. By the time we got to Glenn's mission, we started bringing families down to stay during the time we were deployed at the Cape. It worked out pretty well; we would work through the night and return to our motel and go to bed in the early morning hours. The wives would take the kids to the pool for the day while we slept. Then we would get up late in the afternoon, find some food, go to work, and the wives and kids would take our places in the motel rooms. The only trouble was that the per diem we got for deployment to the Cape was so meager that after our families left and we were just down there on our own we slept two guys to a room to make ends meet.

After all this, the countdown had to be scrubbed again, this time for repair of a leaking tank bulkhead on the Atlas. It seemed that we would never get Glenn's rocket off the launch pad.

Back to Langley. Next deployment to the Cape was for a Valentine's Day launch, and, once more, we scrubbed it, this time for weather. At the Cape, we were focused and on top of the job, but the ups and downs of the launch preparation were nibbling at us. Tempers started to fray. The delays were hard to endure; we knew the stakes were enormous, both for the country and for John Glenn, but we were eager to move to launch and ready to go at a moment's notice.

February 20, 1962, Mercury-Atlas 6

When I look back, I find it hard to believe that when we launched John Glenn we had had a total of three orbits' worth of experience during the two preceding missions. Two of the Mercury-Atlas rockets had failed. As formidable as the Atlas appeared, it was essentially a pressurized metal balloon. If pressurization was lost while the rocket was on the pad, the rocket would have collapsed. We were rolling the dice in a way that would not be allowed in today's space program.

After ten postponements and eighty-two days of delays, this countdown had the usual glitches. The pad crew broke a bolt in the hatch while inserting Glenn and had to scurry off to get a new one. Watching from Kraft's TV, I was amazed as always by the close quarters of the Mercury capsule. The cockpit was smaller than any fighter aircraft I had flown. Glenn entered the capsule feet first, through a hatch less than two feet square, assisted by the pad crew, then ducked under the instrument panel. In the silver space suit, gloves, and helmet, Glenn was the modern-day explorer ready to embark to a new world.

During the hold, Kraft had his customary pint

of milk to settle his stomach, and Williams made increasingly frequent trips outside to look at the weather. After the hatch was repaired, the countdown resumed.

Williams polled the mission team, receiving a solid Go from Kraft and the Mercury capsule and Atlas booster test conductors. With each succeeding Go the energy level in the control room threatened to burst through the doors. Scott Carpenter called from the blockhouse, "God speed, John Glenn." Then he counted down: "Three seconds . . . two . . . one . . . zero."

As the umbilical cables were ejected, I filled in the liftoff message and started a continuous voice briefing for the Bermuda site team. My hand was trembling as I jotted 14:14:39Z on my notepad. When I heard Glenn's report, "Roger, the clock is running, we are underway," I took off for the Teletype room. This time, I remembered to remove my headset.

Returning, I continued to brief Bermuda via the Teletype order wire until they acquired data at three minutes. Other than during the high-G periods, Glenn's reporting was just like his training runs. I consciously had to fight to keep doing my job and avoid being mesmerized by the words.

Shortly after five minutes, Glenn calmly said, "Sustainer engine cutoff, and the posigrades have fired." The posigrades provided the thrust to separate the capsule from the Atlas booster.

The room hushed as Kraft and Tec Roberts engaged in the Go NoGo dialogue. The answer came swiftly and, after checking with John

Hodge, the Bermuda flight director, Kraft nodded to Shepard, who gave Glenn a resounding, "Go! You are Go for at least seven orbits." (The orbital trajectory was designed to provide more orbit lifetime than needed for the mission plan in case we had problems and could not come down at the planned end of mission time.)

Glenn, now in the element of the test pilot, was reporting the control system status, closing out the checklist items, and continually reassuring the doctors that he felt fine. It was a jam-packed ten minutes, but everything had gotten done as *Friendship 7* left Bermuda for the Canary Islands site, the next in our chain of tracking stations girdling the Earth.

For the first time I was nervous. My site controllers were going to see a manned spacecraft for the first time, and now they would have to communicate with an astronaut, run the site, and direct their team. This was it, real-time and pretty heady work for a bunch of young guys. Communications were clear and I listened intently as Glenn reported his status to Llewellyn: "Control check complete. I have the booster in sight out the window. It's probably about one mile away, going down under my position and a bit to the left."

Llewellyn stammered a bit in his first communication, "*Friendship 7*, what is your space . . . spacecraft station . . . status report?" I had to smile that even a tough, hard-ass Marine gets clanked up at times.

After all of the attempts to get Glenn off the

ground, this one seemed unbelievably easy. When it finally happened, it was smooth as apple butter. John kept clipping through the first orbit without a glitch. Teletype messages and occasional voice contacts indicated that the controllers' adrenaline was pumping. It was difficult at such times to maintain my focus. I felt a strong urge to yell out, "We've got an American in orbit!" My remote site controllers were working well and the lines were humming. I felt a burst of pride for our team and how it had progressed in the fifteen months since we had bungled our Four-Inch Flight back in November of 1960.

I had sat next to Chris Kraft since the first mission and was amazed at his aplomb now that he finally had an American in orbit. High-risk leadership beckons many, but few accept the call.

The Teletype post-pass reports filled the gaps when we could not listen to Glenn. The messages, when pieced together, indicated he was on the flight plan and the capsule was performing well. At the completion of each site pass, controllers prepared a systems summary message containing the values of sixteen key systems measurements, recorded during the pass. The short time intervals between tracking sites did not allow the MCC and site controllers to plot the measurements, so the Teletype messages were cut into strips and aligned with the measurements from the preceding sites, then taped together. Trend predictions from this data were rough but this was the best we could do.

The control team at Canton Island in the

South Pacific received Glenn's report of a brilliant "bright red" sunrise, then were startled as he continued, "I am in the middle of a mass of thousands of very small particles that are brilliantly lit up like they are luminescent. They are a bright yellowish green, about the size and intensity of a firefly on a real dark night. I have never seen anything like it. They look like little stars. They swirl around the capsule and go in front of the window."

I showed the Teletype of Glenn's finding to Kraft. He nodded, then said, "Keep me advised."

Other than modest increases in temperature, there were no concerns as we approached the end of the first orbit. The only unusual occurrence was Glenn's report of the "fireflies" he spotted at sunrise over the Pacific.

Before the Cape pass, Kraft walked over to me and said, "Shorty [Powers] has confirmed that President Kennedy will make a call through the Cape at the end of the first orbit. Get with the communications people and make sure everything is set up." When caught off guard I tend to be inflexible, particularly when I am distracted from the business at hand by details that could have been worked out earlier. This was one of those times. Kraft saw the expression on my face, frowned, and then chided, "The President is the boss!"

The presidential call did not surprise the audio technicians; they had been advised of the call by the White House switchboard the night before launch, and I was assured everything was already checked out. I had a lot to learn about the politics of space.

As the capsule passed over the Cape, Shepard, the Cape CapCom, called, "*Seven*, this is the Cape. The President will be talking to you." Caught by surprise himself, Glenn stammered, "Ah . . . the President? This is *Friendship 7*, standing by."

Shepard said, "Go ahead, Mr. President." The communications loop was dead; the call came early and the phone line was not yet patched in.

George Metcalf picked up the phone when it rang at his backroom console. He thought it was a gag when the voice at the other end said, "This is the White House, stand by for the President." Attending other duties and unaware of the planned phone call, George stuttered, "Hello, hello, Mr. President!" Then Metcalf stood up, wildly gesturing for other technicians to come help him set up the patch.

In the control room, a more crucial event now intruded on the team's attention. A warning light had flashed on the instrument panels in Mercury Control. Moments later, Don Arabian, the systems monitor, called out, "Chris, I don't know what to make of this, but I am showing an indication on Segment 51."

Kraft looked perplexed. I overheard the call and immediately pulled out the telemetry listing from my console drawer. "Chris," Arabian went on, "Segment 51 is the impact [landing] bag deploy."

Quickly, I called John Hatcher, the facility

team boss. "John, forget the [President's] phone call and verify the patching of Segment 51." Then I broadcast an all-site Teletype message: "Confirm patching instructions for spacecraft telemetry and report the readings on Segment 51, ASAP." The chilling implication of the telemetry that Don Arabian was reporting was that the impact bag had deployed — which in turn meant that the heat shield had somehow come loose. The heat shield protects the capsule from the fire of reentry. After reentry is complete and the parachute is deployed, the heat shield is released. A rubberized bag which is attached to the capsule structure is stowed behind the heat shield. After the capsule's parachute deploys the heat shield is automatically released and the rubber landing bag extends to cushion the landing impact. On the water the landing bag acts like a sea anchor to stabilize the capsule in the upright position and minimize drift.

If the telemetry indication being reported was correct and the heat shield had come loose in orbit John Glenn would have no protection from the 3,000-degree F reentry temperatures. The capsule would become a meteor that flashed for but a few brief seconds during reentry before burning up.

Kraft and the team were now faced with a grim set of choices. Controllers, distrustful of solitary measurements, immediately started digging out the details of the switch and how it was rigged. The phone system came alive as the problem was pursued. Precious time was lost trying to track down engineers at the blockhouse and Hangar S.

Kraft's controllers had no provisions for emergency access to the total design, manufacturing, and assembly team. I jotted a note to myself to set up a hot line to McDonnell Aircraft for the next mission — if there was one.

Unaware of the crisis unfolding around him, John Glenn coasted over the Atlantic. He was oblivious to the uncertainty over Segment 51, but he was now having an unrelated problem with the attitude control. The capsule was drifting sideways to the right until it hit the attitude limit, then the big yaw thruster would kick in.

The mood inside Mercury Control had changed with the suddenness of a thunderclap. Metcalf, still holding the line to the White House, found no easy way to disengage. "Mr. President," he blurted, "we've gotten pretty busy down here now. I don't think we've got time to talk." The President responded, "Give me a call if you get a chance." George hung up and turned to more urgent problems.

As the spacecraft passed over Central Africa, Glenn reported that he was departing from the flight plan and was troubleshooting the attitude control problems. It appeared to him and those on the ground that the problem was a random movement from left to right, possibly caused by one of the small thrusters working intermittently.

Glenn interrupted his troubleshooting with reports on the flight plan as he moved across the Indian Ocean, continuing his assurances to the doctors that he felt fine. The medical community's anxieties, although reduced by Yuri Gagarin's one-orbit mission, persisted. Kraft and the rest of the control team were elated with Glenn's performance. It was obvious that he was on schedule and had no problems adapting. As John passed over Australia, he embraced the attitude control problem with an energized can-do spirit.

On the ground the story was far different. The blockhouse had reviewed their measurements and verified that they had had no Segment 51 indications during launch. But when the spacecraft passed overhead at the end of the first orbit, the blockhouse had seen the indication also. The data I was receiving from the remote sites was not much help. Half of the CapCom reports indicated they were seeing Segment 51, and the others were not.

The limitations of the Mercury communications system now became evident. The response to each of the Teletype queries took ten to fifteen minutes. The answers often prompted another query.

Chris was now on the phone with the blockhouse, trying to get answers, while I had gone to a small office behind the Teletype room seeking out Ed Nieman and Dana Boatman, the two McDonnell engineers who were developing the system schematics. Both of us were seeking answers to the same question: if Segment 51 is valid, what are we going to do about it?

Chris was not having much luck and his frustration level was rising. The design engineers, in the heat of real-time crisis, weren't quick at coming up with options. One of them suggested, "Let's make sure Glenn keeps the landing bag switch off and we should ask him if he hears any banging noises when he maneuvers at a high rate."

Kraft relayed the messages to Gordo Cooper in Australia, who first verified the landing bag switch position. Cooper asked, "John, you haven't heard any banging noises or anything of that kind, have you? Maybe when you maneuver at a high rate?" Glenn replied, "Negative," and let the question drop. If he found it strange, he gave no indication.

Walt Williams, Max Faget, the capsule designer, and John Yardley from McDonnell now joined Kraft at the console. In short order, the words came tumbling out, like objects falling from a filled closet. Faget proposed to hold the heat shield in place for the early part of reentry with the retropack.

The retropack is a cluster of three solid rockets used to reduce the orbital velocity sufficiently to allow gravity to pull the spacecraft into a safe reentry trajectory. The timing of the retrorocket firing and the spacecraft attitudes are key to hitting the planned landing zone. The retropack is located behind and in the center of the heat shield and is attached to the capsule by three metal straps. After the retrorockets are fired, an electrical signal pyrotechnically cuts the three straps and a small spring pushes the empty retropack away from the heat shield. With the capsule oriented blunt end forward during reentry, the heat shield protects the capsule by dissipating the heat as the capsule enters the atmosphere. The heat shield literally melts away (ablates) on reentry. If the heat shield was not firmly attached to the capsule the aerodynamic forces would tear it off during reentry, leaving the capsule unprotected. If after retrofire, however, the retropack was not jettisoned, it was in a position to hold the heat shield during the reentry until the retropack melted away. When the retropack melted it was believed the aerodynamic forces would hold the heatshield in position during the remainder of the reentry.

I looked around the room and saw faces drained of blood. John Glenn's life was in peril. We were desperate to find a solution, without being sure we knew the problem.

Kraft said, "This is the wrong way to go. It's too damn risky for something that is probably an instrument error." Fighting to avoid a premature decision, Kraft fired off a barrage of questions: "Does anyone know the aerodynamic effects of reentering with the retropackage attached? Do we have sufficient control with the attitude jets to keep us oriented in reentry attitude? Will we damage the heat shield to the point where it cannot protect the capsule as the retropack melts away during reentry?" Faget and Yardley scratched their heads. They were well beyond the bounds of their design knowledge.

Faget, quietly muttering in his Cajun accent, tried to reassure Kraft. "Chris, it should be okay. We designed the heat shield with plenty of margin." His words did not sound convincing. Kraft's gut feeling indicated just the opposite.

He believed Segment 51 was a false telemetry indication and the risks of an unproven, untested entry technique with the retropack were too high. Kraft exclaimed, "Dammit, we've got to find other pieces of data to confirm this before we jump to the conclusion to enter with the retropack."

Slayton and Shepard had plugged into Kraft's console for the telephone conversation with the blockhouse. Returning to his console, Williams leaned over and asked, "What should we tell John?" Kraft ignored the question for the moment.

Faget, the consummate engineer, could not ignore the telemetry. Engineers like Max live by their data. Asking one of them to consider that the telemetry information may be wrong borders on the heretical. Kraft was at the opposite end of the spectrum. "Max, dammit," he barked, "we only got one piece of instrumentation. My guys tell me it would take a dual electrical failure for the heat shield to come loose. The way it is stacked, a mechanical failure is out of the question."

Yardley, who had been conferring with Williams, posed the fatal question: "Chris, what happens if it is valid?"

With only a single data point, the discussion was at an impasse. Williams joined in: "If we come in with the retropack attached, what is the worst thing that can happen to us?" (I could imagine Glenn's reaction, if he only knew: "What's this us crap?") No new information was forthcoming, but at least there was now a grudging acceptance for Kraft's position that the telemetry indicator could be wrong.

The debate at the console broke off again as Cooper came on the conference loop to offer his impression of the problems reported by Glenn over Australia. Kraft knew the discussions were bothering his controllers on the floor and motioned the entourage to return to their seats. I noted the clock. The mission was half over and the team needed to make a decision in less than two hours. I made one more trip to my McDonnell engineers, who had been on the phone back to the plant in St. Louis. I doubted they would have any answers prior to entry.

Kraft was again seated at his console when I returned. For a few moments, it seemed that everything had settled down. My console was flooded with Teletype messages and I scanned the most recent arrivals first. The message from the Canton site brought me to my feet and I handed it to Kraft. The message said:

> ALL
> DE CTN
> FM CAPCOM
> CTN ADVISED S/C NO INDICATIONS LDG BAG DEPLOY.
> S/C RESP DID SOMEONE ASK IF THE LDG BAG WAS DOWN?
> CTN RESP WE HAVE BEEN ASKED TO MONITOR LDG BAG AND ASK IF YOU HEARD NOISES WHEN MANEUVERING.
> S/C RESPONSE NEG.

Teletype messages between the tracking stations and MCC are always abbreviated and terse to save transmission time. This message was painfully clear. The Canton Island CapCom (CTN) had inadvertently informed Glenn (S/C or spacecraft) of our concerns.

Kraft had staked out his position long ago: "I don't worry about things I can't do something about." But in this case, he was worrying a great deal. The impasse on entry techniques bothered him. Left to his own he would press on with a normal entry, ignoring the alarm, but Yardley and Faget were two damned good engineers arrayed against him. Williams also had flight test savvy. They were all telling him he should take a different path.

Still, Kraft had picked up the scent. He believed he was on the right track and wanted to buy more time. "I want to give John a complete story," he explained, "and I need more answers."

Over Hawaii, Glenn was given the Go for the final orbit. After the discussion with Canton, I was surprised John did not mention his conversation with the CapCom. During his last stateside pass, he continued discussing his attitude control problems, and we provided him with a recommendation for backing up the automatic attitude control.

My roommate, Carl Huss, updated the capsule retrosequence clock to the correct retrofire time for the planned end of the mission landing area. The clocks would automatically initiate the retrofire sequence if the spacecraft attitude was correct when the clock timed out. The mission now coasted into the third and final orbit. Under other circumstances, Huss and Tec Roberts would have been delighted. Their tracking data was solid and the planned retrofire times had not changed by even a second during the last orbit. With the decision time now down to less than an hour, I believed we had all the data we were going to get. Kraft and Williams were facing the lonesome task of deciding what to do. John Glenn's life, the Mercury program, and America's future in space were in the balance.

Looking back on this episode, and the other Mercury missions, I find it hard to believe that we did so well. The systems operators did not have the benefit of the massive analytical tools available today. The only computing resources available during the mission were used to process radar tracking data. Compared to the present technology, our computers were the equivalent of a rusty adding machine.

A controller lives or dies based on the information he has at his console. If you lack what you need at liftoff, there is little hope that you will get new information that you would trust during a mission. This realization was the most profound impression branded on me from the Glenn mission. During the final orbit I witnessed the agony and the frustration faced by the controllers and engineers wrestling to help Kraft make the best decision.

There was no right decision that day, nothing in black and white. We could only try to obtain the best answer. At that moment, I also realized that learning by doing was the only way a controller could ever become smart enough to succeed in the tough and unforgiving environment of spaceflight operations.

The last orbit was a stalemate. No more data was coming. The best judgment of the engineers was that there was sufficient attitude control for reentry with the retropack attached. The straps would burn off during entry and should not induce any landing position errors. Kraft restated his position: "It is an instrumentation problem. The heat shield is still attached. If we burn a hole in the damned heat shield we are going to kill Glenn!" Williams, rising to the emotion of the decision, chimed in: "Chris, if you're wrong we are going to kill him, too."

The engineers cautioned that if we kept the retropack attached, we needed to confirm that all three retros had fired. If one did not, there was a good chance it would detonate during reentry. Kraft did not comment on this last prediction.

Kraft was still holding out until the last moment, so that he had a complete understanding of the final instructions before he radioed up to John Glenn. The mission was turning into a horse race. Kraft wanted answers from one final test to be performed over Hawaii before he turned the discussion to the entry procedures modifications.

At capsule acquisition, the Hawaii CapCom advised Glenn, "*Friendship 7*, we have been reading an indication on the ground of Segment 51, which is heat shield deploy. We suspect this is an erroneous signal and would like to check it out. Place the landing bag switch in auto and see if you get a light."

Glenn responded, "Negative. In automatic position did not get a light and I'm back in the off position now, over."

Kraft turned again to Williams. "Walt, this is the best damned data we can get. The test was negative. We should go ahead with the normal reentry sequence."

Without waiting for a response, Kraft advised Hawaii, "Tell Glenn we will go ahead with the normal reentry sequence."

Kraft's instruction to the Hawaii CapCom surprised Williams. Still not satisfied that the test was valid, Walt continued to question the engineers over the telephone. He was getting a mixed input — the design engineers had conflicting feelings. Stormily closing off the final conference call, he said, "If reentry with the retropack is safe, what do we lose by coming in with the pack on?"

While the debate continued, Glenn had now made contact with the California site, and Huss started the countdown to retro sequence. The count was relayed to the capsule at California by Wally Schirra: "Five . . . four . . . three . . . two . . . one . . . MARK."

Glenn responded, "Retro sequence green."

Thirty seconds later the retros fired and John Glenn was heading back to Earth. All that was needed was the final decision on whether to retain the retropackage to keep the heat shield in place during the reentry. Schirra said, "Attitude looked good, John. Keep your retros on until you pass Texas."

Glenn asked several times during the pass about the retropack jettison time. Schirra advised him, "You'll get the final word over Texas."

Kraft called Schirra: "California, can you confirm that all three retros fired?" Wally: "Affirmative, Chris."

In his role as the operations director, Williams leaned toward Kraft, and quietly but firmly said, "That settles it, we're coming in with the pack on." Kraft nodded. Williams was the boss and the final decision was made.

Chris got on the voice loop to the Texas CapCom: "Tell John to keep the retros on through entry."

I had anticipated the decision and had a Teletype message with the procedures already at the site. Glenn would have to override the .05G signal, which changed capsule attitude control modes when capsule reentry accelerations were sensed, and he would need to retract the periscope manually. I handed Kraft the message. Alan Shepard, the MCC CapCom, standing in front of Kraft's console, nodded affirmatively at its content. Kraft then asked the Texas CapCom, "Texas, do you have the message on entry procedures?"

At acquisition, Texas called the capsule. "This is Texas CapCom, *Friendship 7*. We are recommending that you leave the retropackage on through the entire reentry. This means that you will have to override the .05G switch expected at

04:43:53. This also means you will have to manually retract the scope. Do you read?"

Glenn's response was tart. He wanted answers. "Texas, *Friendship 7*. What is the reason for this? Do you have a reason? Over."

Caught in the middle, and without the benefit of the discussions in Mercury Control, Texas passed the buck to the Cape. "*Friendship 7*, Texas. Cape Flight will give you the reasons for this action when you are in view."

When Glenn passed over the Cape during reentry, Shepard calmly recommended the periscope retraction. He then added: "John, while you're doing that, we are not sure whether or not your landing bag has deployed. We feel it is possible to reenter with the retropackage on. We see no difficulty at this time with this type of reentry. Over." Glenn's response on hearing it from Shepard was simply, "Roger, understand."

As the capsule plunges toward the Earth, a sheath of superheated ionized particles surrounds it, causing a communications blackout with the ground stations. This blackout is accompanied by a rapid increase in external temperatures and G (gravity) buildup. The astronaut is literally in the center of a fireball.

Glenn watched as his world turned to a very bright orange as the external temperatures reached toward 3,000 degrees Fahrenheit. Flaming pieces of metal broke off and passed behind him, and for a moment Glenn had visions that they were chunks of the heat shield, but he could only wait to know for sure. The reentry Gs had him virtually immobilized, his body now weighing seven times his Earth weight.

Exiting communications blackout, the spacecraft started to oscillate, and Glenn tried every control mode. As the oscillations started to diverge, and he could finally lift his arm against the G forces, he reached up to deploy the drogue parachute just as the automatic system sent the command.

We listened in Mercury Control as the final events unfolded. I continued the Teletype and voice briefings for the control teams around the world. It was hard to contain my glee. Today we had put an American in orbit and returned him safely, in spite of a grave and, at the time, a life-threatening uncertainty.

The destroyer *Noa* sighted the spacecraft as it descended through a broken cloud layer at 5,000 feet for a landing five miles off the bow. The cheering sailors who plucked him out of the sea painted Glenn's footprints on the deck.

There were a hellacious number of rough spots and much to rethink before the next mission, but this was our day. There was no doubt about the team. Kraft's Brotherhood had pulled it off. The joyous chatter among the consoles as the controllers stowed their headsets and documents belied the rip-roaring party we would throw that night. I sent the final message to the remote sites. I was damned proud of my guys. They had kept on top of a spacecraft traveling at five miles a second with a low-speed Teletype network. It isn't equipment that wins the battles;

it is the quality and the determination of the people fighting for a cause in which they believe.

And, of course, it was John Glenn's day.

None of us could have predicted the emotional reaction to John Glenn's flight: parades in Cape Canaveral, New York, and Washington. Miles of ticker tape. An invitation to speak to Congress. Dinner and touch football with the Kennedys. Half a million letters and telegrams in the first month after his flight. And Glenn tried to answer them all.

He had left college to join the Marines during World War II. He was a decorated hero in Korea, a jet fighter pilot nicknamed "Ol' Magnet Tail" by his buddies because his plane took so many hits. In 1957 he set a transcontinental speed record for jet aircraft.

Glenn was simply an old-fashioned, star-spangled hero. He spoke of God and country and the flag and the bravery of his fellow astronauts, and he actually meant what he said. Even a cynic like Shorty Powers was moved to say, "This guy is for real. I'd say he's the most decent human being I've ever met."

The post-mission analysis confirmed that the telemetry reading had been invalid.

John Glenn's mission was the turning point in Flight Control and in Kraft's evolution as a flight director. Walt's direction rankled Kraft, and Kraft vowed never to be placed in a similar position again. Kraft believed his neophyte team was superior to the designers at real-time integrated spacecraft systems analysis. Learning by doing equipped the controllers with a gut-level knowledge of spacecraft design and operations. When this knowledge was combined with the multidisciplinary skills of the mission team and the integrated risk assessments developed through the mission rules, the Flight Control team had the foundation needed to succeed in the new environment of space. Flight Control rapidly became the dominant systems engineering cadre in the U.S. space program.

There was, of course, a remarkable sequel to Glenn's flight. He retired from the astronaut corps to run for the Senate, withdrawing once after an accident, losing a second race to big money, and getting elected on his third try. No one ever accused John Glenn of being a quitter. He served in the Senate, representing Ohio, then retired after four terms. Thirty-six years after he had made space history he flew again, this time on the Space Shuttle in October 1998. His second flight helped close the chapters in history books that cover the first four decades of America's space program.

4

THE BROTHERHOOD

Langley Air Force Base, Virginia

By the time of John Glenn's flight, time was our greatest enemy at NASA. We had to move ahead as quickly as possible because of President Kennedy's pledge to land a man on the Moon in the 1960s; we also knew that nasty surprises awaited us at every stage in development, no matter how hard we tried to anticipate the most remote contingencies.

Deke Slayton had been selected to fly the second orbital Mercury mission. I had quickly updated the rules based on the Glenn mission and, with the flight planners, was set to give Deke an advance look at the flight rules and the five experiments that had been added. The same scientists who had believed a man's performance in space would be severely degraded by zero gravity and other factors were now eager to have astronauts perform as many experiments as possible, within the limits of a three-orbit mission. Deke had heard the experiments were coming and voiced his well-founded objections to anyone who would listen. I believed he was right to do so. With only three orbits' worth of manned experience, most, if not all, of the team felt it was too soon to distract the astronaut with tethered balloons, fluid studies, and a variety of other observations.

I was pulling together a one-page cheat sheet for Slayton on the flight rules, when the word came down that he would be replaced by Scott Carpenter. This news shocked us all, although not nearly as much as it did Deke. I assumed the switch was because he had raised hell about all the added experiments.

But it turned out that Slayton had been scratched because of an irregular heartbeat. The problem, known as idiopathic atrial fibrillation, had been noted when Deke was being tested on the centrifuge (which simulated increased gravity and other stress factors). After an analysis of his data by NASA and Air Force specialists, he was accepted for flight. When it was his turn to fly, the NASA Administrator, James Webb, had his records reviewed for a final time. Three different groups of medical specialists gave their okay — then Webb got three civilian cardiologists from Georgetown University, Washington Hospital Center, and the National Institutes of Health to review his records and give him a brief exam. They recommended his removal from flight status. Moral: if you ask enough people, you'll find someone who will disagree with the majority and give those nervous about risk a way out. No one doubted Deke's heart when he was one of the hot test pilots at Edwards Air Force Base, pushing the

F-105 to its limits.

Slayton didn't quit the program. Few expected that he would. His initial assignment placed him as coordinator of the astronaut corps, and his first task involved the selection of the second class of astronauts. In October 1963 he was named the deputy for flight operations, putting him in charge of just about everything that concerned his fellow astronauts. It was a legitimate job and a big one, but you could not avoid suspecting that Deke had been given a consolation prize.

In the trenches at Mercury Control, we probably felt the strongest empathy for him, and for the time and energy he had spent training for a day that was now gone from the calendar and might never come again. He would be cleared ten years later and would finally make it into space as part of NASA's joint venture with the Russians. But for now, he was the first of the Mercury astronauts to be washed out, and the controllers could not view that setback with indifference.

Scott Carpenter, the backup for Glenn's mission, was a virtual unknown to most of the controllers. When Scott was given the nod, we were surprised because we had been expecting Wally Schirra, Slayton's backup, to replace Deke. The mission was rescheduled to accommodate the change in crewman, and additional attitude control tests were added to the flight plan. Glenn's mission had finally cleared the way for astronauts having hands-on control of the spacecraft in flight. John had not experienced any disorientation, and his troubleshooting of the attitude control problems demonstrated the value of having a human in control of the spacecraft at critical moments. Carpenter's flight plan was expanded to permit him to perform maneuvers to observe sunrise and sunset, fly upside down to test pilot disorientation, and conduct visual observations of Earth and space phenomena.

We had reached another milestone; two teams were working in concert, more or less, the man in the capsule and the crews in the control room. Both sides were — understandably — a little wary of the scientists.

Now the earthbound Slayton had nearly as much in common with our Flight Control team as his own astronaut corps. We in Mercury Control were like the second-string team in football, who scrimmaged all week and took the banging, but didn't get to make the road trips. Each time a rocket lifted off the pad, we felt pride and elation . . . and a little envy.

We did not mingle socially with the astronauts. Even if that had seemed a desirable thing to do, none of us had the time. When the astronauts were not in nearly nonstop training, they were flying or racing their sports cars or making public appearances to promote the space program. But as our Mercury Control team acquired as much, if not more, knowledge about the spacecraft as those who would fly them, each mission brought both sides closer together in mutual confidence — and we felt a more personal link with the crews.

The first Mercury orbital mission had been at

the core of our lives in the winter of 1961. Now, in the early months of 1962, Gilruth's newly designated Manned Spacecraft Center was moving to Houston amid a massive reorganization. This new NASA center was charged with the design, development, and flight operations for the newly formed Gemini and Apollo programs. In July 1960 NASA had announced plans to follow Mercury with a program to fly to the Moon. The program was subsequently called Apollo. The Gemini Program, which started in 1961, would bridge the technology gap between the Mercury missions and the far more ambitious Apollo lunar program. In 1962 the MSC was in a period of unprecedented growth and change. It departed the Langley Field facilities with a staff of 750. With the staff more than doubling each year it would increase to 6,000 at the beginning of Gemini operations in 1964. This rapid growth necessitated corresponding changes in the Flight Operations Division. Chuck Mathews, the FOD division chief, was reassigned by Gilruth to form a Spacecraft Research Division to develop the design requirements and the technologies needed for the Gemini and Apollo spacecraft. Chris Kraft replaced Mathews as the FOD chief. John Hodge, the Bermuda flight director, then formed a Flight Control Operations Branch. This branch had the responsibility for the mission rules development and the remote site teams and the MCC systems controllers. Hodge selected me as deputy chief. My role as his deputy did not last long. Two months later Kraft selected Hodge as his assistant, and I became the branch chief for Flight Control Operations. I now had the resources I needed to develop an operations team that was fully capable of taking any actions needed during the course of a mission.

As we were preparing for Carpenter's flight, Chris Kraft relocated to his temporary offices at the Houston Petroleum Center on the Gulf Freeway. His staff, however, remained at Langley, starting their relocation in the summer of 1962, during the interval between the flights now assigned to Carpenter and Schirra. Many drove with their families from Virginia to Cape Canaveral for Carpenter's flight, towing rental trailers containing all they possessed. After the mission, they continued on to Houston.

As the space program was rapidly expanding, more land was needed to house people and test the systems being developed. In and around Houston, we had access to water — in the Gulf of Mexico and even at Clear Lake — so we could do drop-testing of the capsules. A key factor in determining the new site was the proximity of colleges and universities, a talent pool from which we could recruit newly graduated engineers and scientists for the rapidly expanding program. (It should also be noted that Houston was in the congressional district of Albert Thomas, the chairman of the House Appropriations Subcommittee that oversaw NASA's budget. He and Vice President Johnson were loyal sons of Texas and highly effective advocates for Houston's suitability as the location for the new Manned Spacecraft Center.)

As a result, in the early years of space, many of my controllers were out of educational institutions in Texas and the Southwest, not the colleges in the Northeast that supplied many of those in the original Space Task Group. We had a few from as far north as Purdue, but they came in waves off the campuses of Texas A&M, the University of Texas, Rice University, and Lamar Tech (in nearby Beaumont).

Tec Roberts, already in Houston, was designing the new Mission Control Center for Gemini and Apollo. Glynn Lunney had stepped in as his replacement as the flight dynamics officer for the remaining Mercury missions. Glynn was the pioneer leader of trajectory operations, who turned his craft from an art practiced by few into a pure science. In the early years, I envied him for his ability to rapidly absorb complex materials and find alternatives. We competed for the leadership role, Glynn pointing the way through his remarkable grasp of the entire complex picture, while I focused on structure and team building.

Carl Huss, one of our math wizards, was training the remote site controller John Llewellyn as his replacement, and Arnold Aldrich was brought in from remote site systems to relieve Walt Kapryan at the systems console. Eighteen months after our first baby step into the world of space flight, the Mercury pioneers were sliding into new jobs and their successors were entering the fray.

In the midst of the preparation for the launch of Carpenter's *Aurora 7* mission, we were advised to take a trip to find housing for our families. We would be given thirty dollars per diem for thirty days for all expenses. By the time the allowance ran out we had to be relocated.

Manfred (Dutch) Von Ehrenfried was a new recruit who joined us as a procedures officer in time for the Glenn mission. He had been teaching high school physics when President Kennedy set the lunar goal and was itching for a piece of the action. Trying to avoid an unnecessary trip to Houston, I called him into my office and decided to give him a real test.

"Dutch, we have to get settled quickly in Houston," I advised him. "We need good, cheap housing with low down payments. Scout around and find the best place to live. We can't afford more than a $250 down payment."

Dutch did not own a home in Virginia and was as eager as any of us to get his family resettled. He did well as a real estate scout and, during the interval between missions, ten families moved into houses he picked out on Welk and Regal drives, an area in southeast Houston that came to be known in the early 1960s as Flight Controller Alley.

March 1962

President Kennedy had challenged us to go to the Moon and dispel any doubts of America's leadership, technology, and spirit. The colleges and universities responded. By the spring of 1962, we were flooded with job applications from a generation of young people drawn to the cause.

The newly created Manned Spacecraft Center more than doubled in size, from 750 staffers to 1,800, in three months. Mel Brooks and Jim Hannigan were the first two engineers I hired. Slightly older than the average controller, they had the savvy I needed to lead the young graduates through Mercury and into Gemini.

Hannigan had been a flight test engineer for the Air Force. Brooks, an infantry veteran of the Korean War, had worked with the satellite control of the Air Force's Agena upper-stage rocket, which had been selected as the Gemini rendezvous target. They were the first to relocate to Flight Controller Alley and were pressed immediately into service. Hannigan was selected as a CapCom assigned to the Kano, Nigeria, site, while Brooks led the training section for the final Mercury missions.

The training classes set up by the Philco monitors were expanded and structured to accommodate the new college graduates being recruited as flight controllers. The two-week, twelve-hour-a-day training program was a crash course in remote site flight control, providing only the most basic background for the work. At the completion of their twelve-hour day, the controllers practiced their Morse code, the last, desperate fallback for communicating. The astronauts and controllers were trained to use their mike switch to transmit the code. Then, when Morse code training was completed, they were taught speed printing so that their Teletype messages might get to the next controller's site before the spacecraft. This first formal flight controller training session was designated Class 101. I passed out the certificates to the first six graduates, one of whom was Charles (Skinny) Lewis.

One month after graduating from Class 101 and getting a crash training course, Lewis and his team were sent out to man our remote (and boy, was it remote) site in Zanzibar, the same site where John Llewellyn and his guys had earlier been surrounded by local citizens engaged in "civil disorder" (i.e., rioting). Like many of his fellow accidental tourists, Lewis had never been out of the country before and now was in a place where his military training as a tank commander in the Army Reserve came in handy. Having been briefed on the dangerous conditions in the area, Lewis, along with his surgeon and his systems controller, was returning from the site to his quarters late one night when he saw a roadblock made up of fires burning in oil drums and manned by natives not in uniform. He floored the gas and drove his Volkswagen right through a gap in the blazing obstruction. He survived that and other adventures to eventually become an Apollo flight director.

May 24, 1962, Mercury-Atlas 7

The launch preparation for Scott Carpenter's mission unfolded with the usual glitches that we had come to expect. We used three launch scrubs wisely in local training. I updated the procedures and flight rules for the next mission, reviewing them with Kraft, the controllers, and the astronauts during slack periods at the Cape.

Because of the relocation of our operations to Houston, we lost almost a month of preparation time and we had to get ahead of the power curve. But finally the day came when we were ready to light the fire on Mercury-Atlas 7. (The Mercury spacecraft were given names by the astronauts, but the Mercury-Atlas or Mercury-Redstone designation was used by the launch and flight teams. The word "flight" referred to the spacecraft and booster events from liftoff to landing in the Mercury program. During the Gemini and Apollo programs, the term "mission" was used by the program, launch, and mission teams. "Mission" carried the collective connotation needed when dealing with two or more spacecraft or launch vehicles. Although this was the standard, the terms "mission" and "flight" were often used interchangeably.)

The launch day countdown moved along smartly with few glitches. If you counted orbits, the teams were virtually doubling their experience with each mission. The cumulative lessons of the missions and the training were bringing the controllers to instant adulthood, but as soon as one group made it they moved into new jobs in the rapidly expanding organization and another new team took their place.

In the design of Mercury Control a spare console was located to the left of the flight director. The console was assigned to an assistant flight director (AFD) and the Mercury Control Center procedures manual stated that "The AFD is responsible to the Flight Director for assistance in the detailed control of the mission, and assumes the duties of the Flight Director in his absence." During the early missions no individual had been designated to that position and the console was primarily used during training by the simulation supervisor, SimSup, or a member of his team. At the conclusion of the training for Scott Carpenter's flight, Kraft walked over to me at the console saying, "You've trained Dutch [Von Ehrenfried] well, and I think he is ready to go at the procedures console. I'd like you to sit here next to me from now on." Having become Kraft's assistant flight director, I had to figure out what this new job should cover. In Kraft's traditional fashion, while he allowed us to develop our own unique style and identity, he was cloning his team with a common genetic code. On that day I started along the difficult road to becoming a flight director. Kraft's fire and passion for his work had inspired me just as it had inspired every other member of his team. His cause and his victories were mine, and the emotional content of each day carried into the night. It was impossible to slow down.

To the right of the control room, Carl Huss coached Llewellyn through the myriad checks designed to prove the integrity of the launch trajectory system. It was fun and somehow reassuring to see the two of them at work. Llewellyn drove Huss crazy at the console during training. Huss, orderly to the point of distraction, spent his time checking and rechecking every step of every procedure. Llewellyn, in contrast, trusted his instincts, and when a procedure was completed, he left it alone. Each night in our Cape motel room at the Holiday Inn, I would listen to Huss critique Llewellyn's daily performance. Dancing Bear, facing the end of his career as the "World's First Retro" (retrofire controller), was convinced that John would never succeed in the position he, Carl, had invented.

Huss, now with real-time trajectory experience under his belt, became a section chief in the Mission Analysis branch, designing lunar trajectories.

The generation of controllers who grew up at the remote sites and were now in Mercury Control were about to be tested — and win their spurs.

The launch came off without a hitch and Scott Carpenter became the second American in orbit. His mission, like John Glenn's, would be three orbits. The first orbit went by the book; the controllers and teams were sharp. The training program had paid off in solid data flow, and the Teletype messages told the complete story of each site pass. Class 101 was doing well, damned well.

Except for a minor variation in his suit temperature, Scott reported no disorientation as he moved through the visual observations, and was thoroughly enjoying the zero gravity environment. During the sunrise and sunset periods, he described the same "firefly" phenomenon reported by John Glenn on his flight and tried to determine the source of this phenomenon.

Each flight, however, would present us with a new headache. Starting at the Canary Islands on the second orbit, Carpenter indicated that his out-the-window attitude did not agree with his instruments. The Teletype messages I received from the African, Australian, and Hawaii sites indicated errors between the horizon sensor and capsule pitch attitude.

Carpenter was advised to realign his gyros over the Canary Islands and then attempt to correlate his readings with Skinny Lewis's data over Zanzibar. The site data indicated significant errors.

Horizon sensors detect the difference in infrared radiation between space and Earth. The sensors provide signals to update the gyros that control the pitch and roll, and *Aurora 7*'s pitch sensor was varying by as much as plus 50 to minus 20 degrees. Carpenter was the only one who could put all of the pieces together by comparing his instrument readings with the spacecraft periscope and the view of the horizon from the capsule window. If the readings were off, like a pilot in an aircraft, he could realign the gyros to the correct position.

The Mercury capsule design provided the astronaut with two attitude control systems — an automatic system containing 12 thrusters and 32 pounds of fuel (hydrogen peroxide), and a less capable manual system with 6 thrusters and 23 pounds of fuel. In Mercury control, Arnie Aldrich was watching the drain of fuel from the tanks. Whether on the manual or automatic attitude control system, the high usage continued and, given the infrequent site contacts with the spacecraft, Aldrich was unable to identify the cause. Carpenter was repeatedly advised to conserve fuel by turning off all control and going into drifting flight. At the start of the third and final orbit the propellants were down to 45 percent remaining in both systems. Kraft's concern at this time was not the fuel level as much as the control techniques used by Carpenter. Every time he maneuvered the capsule the fuel quantities plummeted. If the trend we were observing continued, Carpenter would run out of attitude control fuel before reentry. Scott was again told to go to drifting flight and conserve his fuel for retrofire and entry. The site reports across Africa and Australia indicated that the fuel usage had stopped.

The news from Hawaii on the final orbit was not good. Their job was to get Carpenter into retro attitude and complete the pre-retro checklist. As Carpenter maneuvered the spacecraft into deorbit attitude with the automatic control system, the controllers were alarmed by the system's rate of fuel usage. We didn't have the full picture back at the Cape; only Hawaii could look at the telemetry and communicate with Carpenter at the same time.

We could hear that he was running behind in his preparations for deorbit and, given the low fuel level, we had to bring him home quickly. Then came more bad news. The automatic attitude control system was not functioning properly; Carpenter had to go to manual while the fuel continued to bleed away. With three minutes of contact remaining between Hawaii and the spacecraft, and five minutes left to complete the retrofire checklist, it was a horse race to get into deorbiting attitude if we were to land anywhere near the planned splashdown point.

Each of the Mercury astronauts was made of the right stuff, but all were vastly different. While others would be making last compulsively detailed checks on the spacecraft, Scott might be found sitting on the beach, strumming a guitar. What I knew for sure at this point was that we had an astronaut in trouble, who was way behind the power curve, and if he wasn't careful he was going to die. Whatever the problem, the real-time California tracking station pass would determine if Scotty lived or died.

("Real time" is a term used when the capsule is electronically in view of a tracking station. It generally stretches from horizon to horizon as the station antennas track the capsule orbit. The initial contact is called acquisition [ACQ] and real time ends at loss of signal [LOS]. At the low orbital altitudes [between 100 and 200 miles] of the Mercury capsule, the real-time period could range from seconds up to eight minutes. To many controllers it seemed a lifetime.)

As Hawaii was losing the signal from the spacecraft, acting on sheer hope (and maybe a bit of prayer) they continued to read out the checklist. Their hope was rewarded when, well after Scott had passed over the horizon, he contacted Hawaii and told them that he had copied all of their message on the deorbit checklist.

For the first time in what seemed like hours, the controllers at the Cape took a deep breath, wondering what the hell was going to happen next.

Kraft leaned over and punched up the loop to California. "California CapCom, have you been tracking this?" Alan Shepard responded tersely, "Rog, Chris." Shepard and Ted White had been listening to the Hawaii pass and knew what they needed to do. There was nothing we could say or do at the control center; it was now up to the California team to get him down.

Sixty seconds after Hawaii lost the signal, Shepard made his first call from the California site. "Seven, this is CapCom, are you in retro attitude?" Carpenter replied, "Yes, but I don't have agreement with the window, Al. I think I'm going to have to do it manually, with the window and the scope."

Ted White, the systems monitor, scanned the console meters for the attitude indications and exclaimed to Shepard, "Dammit, Al, if he is in retro attitude his gyros are way off. I don't think he ever did a gyro alignment. His pitch gyro is off by at least 25 degrees."

Reaching forward, Ted picked up a grease pencil and slashed a line on each of his meters at the current indication he believed represented the retro pitch attitude of minus 34 degrees, zero yaw. Leaning toward Shepard, he pointed at his meters, saying, "He's about out of manual fuel, automatic is about 35 percent. Attitude is okay in pitch . . . okay, but he's still moving in yaw."

Knowing he had to get Carpenter down, Shepard nodded, gave a thirty-second mark, and then began the countdown to retro sequence, ". . . six, five, four, three, two, one, MARK!" Carpenter confirmed, "Retro sequence Green."

Scott made the final attitude check by looking out his window. White checked his meters; *Aurora 7* was still moving. Retrofire was designed to be accomplished by the automatic attitude control system. The sequence began but immediately hung up.

Shepard had to pack all his verbal instructions to Carpenter into the next half minute to hit the planned retrofire time: "Try the automatic system, quickly. If your gyros are off, you will have to bypass the attitude interlock." After hearing Scott confirm, "Gyros off," Alan continued, "Bypass attitude and use manual."

At the Cape we heard Carpenter indicate he had fired all three rockets. "I had to punch them off manually," he said, "and I have a bit of smoke in the capsule." Ted White confirmed that the retros were firing, but the capsule was well out of attitude in yaw. Seconds later, it seemed to tumble.

Retrofire was only three seconds late, but Llewellyn's concern was the attitude. If Carpenter was far enough out in yaw or pitch, he would be stuck in orbit with no way home. One way or another, the die was now cast. Carpenter reported, "I think the attitudes held well, Al. I think they were good. I can't tell you what was wrong about them since the gyros were not quite right."

Shepard glanced at the fuel quantity meters on White's console; the manual fuel was reading zero. Less than 20 percent of the fuel remained in the malfunctioning automatic system. Al knew it was going to be close.

Carpenter said, "I'm using fly-by-wire to stop tumbling. I'm out of manual fuel, Al." Shepard, speaking with deliberate calmness, pronouncing each word slowly and distinctly, responded, "You've got plenty of time. Take your time on fly-by-wire to get into reentry attitude." If the capsule ran out of fuel before getting into the correct attitude, it could enter the atmosphere nose forward and burn up during entry.

Whatever difficulty Scott was in, he could hardly have found himself in better hands. Shepard, in the opinion of just about everyone, was the most unflappable of all the astronauts. The man at his side, Ted White, was one of the pioneering group of Philco technical reps who had been hired as monitors for the Mercury system. He had no peer. He had dedicated himself to acquiring experience in remote site operation, even if it required him to volunteer for the ships and spend weeks at sea.

Llewellyn, now faced with his first emergency as a retro controller, continued to talk to White to get the specific attitude. "It looked like he was okay in pitch," White explained, "but he was all over the sky in yaw. I thought we lost him when he started to tumble. I'll get you more data after the pass."

Llewellyn alerted the recovery teams that the capsule would land long, and searched for data that would give him a better estimate. Shepard continued coaching Carpenter: "You have plenty of time, about seven minutes until .05G, so take it slow and easy."

At Mercury Control, we heard Carpenter's

response, "Roger. Okay, I can make out very, very small farmland . . . pastureland below. I see individual fields, rivers, lakes, roads, I think. I'll get back to reentry attitude now."

Shepard's response was all business. "Recommend you get as close to reentry attitude as you can, using as little fuel as possible, and stand by on fly-by-wire until rates [capsule motions] develop, over."

Every controller knew it was going to be damn close. If Scotty had not fired the retros in the correct attitude, instead of coming down in the landing zone he'd be stuck in a slowly decaying orbit, circling for a few hours until his power and oxygen ran out.

With the second critical reentry facing him, Kraft was a study in controlled fury. The mission had been close to perfect until these last few minutes and now everything was going to hell. Yelling over to Grissom, the Cape CapCom, Kraft said: "Dammit, Gus, keep his [Carpenter's] mind on the job. I think he's delirious."

I was standing at the console next to Kraft. There was not much we could do except try to keep Scotty focused on the very few remaining things that would keep him alive, nursing the last few drops of attitude control fuel and maintaining reentry attitude. In the years to come I would often feel the absolute frustration at being helpless during the blackout that concludes every mission. It is like watching your wife in labor . . . there is nothing you can do except hold her hand and pray that all goes well.

Llewellyn looked like a guy juggling raw eggs,

trying to interpret Ted White's report and coordinate the recovery forces and radar trackers. For his first mission he had drawn a big one. Using high school geometry, and after consulting a plot under the plexiglass cover on his console, John said: "Chris, with the California data I think we're going to be a couple hundred miles long. We'll get a good hack when we see what time he enters blackout."

Shepard chimed in: "This is California. We're losing you now. Stand by for the Cape."

After contact three minutes later, Grissom's dialogue with Scott centered on the fuel status and the few remaining checklist items. Aldrich said quietly, "Chris, he's down to 15 percent on auto fuel, the manual tank is empty." After relaying the recovery area weather, Grissom reminded Scott of the impending blackout. The voice loop exchanges were crisp, but the overall mood in Mercury Control was somber as Scotty entered blackout. In the next few minutes, America might lose an astronaut.

At the Mercury Control loss of signal, Llewellyn called Kraft, "Flight, we were about thirty seconds late going into the blackout. We're long. I'll give you a better hack in a few minutes."

Within minutes, Llewellyn had the first landing prediction: "Flight, I've got an impact point 250 miles downrange from the predicted target. I have good agreement with all of my tracking radars. This is the best impact point we will get. I'm passing it on to Recovery."

Bob Thompson, the recovery operations chief,

called, indicating that an Air Rescue Service twin engine amphibious aircraft had been launched from Puerto Rico and was flying to the new landing point north of the Virgin Islands.

The wait was agonizing, as it had been on John Glenn's reentry. I muttered a silent prayer to St. Christopher, the patron saint of travelers, and it seemed everyone was on his second cigarette during blackout. Ten minutes later, we got our first call from *Aurora 7*. "I'm on the main chute at 5,000. Status is good." It was if ten tons had been lifted off our backs. We felt like standing and cheering, but that didn't happen in Kraft's control room, only in the movies. Chris's smile was more like the grimace of one receiving a last-minute reprieve from the guillotine. He was not happy about the ending of his second manned orbital mission. A relieved Grissom advised Carpenter, "Your landing point is over 200 miles long. We will jump Air Rescue personnel to you in about an hour."

For nearly thirty minutes after the landing the press was in a panic, thinking an astronaut had been lost in space, or dropped into a watery grave. When the recovery helicopters got to him, a typically laid-back Scott Carpenter was described as quite comfortable in his raft, even a bit irritated because they had interrupted his contemplation. He was eating a candy bar.

In the rescue process Scott was quoted as saying, "I didn't know where I was and they [Mercury Control] didn't, either." Those were fighting words to a new young RETRO who prided himself on perfection in the emerging art of retrofire control.

"Bullshit," Llewellyn exclaimed. "The SOB is damned lucky to be alive."

Carpenter's words would often be remembered. At beer parties, or during debriefings, if we wanted to get John Llewellyn to tell the story of *Aurora 7* and his first mission as RETRO, we would stand and say, "I didn't know where I was and they didn't, either." And the story would commence. Later, Scott Carpenter became entranced by the mysteries of the sea. He lived and worked on the ocean floor for periods of thirty and forty-five days, developing techniques and systems for undersea operations. He retired from NASA in 1967 and was assigned by the Navy to the deep-submergence-systems project.

During mission debriefing, Scott said simply, "I think the reason I got behind at retrofire was because approaching Hawaii at dawn on the third orbit I discovered the source of the fireflies. I took out the camera and tried to get some pictures. I felt that I had time to get that taken care of and prepare for retrofire properly, but time just slipped away."

The "fireflies" reported on Glenn's and Carpenter's missions were simply frozen droplets of water from the evaporators used to cool the cabin and space suits. They were most noticeable at sunrise on each orbit.

Scotty's mission was close, too damn close. A crewman distracted and behind in the flight plan is a risk to the mission and himself. A major component of the ground team's responsibility is to provide a check on the crew. The ground had waited too long in addressing the fuel status and should have been more forceful in getting on with the checklists. I also thought we started too early in introducing such scientific experiments as a balloon drag experiment, zero-gravity fluid studies, and photographic observations. Once again, we were lucky — but luck has no business in spaceflight.

June 1962

During these times of endless hours and heavy stress, keeping a marriage intact was no small achievement. I couldn't have done it without an extraordinary wife. Our newest daughter, Joan, was now almost a year old as we packed up and joined the caravan moving west. As we rolled into Flight Controller Alley in Houston I knew that while the controllers faced many tests, we, too, had "the right stuff" for spaceflight.

The nature of our work made us develop strong loyalties to each other. Most of us were in our late twenties and early thirties. Outside of wartime, I do not believe that young people had ever been given responsibilities so heavy or historic. We were in jobs that appealed to the adventurer, dreamer, and Foreign Legionnaire in each of us. Over half of the original members of the Space Task Group elected to leave flight operations in the first two years. Many had come from working in aeronautics at Langley and wanted to get back to the airplane business; others did not want to relocate to Houston from Virginia. There were some who handled one or two missions, but the responsibility scared the hell out of them and they left. But those who stayed with the program grew in experience and judgment — and provided the hard-core leadership for Gemini and Apollo.

With the selection of the second astronaut group in September 1962, it was obvious that we were approaching the end of the series of missions that made up the Mercury program. While waiting for the construction of our new offices, the Flight Control Division was located in a sporting goods warehouse on the main freeway between Houston and Galveston. The remaining offices of the Manned Spacecraft Center were scattered at eleven other locations, including a bottling plant, a TV station, a bank, and an electric fan company.

The accommodations for my branch were short on windows, rest rooms, and decent lighting. The first floor was a huge open warehouse on which we located our gunmetal gray government desks. The secretaries were in pay booths just like those in a movie theater at the end of the hall on the second floor. These facilities would be our home as we closed out Mercury and prepared for Gemini.

In August, with their dual launch of two Vostok spacecraft, the Russians had shown us they were in the race to stay. Their spaceships came within less than three miles of each other. We believed they had intended to rendezvous but didn't execute the final maneuvers needed to fly formation in the same orbit. For the first time, I thought we might have overestimated their technical capabilities. I respected their accomplishments, but the Russians no longer seemed invincible.

Kennedy's bold commitment to put a man on the Moon had set America on a faster track. Daily we could see progress in developing the computing, communications, and precision guidance and control technologies we needed. I was confident the tools would be there, but I was concerned that we wouldn't be smart enough, or at least have sufficient hands-on experience to use them well operationally.

September 1962

It was in the interval between the first scrub of Wally Schirra's launch and the second attempt that President Kennedy made his speech at Rice University that confirmed his commitment. This time I was more attuned to his words. On a makeshift stage erected on the fifty-yard line at Rice Stadium, Kennedy repeated the question that many had raised: "Some have asked, why go to the Moon? One might as well ask, why climb the highest mountain? Why sail the widest ocean?"

Kennedy's words were to echo across the decades, and we, along with the rest of the country, found out that a noble cause brings out a nation's best qualities. I now believe that Kennedy fully understood the difficulties before us. I listened closely to his speech, feeling that I was ready to do whatever it took to turn his great dream into reality.

Next up for a Mercury mission was Schirra, a known quantity to the controllers, the astronaut with the best feel for the emerging relationship between the flight crews and ground control. Schirra's six orbits would be the bridge needed to go from three orbits to a full twenty-four-hour mission. (The medical community's concerns about man's ability to adapt to zero G led to conservatively planned incremental increases in mission duration during Mercury and early Gemini. In general, we attempted to double the flight duration of each mission.) I believed that the name of his capsule, *Sigma 7*, symbolized the sum of all the efforts of design, test, and operations necessary for success in space. For us, it signified teamwork.

Since I had hired Mel Brooks at Houston, he had been quietly transforming the training team. Previous training exercises had focused on the controllers' knowledge of the mission sequence and had tested simple applications of their judgment on the flight rules.

With the perspective he gained from his satellite operations, Brooks had taken over the training of the flight control team with a single-minded zeal. Think of Alec Guinness in *Bridge on the River Kwai*. Recognizing their youthful naïveté and technical limitations, Brooks's training staff attacked us individually and as a team. He went after those he felt were short on theory. He picked apart their communications and forced them to own up and say, "I don't know."

As we prepared for the final day of training,

Brooks decided Kraft was too dependent on my backing up his every move and anticipating his command and data needs. So he took me by the shoulder, walked up to Kraft, and said, "Kranz is in the hospital. He was injured in an automobile accident coming to the MCC today." My response was quick. "Dammit, Brooks, what the hell do you think you are doing! This is our last training run." Kraft, amused by the byplay, said, "It looks like you've been benched." Brooks believed that Kraft should conduct Schirra's final training simulation without his regular wingman. Von Ehrenfried filled in and did well.

Schirra's was a textbook flight. But within days of its conclusion the world was again on the brink of war, this time over the discovery of Soviet missiles in Cuba. I was glad we were not in Florida. My Air Force Reserve unit, among many others, was put on standby status as a showdown between the Soviets and the United States developed.

For almost two weeks, we in the space program were understandably preoccupied by the blockade and possible invasion of Cuba, which could presage an all-out nuclear conflict with Russia. With the two countries "eyeball-to-eyeball," in the apt phrase of the Secretary of State, Dean Rusk, the Russians backed down, turned their ships around, and removed the missiles.

The textbook flight of Schirra cleared the way for Gordo Cooper's one-day mission to conclude the Mercury program. Cooper's flight required two MCC control teams and a relocation of the ships to plug the gaps in the network coverage. The length of the flight represented a different test for all of us. The capsule would be out of communication on several orbits for over an hour as Cooper slept. We had no option but to trust the capsule systems. Still, I was less concerned about this mission than any of the preceding ones. The Mercury spacecraft, if used well, was getting the job done. The knowledge level of the controllers was at an all-time high and the remote site teams had proven that they could respond rapidly. The technology was advancing so rapidly that we could now reliably bring the tracking and telemetry data from the Bermuda, California, and Texas tracking stations to the MCC. Now we had almost twenty minutes of continuous data every time the capsule passed over these stations. Since we no longer had to staff these stations with controllers, we had the resources to make decisions in a much more focused and efficient fashion at Mercury Control.

Early 1963

With the New Year, more changes came. Bob Gilruth, director of the Space Task Group, became the director of the Manned Spacecraft Center in Houston. He was shifting resources to Gemini and Apollo, so the Mercury office was down to less than fifty people.

Marta was in the hospital in labor with our fourth child and for a change I was holding her hand. I had been in Korea for Carmen's birth, out to lunch during Lucy's, and just made it back to Virginia for Joan's. Now the women in Flight Controller Alley decided it was time for me to do my duty. They selected Jim Strickland, a neighbor and systems controller, to carry the message. "Gene, we can carry on here at work. The Alley has decided you should be with Marta, so get home and take care of your wife!" The handholding was an experience in total helplessness. All I could think was to thank God for the courage women have to go through childbirth. Mark, our son and fourth child, was born in Houston in January 1963. We were staying with the Von Ehrenfrieds at the time we think he was conceived, so we have since concluded that drinking a lot of good German beer and living with a family that had two boys in it had a decisive effect. (Twenty-seven years later I again learned how powerless one can be. Mark was hit by a drunk driver. For an entire night Marta and I held his hand while doctors tested for neurological damage. Mark had to endure without medication. He spent the best part of two years in multiple surgeries and therapy before he recovered.)

With the organization in high gear, I prepared for the final Mercury mission. The Air Force Atlas program had suffered two unexplained flight failures. We could not move ahead until we figured out what had gone wrong. Then the Atlas booster to be used for Cooper's liftoff failed its rollout inspection in San Diego in January. These setbacks had emphasized the need for support from inside the plant to track design changes.

The second Atlas rollout in March was successful and finally we had the flight elements for the final mission. Gordo Cooper's *Faith 7* mission was the first where we would literally fly beyond the coverage of the ground network. The ships *Coastal Sentry* and *Rose Knot* were moved to the Pacific Ocean to provide the mid-orbit communications coverage and to support the capsule retrofire sequence preparation. The two ships provided the majority of the controller support during the middle section of the mission while Gordo Cooper was sleeping.

Chuck Lewis led the Australian team and Ted White had California. At least one team would spot the capsule during each orbit. Cooper's mission would involve a global effort of twenty-eight ships, 171 aircraft, and 18,000 military personnel, in addition to the support of the ground control crews.

Mel Brooks, in his capacity as SimSup, had achieved a degree of notoriety among the flight control teams as a result of the training exercises for Schirra's flight. He was forming a new organization for Gemini, and for his farewell training run he unleashed an exercise that tied the control teams and especially the doctors in knots.

Most of the doctors assigned to the remote site teams were military personnel with broad experience; some were Army Airborne and others were Air Force and Navy flight surgeons. Those in Mission Control came from research backgrounds. Brooks was determined to build on their uncommon skills and motivation to create a real-time ready team. While medicine often has occasion for differential diagnoses (i.e., doctors who disagree), Brooks felt that the control teams didn't need any doctors telling them, "on the one hand . . . but on the other hand." So he did things like obtain electrocardiograms of people having actual heart attacks and patch those into training tapes. In the final simulation session Brooks had trouble keeping a straight face as tapes peppered with this type of material and instructors simulating astronauts in medical distress ratcheted up the anxiety level in their voices, giving our doctors plenty to think about — and react to. Fast. They quickly learned there was no time to wait until the next pass, and they learned to resolve differing diagnoses very quickly.

Brooks had a field day debriefing the medics and was in a merry mood. His final training run had been a success and now it was time to put on his other hat and become a capsule systems engineer on Hodge's team for the final Mercury launch.

Shortly before launch, Cooper's mission was expanded upward from one day to twenty-two orbits, placing the planned landing point near Midway Island in the Pacific. Since Gordo Cooper had been the first astronaut I had met, I was happy to be working his mission and even a tad sentimental about it. I was sure that my remote site controllers were in peak form. With the longer mission duration Kraft and Hodge formed two teams to provide twenty-four-hour coverage at MSC. Only Kraft, Hodge, Huss, Lunney, Aldrich, Llewellyn, and I remained from Kraft's original Mercury Control team. The new teams were young, but we all were young together in the early years of space.

May 15, 1963, Mercury-Atlas 9

Thirteen seconds after 8:04 in the morning EST, *Faith 7* lumbered skyward, and the final mission of the Mercury program was bound for orbit. The early orbits went rapidly. Cooper precisely checked off the procedures for the eleven experiments his capsule carried. On the fourth orbit, he launched a six-inch satellite with a flashing light to test an astronaut's visual acuity in space. The orbits clicked off, the systems performing perfectly and the early Go NoGo decisions easily made. Mercury Control operated on a two-shift basis for the first time. After the Go NoGo on orbit six, Kraft's team handed over to John Hodge's team. In my newly assigned role as assistant flight director Kraft had assigned me to support Hodge's team. My vision of the flight director's role in Mercury Control was based on associating with Chris during the pre-mission planning periods and watching him in action during missions. Kraft was called the "Teacher" because of his hands-on mentoring of his young charges. The longer mission duration and infrequent contacts during the mid-orbits of Cooper's mission allowed me to study another flight director at work, John Hodge. Hodge was different from Kraft, less on edge and more dependent on his team members. I quickly sized up John as a flight director who sought to obtain consensus in selecting a direction and one totally at ease with his people — but not with his role as their leader. John was a damn nice guy.

Cooper, a celestial observer, enjoyed the middle period in drifting flight, with all the capsule systems powered down. His thirty-four-hour mission would take him through twenty-two orbits. With the control center, network, and capsule systems reacting flawlessly, it was easy to get lost in Cooper's descriptions of the view from orbit.

When things are going well, the controllers slip into a mode of relaxed awareness. You tune your senses keenly to pick up even the slightest departure from the norm; it seems that you have a second sense running in the background, almost subliminal, that can pick up the slightest deviation. It could be a minor glitch in a telemetry measurement, a procedural step overlooked, or an unexpected observation from the astronaut. It can be the tone of voice of another controller. When you are well tuned, a second sense kicks in, looking for something out of order while you proceed with the normal or routine. The mission continued in this fashion through the nineteenth orbit.

Kraft's team was in charge as the mission entered the gate for the final three orbits. The recovery forces were calling in, and Llewellyn, checking the retro times, anticipated a perfect finish for the final mission.

Emerging from the network coverage gap at Hawaii, Cooper began reporting to Scott Carpenter, his CapCom: "Scott, I wonder if you would relay to the Cape a little situation I had happen and see what they think of it. My .05G telelight came on after the light check. I have turned off both the .05G normal and emergency fuses. Relay it to them and get their idea on it, over." The telelight was indicating that the capsule was sensing the onset of reentry gravity.

Mel Brooks, seated next to Arnie Aldrich at the systems console, unfolded his schematics and began tracing the .05G circuitry. After a brief consultation with Mel, Arnie reported to Kraft, "Chris, the signal that illuminated on Cooper's panel changes the operation of the automatic control system. It is used to provide a steady roll rate and dampen the capsule motions during entry." After giving Kraft a few seconds to digest his input, Arnie continued, "With the .05G indication, the automatic system cannot be used for retrofire. I'd like to do a few tests with Gordo."

The joint testing began over the Cape and continued through the next orbit, with Cooper reporting the test results to John Glenn, the CapCom on board the *Coastal Sentry*. The testing confirmed that the automatic system had malfunctioned and was in reentry mode. Cooper was informed that retrofire would have to be performed manually, but after retrofire was completed the automatic system would be usable for the final phases of entry.

So far everything had gone well. The team had responded to the glitch, the controllers smoothly regrouping and updating the retrofire and entry procedures. I kicked into gear, getting the new reentry plans out on the Teletype and setting up the backup communications to the *Coastal Sentry*.

You can feel the atmosphere of a crisis center; it is almost like sensing the change in pressure when a storm moves in. In Mercury Control it consists of the noise level of the room — conversations change from an informal banter to crisp dialogue, thick manuals thump open, small huddles form at the consoles. The feeling is unmistakable, and when problems developed for *Faith 7* I nonetheless felt secure: the teams were poised, professional, and competent.

I was confident that whatever the problem, Mercury Control and the remote site teams could handle it. The last-minute decision to expand from sixteen orbits to twenty-two got Cooper's mission into the record books as the third longest manned spaceflight. The Russians had flown sixty-four orbits with Andrian Nikolayev, and forty-eight orbits with Pavel Popovich. Cooper's flight now edged out Gherman Titov's seventeen orbits. The increased duration, however, moved the landing point from the Atlantic Ocean to the Pacific, resulting in reduced coverage at the time of retrofire and reentry. As the capsule moved toward its de-orbit point, only three sites remained on the ground track. Cooper had worked around the small glitch in the .05G circuit by the twenty-first orbit and everything was returning to normal operation. Then all hell broke loose.

Jim Tomberlin's job at Zanzibar was to complete the stowage and pre-retro checklists so that Gordo had an "all green" capsule, as he coasted to the *Coastal Sentry* site in the Pacific Ocean south of Japan, where John Glenn was CapCom.

Jim started reading the checklist only to be interrupted by Cooper. "Zanzibar, I have one item for you. My automatic control system inverter has failed, so I will be making a manual entry." Tomberlin, momentarily startled, asked, "Has the automatic system inverter failed?" Cooper responded, "That is affirmative."

Like a tennis match, Tomberlin again volleyed, "Have you tried the standby inverter?" Cooper's response came like a firecracker, "Roger, it would not start." Mystified, Tomberlin replied, "Roger. Then we better get on with the checklist now. Attitude permission to bypass."

"Roger, retro rocket arm to manual," Cooper replied.

The checklist dialogue continued for the next four minutes and when completed, Tomberlin switched gears and started a review of the backup procedures. During the six-minute pass, Gordo and the Mercury team at Zanzibar had prepared *Faith 7* for retrofire and reviewed the backup procedure, and a potentially serious problem had been quickly addressed. We had come a long way in the two years and ten days since Alan Shepard's launch.

Seventeen minutes later, Cooper was holding attitude, using the window horizon as he reported to Glenn on the *Coastal Sentry*, "Checklist complete except for the pyros." After a few moments of banter, Cooper continued, "Oh, my automatic inverter failed along with a few other odds and ends. I will shoot the retros on manual and then reenter manually also."

While waiting for the retrofire time, he added, "I'm looking to get a lot of experience on this flight." Glenn's response was laconic: "You're going to get it."

Moments later, John issued the retrofire countdown to *Faith 7*. The Mercury program was coming to a close in the sky over the Pacific. The manual retrofire and reentry were virtually flawless. Forty minutes later, Cooper emerged from the capsule and stood on the deck of the aircraft carrier *Kearsarge*.

Cooper's post-mission comments said it all. "My analysis of the malfunctions," he said, "illustrated that the entire Mercury network had developed a concept of teamwork that culminated in an almost perfect example of cooperation between the ground and spacecraft. Almost everyone followed the prestated ground rules exactly, and the radio discipline was excellent."

Cooper, the loner and rebel against the spaceflight bureaucracy, had pulled off a great mission and a picture-perfect entry. Gordo's test pilot mentality, coupled with the superb performance of the ground team, was a fitting finale to America's first manned venture into space.

The final Mercury program party was held at the Old Governor's Mansion on Galveston Bay on July 27, 1963. It was more formal than most MSC parties and was complete with an invitation worthy of a scrapbook. After a few drinks,

Walt Williams began speaking about Mercury. Remembering the first Atlas launch failure and his decision to launch into an overcast sky, he started talking about making risk judgments. Then, in a melancholy tone, he concluded, "You will never remember the many times the launch slipped, but the on-time failures are with you always." In the years ahead we would have occasion to remember those words.

The Sabre Jet was the goal of every fighter pilot of the 1950s. *My Darling Marta* was my assigned aircraft in the 69th Fighter Bomber Squadron in Korea. The lessons in leadership, trust, and teamwork learned as a fighter pilot served me well in Mission Control.

NASA
S-62-303

Mercury 6, February 20, 1962. The Mercury-Atlas 6 lifts off with John Glenn for a three-orbit mission in the *Friendship 7* capsule, ending Russian dominance of space.

My rocketlike ride in Gordo Cooper's Chevy convertible was my introduction to the Mercury astronauts. Cooper became my favorite astronaut for his boyish enthusiasm and go-for-broke approach to everything he did.

Two pioneers of early spaceflight and a rookie await word that Glenn's spacecraft survived reentry. Walt Williams was the tough Mercury operations boss while Chris Kraft wrote the book on the flight director's job. The Mercury missions prepared me for my later role as a flight director.

With Neil Armstrong and Dave Scott critically low on fuel after a thruster malfunction, John Hodge decides to bring the crew down into an emergency landing area in the West Pacific Ocean on Gemini 8.

Traditionally the entire management chain would take the heat during a press conference, winning the respect and admiration of the troops. From left to right: Dr. Chuck Berry; Deke Slayton; me; Program Manager Chuck Mathews; Mission Manager Bill Schneider; Lieutenant General Leighton Davis; MSC Director Dr. Robert Gilruth; and NASA Associate Administrator, Manned Space Flight, George Mueller.

The first American EVA or space walk: astronaut Ed White on Gemini 4. The success of this EVA caused us to underestimate the complexity and difficulty of working outside the spacecraft.

Two men who shaped the future of manned space operations: Chris Kraft, director of Flight Control, and Deke Slayton, the astronauts' taskmaster. Slayton, grounded for heart irregularities in Mercury, ran the astronaut corps, eventually regaining flight status during the July 1975 Apollo-Soyuz mission. Kraft, the founder of the Brotherhood of Flight Control, became the director of Johnson Space Center.

After achieving Earth orbit on Apollo 9, the crew prepares to dock with the LM (Lunar Module) on top of the Saturn IVB stage. The docking cone is at the center of the picture. The target used by the crew when docking is below the cone.

The prime crew for the first Apollo manned launch is named by Dr. Gilruth at a press conference on March 21, 1966. Seated left to right: astronauts Roger Chaffee, Ed White, and Gus Grissom, and Gilruth.

The Apollo 11 LM ascent stage as seen by command module pilot Mike Collins during the rendezvous after the first lunar surface EVA. The spacecraft is upside down with the rendezvous radar antenna at the bottom of the picture.

Apollo 11 lunar landing team — Front row, from left to right: Don Puddy, Bob Carlton, me, astronaut Charlie Duke, Dr. John Zieglschmid, Captain George Ojalehto, Spencer Gardner, Frank Edelin, Arnie Aldrich, and Buck Willoughby. Back row, from left to right: John Aaron, Dick Brown, Chuck Lewis, Larry Armstrong, Bill Blair, Ed Fendell, Jim Hannigan, Jerry Bostick, Jay Greene, Gran Paules, Steve Bales, Chuck Deiterich, and Doug Wilson.

August 13, 1969, outside the Century Plaza Hotel in Los Angeles. The presidential party celebrating the Apollo 11 lunar landing was the first opportunity Marta had to share in the benefits of my work. (Courtesy of the White House)

Apollo 11 cigars and celebration at the flight director console. Left to right: Flight Directors Cliff Charlesworth, Glynn Lunney, and me; Director of Flight Operations Chris Kraft; Associate Administrator George Mueller; MSC Center Director Robert Gilruth; Apollo Program Manager George Low; and Charles Mathews, NASA headquarters.

The awful uncertainty during the extended blackout period as Apollo 13 reentered Earth's atmosphere is captured in the expressions of the flight directors. The flight directors who led their teams through their finest hour are Gerry Griffin, me, Lunney, and Milt Windler. Flight Director Pete Frank stands with jacket on at the end.

President Richard Nixon presents the Presidential Medal of Freedom, the nation's highest civilian honor, to the Apollo 13 Mission Operations Team. Seated at left is Dr. Thomas O. Paine, NASA Administrator; standing are President Nixon, Lunney, me, Griffin, Windler, and Sig Sjoberg, the director of Flight Operations.

Apollo 15 liftoff from Launch Complex 39, July 26, 1971. The 363-foot-high Saturn was the world's most powerful machine. With 7.65 million pounds of thrust provided by the five engines at liftoff, the engine start alone used twenty-three tons of propellant.

The new and old Apollo flight directors. Four new flight directors were broken in during the final three Apollo missions to continue the leadership role in the MCC. First row, left to right: Chuck Lewis (Bronze), Don Puddy (Crimson), Neil Hutchinson (Silver), and Phil Shaffer (Purple). Standing, left to right: Windler (Maroon), Griffin (Gold), Frank (Orange), Lunney (Black), and me (White).

The Apollo 15 CSM (Command and Service Module) displaying the eight scientific instruments used by the command module pilot Al Worden to photograph and study the lunar surface. A bay similar to this was destroyed when oxygen tank two exploded on the CSM on Apollo 13.

Jim Irwin prepares the Lunar Rover for the first EVA of Apollo 15. Mount Hadley is in the background. The Rover weighed 400 pounds and could carry 970 pounds a distance of 75 miles on its batteries.

5

THE MAKING OF A ROCKET MAN

Toledo, Ohio, 1940s

I always wanted to fly. As a boy in Toledo, Ohio, I had my head in the clouds and my heart followed. The cottonwood tree in our backyard was my telescope to the world. On a windy day, I was on the tall mast of a ship plowing through stormy seas, calling out commands to my crew below. On calm days, I was an eagle, lifting and soaring silently, searching for my prey.

The tree was so tall I could see the Willys Overland plant where Jeeps were made. If I stretched out, I could see an occasional airplane to the far north, over Franklin Field.

My grandfather was a German immigrant, Peter Joseph Kranz, a remarkable man who came to America in his early twenties. He founded a savings and loan office in Toledo, was elected city treasurer, and became famous as a trout fisherman. He was wiped out in the Depression. His five sons assumed the responsibility of repaying his creditors.

The oldest son, my father, Leo Peter Kranz, died in 1940 of a bleeding ulcer, misdiagnosed as a heart problem. He was forty-nine; I was seven. One of my greatest losses is that I never got to know him. Of the few recollections I have, the most vivid is of my dad listening to the radio when the Germans invaded Poland. He commented, sadly, that this was the start of a "world war." I didn't know what he meant, but in later years I learned he had been a medic in World War I — the so-called war to end all wars.

I'm fond of his picture in uniform, erect, proud, crew-cut, and blond. I closely resemble him and for most of my life I have worn my hair in a crew cut. Like many others of his time, my father had no life insurance, so things got tight for the Kranz family after he died. My mother moved us — me and my two older sisters, Louise and Helen — to West Toledo so that she could raise her kids in a better environment. A staunch Catholic and Republican, her German side was evident when she was faced by a challenge; she never backed down, was stubbornly self-reliant, and expected others, particularly her children, to live up to their beliefs and stick to their guns. She was loving, nurturing — and tough. She turned our home into a boarding house to pay the bills and she ran it like a drill sergeant. Among the boarders who stayed there, a few days at a time, were the servicemen who became my heroes, soon to be fighting the land, sea, and air battles in faraway places — some of whom would give their lives in combat.

I retain vivid memories of World War II

because of those young men who lived in our house, flirted with my sisters, and wrote them letters from the various fronts on which they fought. As a paperboy delivering both the morning and evening editions, I would walk past the houses of my customers, hurling the paper and shouting, "Extra! Extra! Read all about it" as the first reports on the Doolittle Raid or Midway came in, right through D-Day, the Battle of the Bulge, and, finally, the capitulation of the Germans and the Japanese. I plotted the major battles on maps I had pinned up on my wall at home.

I also was obsessed with aircraft — just by looking at pictures of fighters and bombers, I could construct models of them the old-fashioned way, gluing them together out of balsa wood and tissue paper. This led me to experiment with powered model aircraft and, later, rockets. I dreamed of the day when I would climb aboard the real thing. I hitchhiked to Cleveland and Detroit for the air races. Aviation magazines flourished after the war and I hoarded nickels and dimes to buy them. The writings of Willy Ley, Wernher von Braun, and David Anderton inspired me.

My high school thesis was entitled, "The Design and Possibilities of the Interplanetary Rocket." I proposed a gigantic two-stage rocket shaped like the German V-2. I rendezvoused it in Earth orbit with another stage for propellant resupply. Once the ship reached the Moon, I had to establish a base to manufacture the propellants needed to return home. Writing in

1950, I made this forecast:

> An examination of the current technical and industrial development demonstrates the high probability that the Moon will shortly be conquered by man. The base will probably be established in five years and completed in ten.

I scored a 98 on the paper, never imagining that someday I would be a member of the team that would place an American flag on the lunar surface. But, then, my thesis was not your typical high school junior's idea of light reading.

Work was a way of life in our home. I spent eight years absorbing the lessons and discipline of the Ursuline nuns. I can't remember not working two jobs, even while cramming and crashing to keep up my grades. I earned a Naval ROTC scholarship to Notre Dame and a congressional appointment to the Naval Academy, then received the crushing news that I had failed the physical. I had shown signs of diabetes. I had high blood sugar because of my diet — heavy on sweets — while working at the A&P grocery store. The condition was temporary, a small consolation. I believed that my world had ended.

My mother, and Sister Mark, my history teacher, would not allow me to surrender. They found a $500 scholarship for the children of deceased veterans of World War I. We sold my father's stamp albums and with every resource on the table I enrolled at Parks Air College in East St. Louis. The Korean air war had focused my goal to become a fighter pilot. Parks offered training in aeronautical engineering and a no-frills flight program. The dormitories were Army barracks at the edge of crossed cinder runways. Classes were punctuated by the roars of the PT-17 Stearman biplanes, the primary trainer for pilots in World War II, passing overhead. This was the first aircraft I ever flew, feet on the rudder pedals and my hand on the stick. The emotions of that first flight literally brought tears to my eyes. I was no longer swaying in the top of my tree; I was now truly an eagle, climbing, diving, and practicing turns with the wind singing in my ear. At Parks my dream came to life.

I intended to stay for one year, then apply for pilot training. But with the Korean War winding down, I decided to finish college. Parks turned out to be my field of dreams. The college had opened in August of 1927 in a rented hangar at Lambert Field in St. Louis. Oliver Parks was the owner and only flight instructor, and he started with two aircraft. During World War II, 10 percent of America's pilots had received their primary flight training at one of his schools.

I graduated on a hot, steamy day in July of 1954 with a commission in the Air Force Reserve. Present were my mother and one of my sisters, Louise, as well as my Uncle Albert, who had helped us through some difficult years and was always there when I needed him. The entire family had pooled resources and bought me a green 1954 Plymouth coupé as a graduation present. It was a wonderful surprise — I had planned on saving my money to buy a car before I entered Air Force flight training. There was one problem: I could fly airplanes but had not yet driven a car!

The fifty-two members of my class consisted of Korean War veterans getting their schooling on the GI Bill, six Israelis, and thirty-eight rookies like myself. I was one month short of my twenty-first birthday and the gold bars on my shoulders were more meaningful to me than my college diploma.

While waiting for a training slot, I applied to McDonnell Aircraft Company in St. Louis and was grateful for the chance to continue my transition from student to pilot. Graduates entering the aircraft industry in the 1950s were generally given options to work in drafting or reading and plotting the data records from flight tests. I chose the latter because it was attached to the flight test department, which was where I wanted to be. The roar of jet engines, the smell of jet fuel, and the constant rumble of the factory permeated the buildings. It was an exciting, busy place filled with high-energy people.

As I walked through the office maze on my first day, I heard a gruff voice ask, "Are you Kranz?" I stopped and the voice continued, "Where have you been? I've been waiting for you." I turned to face a balding cherub, with a red nose and forehead, as if he had just emerged from a sauna. He looked like one of Santa's helpers. He was my height, but he leaned forward, neck bent in a questioning attitude. He had clear, piercing eyes under shaggy brows. I noted his bow tie and suspenders as he said, "Hi, I'm Harry Carroll. Follow me. I'm your new boss."

He moved out quickly and I followed him to a desk covered with rolls of paper. He shoved them against the wall, pushed me into the chair, and said, "When you reduce these rolls of oscillograph readouts and learn to read the data, you will know more about what happened during a flight test than the pilot, the engineer, and the designer. These rolls of paper are like novels. It is up to you to get the meaning, then sense the plot and determine whether flight objectives were satisfied. You must watch to see if we are getting too close to the flight limits." Then he stepped back, chuckled, and said, "This is the best job in flight test! Get started."

His enthusiasm was my enthusiasm; his passion for work was my passion. I had to learn from others that he had flown eighty-six combat missions over Italy and Germany in the B-17 Flying Fortress and over Japan in the B-29 Superfortress. He had many inventions related to data reduction and aviation safety. He was also a poet, actor, and scoutmaster and he led the rugged and difficult grand portage canoe trips across northern Minnesota to Lake Superior. After retirement, he served as a deckhand for barefoot cruises in the Caribbean, and became the oldest individual to complete the Outward Bound mountain survival program.

Each day there was a new discovery under his guidance. No work was insignificant, no job unimportant. The standards had to be the highest if you were to meet with his approval. Harry Carroll was the first in a string of mentors who changed my life.

By the end of my third month on the job, I was sitting with the test pilots and flight test engineers during debriefings, reviewing flight cards (a pilot's checklist for the test sequences), transcribing pilot's notes, and validating flight test objectives. Heady stuff for a recent college graduate and the next best thing to being in the cockpit. This experience would serve me well later when I sat with the backroom guys and reviewed data on our space missions. At a glance I learned to identify the essentials and put the story together.

The months passed rapidly, and then it was time to pack up and report to Lackland Air Force Base in San Antonio, Texas, for preflight training. As of March 1955, I was now on active duty and assigned to pilot class 56M.

Other than St. Louis, I had never been west of the Mississippi River, and I soaked in the scenery as I drove through western Missouri and down into Oklahoma. My vision of Texas was crushed when I crossed the Red River. I had been expecting the sandy desert, cactus, and rattlesnakes of the movies, but I saw rolling hills starting to green up. I was sure that would change as I neared San Antonio.

I was wrong. This wasn't the lonesome prairie, the Texas of parched land and skeletal oil rigs. A scenic river wound its way through San Antonio, and the blend of Mexican and western culture gave the city a gentle and festive character. I was a willing believer in Texas charm and hospitality as I drove through the gates of Lackland Air Force Base. There I would have twelve weeks of preflight training, a good part of which taught you confidence by putting you through some pretty physically demanding exercises out in the boonies. I also learned the essence of leadership through being given responsibility for raw recruits who were wearing a uniform for the first time and were badly in need of understanding why the military demands order in everything from the state of your locker to the crispness of a salute, instant compliance with commands, and other basic military cultural imperatives. As the song puts it, "by your pupils you'll be taught." It would take some twelve weeks for them to transform me from college student to officer, one hell of a speedy transition. It was the NCOs (noncommissioned officers) who taught me the basics — and my respect for the sergeants on the line grew with every passing year.

My travels in the Air Force next took me to Spence Air Base in Moultrie, Georgia, where Jack Coleman, my primary flight instructor, opened the world of flight for me and taught me much more. In the hot steamy air over southern Georgia he tested my skills, but in the briefing rooms and on the ramp he taught teamwork and the belief that "There is no such thing as good enough. You, your team, and your equipment must be the best. That is how you will win victories." The day he turned me loose to solo, he taught me that the teacher's role is to instill the confidence to fly at the edge of peak performance. Your primary flight instructor is the man you never forget. Coleman's lessons helped me in my years at Mission Control. I could empathize with what the controllers felt during the brutally demanding debriefings after a mission and tactfully handle the one-on-one critiques after a simulation. He taught me, by example, how to train my controllers, build their confidence, and turn them loose when they were ready. Coleman also gave me an appreciation of the fundamental importance of teamwork and mutual trust among team members.

Of course, some lessons can only be experienced, not taught. One of these is dealing with fear, which comes to every pilot, like a bolt from the blue. Fliers and fighters alike have referred to this as looking into the eye of the tiger. I looked into mine one night on my first solo cross-country flight, over the blackness of Georgia's Okefenokee swampland.

I had turned to a new course over the town of Alma and looked down to change my radio frequency. Moments later, when I raised my eyes, the lights of the town seemed to fill the cockpit like tiny diamonds. For a few seconds I was mesmerized, then confused, as the sound of my engine and my rapidly accelerating aircraft snapped me out of my reverie.

In whatever part of my brain was still working, I put it together: I had rolled upside down over the city and was diving, inverted, toward the heart of Alma. I rolled visually away from the lights, into total darkness, all sounds diminishing as I approached a stall. Fighting vertigo, I recovered and flew on instruments, not trusting my senses, all the way back to base.

The next night, when it was time for my second solo cross-country, fear was in my bones like a winter chill. I realized it was either conquer the fear and fly, or wash out of training. I delayed going to my aircraft, chain-smoking cigarettes. Then someone started testing the flight line loudspeakers for the Saturday parade. The music of John Philip Sousa's "Stars and Stripes Forever" soaked into the Ready Room and something almost mystical happened.

My desire to fly overcame my fear. I picked up my parachute and walked to my plane and aced my second night solo flight. From that day forward, whenever I found myself looking into the eye of a tiger, the cadence of "Stars and Stripes Forever" — if only in my head — got me through it. I have a record and tape collection of over twenty versions of the march. It became a key element of my way of life.

My training next took me into jets and an assignment farther west in Texas at Laughlin Air Force Base. Flight training is partly about the fear of failing, and partly about firsts. The top graduates got the best assignments. I wanted to fly the hottest fighter in the Air Force — the F-86 Sabre. In order to get the chance to fly fighters I had to be the best. To be the best I had to go all-out, reach into myself for every resource I had to meet the challenge. In the process I determined what my real capacity was and discovered that for much of my life I had just been coasting along.

My competition for the Sabre assignment was First Lieutenant Anthony (Zeke) Zielinski, who had made his way through the ranks. He had won the wings of a navigator as a sergeant flying over Korea in B-29s and was described as a natural-born pilot.

The competition between us extended to the social, as he tried to cut in on the girl I was dating. We got our pilot wings, graduated with identical scores, and both qualified for the Sabre. Later I aced him, however, for I would marry a wonderful Texas girl, Marta Cadena, whom Zeke also wanted to date. Marta pinned the wings on my chest. Zeke, a true wingman, would be my best man.

Zeke and I headed to Nellis AFB, near Las Vegas, to fly the world's best and fastest fighter aircraft. Six weeks after I arrived at Nellis I climbed into the cockpit of an F-86H Sabre. After a takeoff and climb-out that felt like it would never stop, I started getting ahead of the power curve, just a bit, with this beautifully maneuverable airplane. An instructor flying alongside me put me through the flight. We landed, debriefed and refueled, then took off again and climbed to 35,000 feet, did some steep turns and a bit of trail formation. At this point my instructor told me to perform a split-S, roll out into a 45-degree dive, and call out the readings on the Mach meter. (The Mach meters in those days didn't go a hell of a lot higher than Mach 1 — the speed of sound.) I rolled inverted, picked up the dive angle, and was quickly at .95 Mach, where the nose wanted to pitch up. The instructor then told me to trim out the pitch-up stick force and comment on the aileron forces. Wing heaviness increased and the Mach meter rose to 1.0 and hung there. I had the feeling that I was skiing down a steep mountain bowl, pushing a lot of powder, and that it was impossible to go any faster. I throttled back, recovered, and followed the instructor's lead in a high-speed spiral descent. It took me a moment to realize that I had just broken the sound barrier, a big deal with the airplanes we had in the 1950s.

But that was the bright and shiny side of military life. While I was at Nellis I broke my left wrist in a stupid accident (with the help of a goodly amount of beer) and managed to get myself grounded on September 6, 1956. This gave me ample time for reflection, during which I realized I was in love with the girl from Texas who had pinned on my wings. Since leaving Texas, where Marta and I had first met and started to date, I dated a few other girls but none of them measured up to Marta. I still had her phone number and the good sense to call her. She was still interested so I started to commute between Nellis and Marta's home in Texas. The weeks passed too quickly and soon my duty at Nellis was over.

Upon completion of advanced training at Nellis, Zeke and I were stationed at Myrtle Beach, South Carolina. Myrtle Beach was a base in name only, parts of it still under construction. The only good part of the deal was that eventually we would have shiny new (they were still on the production line when I reported in) F-100 Super Sabres to fly. During long-distance phone conversations, Marta could always raise my morale thanks to her good humor and indomitable spirit. I wanted her to meet my family, so she came to Toledo during my Christmas leave. On Christmas Eve I asked Marta to marry me. We set the date for April 27 of the following year at Eagle Pass, Texas, where we had first met — fifty miles from Laughlin AFB but a commute I never minded making! It was one year to the date she had pinned on my wings. It started to snow as we left the chapel where I had proposed and we started to sing Christmas carols, wishing the magic of the moment would go on forever. We were on top of the world, and there was no question that whatever we faced in the future, we would face it together, and we would emerge victorious.

In courting Marta, I had been at a serious disadvantage, unable to communicate with her mother, Pura, who spoke only Spanish. Marta's parents were born in Mexico and came to the United States after the revolution of 1910–1920. Her father became a citizen and opened a drugstore in 1950. While I was dating Marta, her mother was studying English preparing for her citizenship test. After the wedding, the señora was able to welcome me into the family with two words in English: "No givebacks!"

To the true regret of both of us, I soon was forced to do exactly that.

We had been married only three months when my orders arrived in June of 1957 assigning me to the last squadron of F-86Fs on active duty, the Fighting 69th, at Osan, Korea. My squadron commander at Myrtle Beach bent the rules and let me finish F-100 training before I shipped out to Korea. Marta told me that she was pregnant. This was when I started to learn that if you have a good marriage in the military, you will probably wind up with a great marriage, if it lasts. Our eighteen-month courtship consisted of seven dates.

Even though the war had ended in July of 1953, Osan Air Base, thirty miles south of the Korean capital at Seoul, was on a wartime footing. The squadron's purpose was to provide fighters capable of striking northward, supporting the Korean and American units emplaced along the demilitarized zone and achieving air superiority. We had about fifteen minutes to launch defending aircraft in case we were attacked.

Sabre pilots were a rare breed, flying the hottest machine of its time, the greatest sports car ever invented for the air. When flying overseas, you had no regulations. We buzzed everything in sight, flying as fast and as low as our nerves would allow. This was balls-out flying. We did things we would never be permitted to do stateside. Every flight ended in a mock dogfight with one or more pilots diving, rolling, and scissoring for advantage. The fights continued until you ran out of altitude or you had the other guy on your gun camera film. This confidence was essential to aggressive flight performance.

The Soviets' launch of Sputnik on October 4,

1957, while I was on a thirty-day deployment at Tainan Air Base on Formosa gave the Cold War a new dimension. I had seen the Russian fighters pulling contrails over the Korean demilitarized zone and the Straits of Formosa. Now their new prowess in space raised doubts about America's commitment to lead the free world. I could suddenly see shadows of doubt about America's technological superiority in the eyes of the Nationalist Chinese pilots.

I finally got my own airplane, an F-86F, serial 24872. I promptly named it *My Darling Marta* and had that painted on the left side just below the gun ports. My crew chief and I gave that plane every ounce of our attention and like a human it responded to love and care. It never let us down.

I drew a forward air control assignment with the Army 7th Infantry division in January 1957. My job for a month was to lead a small ground team that directed air strikes by Air Force and allied fighter aircraft supporting the Army front-line troops. This FAC experience prepared me to work effectively with the forward air controller as a pilot attacking targets. I do not think any work in my life has ever been as demanding as close air support missions. Every element of mind, body, and soul was working intuitively and perfectly, putting the pieces together, planning ahead, communicating and pressing the attack. With mountains on all sides, poor visibility, and the high aircraft speeds, it was incredible we did not lose more pilots — and we were just training.

The 69th Squadron was counting down the days to decommissioning and returning stateside. I was looking forward to my next assignment. Marta was doing well in the latter part of her pregnancy, counting the days until the arrival of our first child and my return. On my last flight in a Sabre, I ferried the airplane to Taiwan and turned it over to the Nationalist Chinese.

Returning from the ferry flight, I was furious when I found out that my next orders took me to Altus, Oklahoma, to train in KC-135 jet tankers. I could not believe that with all the pilots coming out of flight school, the Air Force would take operational fighter pilots and send them to tankers. I had flown a Sabre at 400 knots, forty feet above the ground, missiles and guns blowing away everything in my wake. This was the environment I craved. And now I was going to fly tankers?

The assignment dimmed the joy of our returning home, and the K-55 Officers Club rang with the sad songs of the 69th fighter pilots. We had all received the same orders. I wrote Marta a long letter that evening telling her I would be returning to civilian life. The next morning I requested my discharge from active duty.

I wrote to four aircraft companies, hoping to get a job that offered a cockpit position. Marta was waiting in San Francisco and, after a joyful reunion, we left for Texas so I could meet my new baby daughter, Carmen. After spending so much time overseas I looked at my country with an even greater love and appreciation.

The only job offer came from McDonnell, and my feelings on returning to St. Louis were mixed. I missed the flying and the camaraderie of the pilots, and the daily adventures we lived in Korea. But Harry Carroll's exuberance helped me through the first few weeks and gradually I adapted to being a civilian.

I was assigned to the F-101 Voodoo flight test. I could touch the aircraft, brief the pilots, check the preflight instruments, and climb into the seat, but I could not fly the planes. That hurt. Fresh from Korea, I was used to doing things for myself and I got crosswise with McDonnell's union mechanics and inspectors on a variety of issues. Union stewards became familiar faces to my bosses and I was directed to follow the rules. I had been on the job only a few weeks and I was already thrashing around, trying to get a job done and unhappy with my new role.

In October of 1958, McDonnell posted a notice for flight test assignments at Holloman Air Force Base, New Mexico. I applied. Holloman is located sixty-five miles north of El Paso, Texas, and is near White Sands. The Sacramento Mountains lie to the east and lava beds to the north. The valley is a giant corridor stretching 150 miles, almost to Albuquerque. The first atomic bomb was detonated at the Trinity site at the midpoint of the corridor. The vast, uninhabited areas provided the remote location needed to test the early developments in rocketry and aircraft missile systems.

When I arrived, every conceivable type of test

was being conducted. High-speed sleds shrieked across the desert, balloon flights took man to the edge of the atmosphere, and parachute jumps were made from altitudes of over 100,000 feet. Aircraft were being tested with new missile systems and the Army was using rockets to shoot B-17s from the sky. The Zero Launch projects strapped pilots into airplanes that were powered by rockets and screaming engines. In a burst of fire and smoke the pilots were sent airborne. The Matador and the Mace, pilotless bombs with wings and jet engine, were zero-launched as well. Occasionally, one would get loose over the town, flying erratically until it crashed into the mountains or drifted back over the test range and was downed by the shotgun aircraft.

I admired the steely raw guts of the pilots, engineers, and doctors who volunteered for these tests, pushing their bodies and minds to find the boundaries of human performance. Holloman ran on the pure energy of the test projects. The air crackled with high-altitude missile firings and the flight line never slept. I felt alive again.

The Quail, powered by a small jet engine, was McDonnell Aircraft's entry into a competition to develop a decoy missile that could be launched from the bomb bays of the B-47 and the eight-jet B-52, the most advanced bombers in the U.S. arsenal. The decoy's purpose was to confuse the Russian air defense systems by replicating the radar signature of a large bomber.

The head of the Quail flight test was Ralph Saylor, known as "The Great White Hunter."

Saylor was six feet tall, lanky but imposing. He had a sun-bleached crew cut, a great bushy mustache, and crystal blue eyes that peered out from beneath a brown Aussie hat. He was formidable; there was never any doubt about who was in charge of the 200 yards of the McDonnell flight line at Holloman. In short order, he assigned me as the lead flight test engineer for the B-52, with authority over the aircraft. No one got a seat without my okay. My job was to plan the mission, install the launch gear and missiles, test the missile and launch system, and hand the flight test data to the engineers. It was my baby, politics and all.

In the meantime our family continued to grow. On July 27, 1959, we were blessed with our second daughter, Lucy. I was nervous and clumsy in my first experience with a new baby, but savored the chance to share Marta's delight. Carmen was walking, not a rare thing, but I would watch for hours mesmerized as she learned to balance, standing uncertainly, then stutter-stepping around the room. We enjoyed our weekends driving through the mountains from Cloudcroft, New Mexico, to the Mescalero Apache Indian Reservation. We formed friendships with the families of the flight test team and, in many ways, it was like being back in the Air Force.

Jack Ernst was the ground test conductor and on hot missions (with live missiles to be air-launched) he operated from King 1, the range control center, plotting the armada of chase, photo, and shotgun aircraft that accompanied a Quail launch. He made sure that we observed the range safety boundaries and that we stood by for any test replanning or flight contingencies. Jack was always trying to get volunteers to join him in the Officers Club annual rattlesnake roundup. Prowling through the lava beds, Jack had collected several over six feet long. He was not easily disturbed by the critters of the desert.

In December, shortly after noon one day, I received a call from Jack. "We have a problem," he said. "Meet me in the operations room at Boeing as soon as possible and bring your schematics!" I took one of the flight line trams and entered the operations room just as the missile launch operator, Milt Norsworthy, was on the radio from the airborne B-52, explaining his problem. "I was extending the lower missile-to-launch position. Everything was normal, and then I had indications that the launch carriage was gone. We've lost all power to the missile, but the chase aircraft says that the missile is still in launch position."

The implications were immediately evident and they were ominous. Normally, in this condition we could jettison the entire launch assembly. In the current situation, the missile was hanging two feet below the B-52's landing gear and we had no control over it. We had only two options: land on top of the missile and hope that nothing went BOOM, or have the crew eject and lose an instrumented B-52 — and maybe the Quail program. If we landed, the possible loss of both the aircraft and crew was very real. We had to figure out how to get the crew and the airplane back safely.

Landing on top of the missile posed obvious problems. The B-52's main fuel tank vents were aft of the bomb bay, as was the rear landing gear. The engine start tank was on the underside of the Quail and was fueled with explosively flammable ethylene oxide. When we landed, there would be one hell of a ball of flame and, if the missile came off the launch shackles, it would hit the rear landing gear, blowing the tires, tearing out the hydraulics, and possibly igniting vapors from the fuel vent system.

The B-52 had plenty of fuel and continued to circle overhead while we discussed the quandary.

As the errant B-52 continued to circle the base, there were some wild suggestions, including one for Norsworthy to climb down to the launch gear in the bomb bay and attempt to reconnect the electrical umbilical. I described the rigging of the umbilical and clearly and unequivocally stated, "We are wasting time. There is no way to reconnect the umbilical in flight. We should start working on things we *can* do."

Without a pause, I turned the discussion to other options, brainstorming the problem with my Pacific Airmotive team of mechanics. Bob Brown, McDonnell's best electrical engineer, suggested that if we landed on the missile "softly" the pins in the carriage drive motor would shear, allowing the missile to be pushed back up into the bomb bay. We decided that was the best option. Ralph Saylor, the Quail flight test boss, concurred with the recommendation and voiced the plan to Al Perssons, the B-52 pilot. We then started to look at landing techniques. We were tuned in to the discussions between the pilots, who thought there would be little difference between a lakebed or concrete runway landing. All agreed that we should land on foam to smother the flame when the ethylene oxide torched off. Perssons made the decision to land at Holloman and after a practice approach came in for a perfect landing. If ever there was a need for a "grease job," it was that day and he pulled it off.

After a brief flash of fire, the missile pushed up along the launch track, the B-52's landing gear touched down, and the aircraft continued rolling down the runway to a safe stop, chased by a fleet of fire and rescue vehicles. My team had its first flight test save.

By the new year, 1960, we could see the end of the Quail competition in sight. We had the winner, and it was time for me to think about moving on. I wanted to return to active duty, citing my B-52 experience and gaining endorsements from both Boeing and McDonnell flight test pilots and management. I was willing to fly anything.

I received a standard form letter from the Air Force, turning down my application. The Air Force did not need any more active-duty pilots. I was devastated. I had been declared surplus by the Air Force at the age of twenty-seven.

During lunch at Holloman I often read *Aviation Week* magazine, searching out news of our competitors and looking for pictures of my B-52. In the spring and summer of 1960, as I worked on the Quail program, the magazine was devoting more and more attention to the man in space project. McDonnell Aircraft had experimented with the concepts for a blunt-body ballistic spaceship as early as 1955, when I was there briefly after my graduation from college. In January of 1959, the company was awarded a contract to develop the one-man Mercury spacecraft, and they were expanding their engineering team rapidly.

With the flight test program concluding at Holloman, it was time to figure out where to go next. Remembering my chagrin when the Soviets grabbed the high ground with the launch of Sputnik, I decided to move to space. After the Air Force turned down my request to return to active duty, I accelerated my plans when I noticed a small ad in *Aviation Week* that said: "The NASA Space Task Group is looking for qualified engineers seeking to work in the space program. Project Mercury positions are available at Langley Field, Virginia, and at Cape Canaveral, Florida."

I took the advertisement home and talked to Marta. That evening I wrote a letter to the Space Task Group requesting an application for employment. My bosses, Ralph Saylor and George Doerner, applauded my choice, but the pilots thought it was a bad call. They derided the civilian space program because the equipment was wingless and fully automatic; the spacecraft were able to function just as well without pilots, they said. They thought man should go into space in winged rocket ships like the X-15.

Test pilots labeled Project Mercury a Mickey Mouse operation, a man in a can. Then they said, "Everything they launch blows up!" The rockets did have an unfortunate tendency at launch to keel over on their side, a scene that reappeared frequently in the newsreels. Of the nineteen unmanned U.S. rockets launched in 1959, nine failed their missions.

In spite of the risk, I felt I had to press on. I had chosen my direction. I believed that space was the future.

I was hired sight unseen, as virtually all of us were in those early days. I mailed my application and a few weeks later received a phone call saying I had been accepted. This was how fast the agency was moving.

Driving into Hampton, Virginia, on a dreary, rainy day in October 1960, I felt lost. My unease was not just a product of changing jobs. I missed the desert, missed flying, and suspected that I had made a disastrous mistake. Maybe the pilots were right. Maybe this *was* a Mickey Mouse deal.

Marta and I checked into a motel on Military Drive and camped out there while we looked for a house with enough room for us and our daughters, Carmen and Lucy. (Little did I know then that I would not be spending much time at home for the next three years.) I was familiar with Langley Air Force Base from my flying days, but the material sent to me by NASA gave a much deeper perspective. Congress had funded the National Advisory Committee for Aeronautics in 1915 "to supervise and direct the scientific study of the problems of flight, with a view to their practical solution." In 1917, the first aeronautical laboratory was built at Langley. This was the cradle of early aviation and, as I drove around the base in search of my new home, I felt a reverence for the early pioneers who worked in the labs.

Most of the two-story red-brick buildings dated to the 1930s and seemed to emit a musty antique smell. The people of the Space Task Group were young, much like the members of my squadrons in the Air Force. But the similarity ended there. Just looking at the people, you sensed they were radically different.

In the military, everyone was cut from a common stencil that allowed only minor variations. Those in the Space Task Group were a rainbow of personalities, mannerisms, and speech. They were friendly, almost collegial, but they did not seem to have the intensity and focus I had experienced elsewhere. They seemed to be dreamers, dealing in ideas rather than actions. I wondered what the hell I had gotten into and how I would fit in.

I reported to the personnel office and was directed to walk across the alley to the operations building to find Chris Critzos, the administrative assistant for operations, who had hired me from Holloman. Critzos, a natty dresser with a nasal inflection to his voice, took me down the corridor to my new office. Over the years, I learned to respect Critzos for his ability to slice through rivers of red tape. He was the mechanic who got things done, and I don't think he ever failed. There were no doors on the offices, so we walked in. The introductions to my office mates were brief and businesslike.

Paul Havenstein was an engineer and a naval officer detailed to work in establishing Mercury Control. Another engineer, a guy named Paul Johnson, was working on the remote sites. A third, Sigurd Sjoberg, was the assistant to the Mercury flight director, Chris Kraft.

Havenstein and Johnson, while pleasant, continued an intense conversation. Only Sjoberg seemed to acknowledge me. I was taken immediately by his friendliness and sincerity. Just talking to him brought a smile, but as I listened to him I saw a depth, a passion, that frequently broke to the surface like a trout taking a fly. Sjoberg handed me off to Johnson to learn about my new job.

This was the first clue I had about the work I would do. Johnson gave me a three-page job description of my position on the control team and an IBM book on Mercury Control. I had been onboard only minutes, and the job description was Greek to me. Johnson said he would get with me later, which, of course, turned out to be those all-too-short two weeks at Cape Canaveral. But even when he wasn't around, he was still my guardian angel in my first, uncertain months in the program.

The people of the group were friendly, but

unlike the Air Force they did not go out of their way to make a stranger welcome. Their reserve, combined with their preoccupations, made it tough to get started. Gradually, I got to know the rest of the office: John Hibbert from Bell Labs, the Englishman John Hodge, and Kraft, who answered to Chuck Mathews, the operations division chief.

Intense and high-energy types from Britain and Canada milled around like Boy Scouts at their first camp trying to figure out where to place the tents and campfire. They filled critical positions in every work area and much of the important midlevel leadership. Fred Matthews, Tecwyn Roberts, Rodney Rose, and John Hodge seemed to be everywhere, covering every base. Months later I found this was the elite Avro flight test and design team. When the Avro Arrow, the world's top performing interceptor aircraft, was canceled by the Canadian government, the engineers came south to the United States and into the Space Task Group, providing much of the instant maturity and leadership needed for Mercury.

I found it difficult to believe that the people in my building were the core of the team that would put an American in space. For the first time in my life I felt lost, unqualified, but no one sensed my confusion. Then I thought, maybe they feel just like me. All I knew was that the clock was ticking down to the next launch and, after the Space Task Group's first Mercury-Atlas launch disaster, this one had better work.

Behind the friendly faces there was an air of

formality and an informal pecking order not represented on any organization chart. The local people talked about Tidewater, their little spot in Virginia, as though it was heaven on earth. After coming from the desert and mountains, I wondered if they had ever been out of their home state. The Tidewater group was like a country club, with a bunch of unwritten rules that only the longtime members knew. I soon found that I had some measuring up to do.

Kraft advised me to dress up. No more sport shirts and khakis. Then a few days later he said, "Let the secretaries do your work." I had been doing my own typing and other office functions and had offended his secretary. I was out of step. Everyone seemed to be busy and moving to some cadence that I did not hear. I wondered whether I had come so far to be that saddest of all figures, an unnecessary man.

Hampton, Virginia, 1960

During the second week, I started to grasp the lines of authority. Kraft's role was like the operations officer in a squadron. He called the shots, assigned the resources. Kraft's leadership style was to state a position that he had thought through and see who would challenge him. Familiar with his technique, Sjoberg and Hodge would rise to the bait. Kraft liked to lead and at times deliberately injected an emotional content into the discussions by overstating a position, just to see how strongly others felt. Chris and Hodge could really get going, but with Sjoberg acting as moderator it stayed friendly.

I was just an observer and, while most of the dialogue went over my head, I slowly came to realize that since there were no books written on spaceflight, these few were writing them as they went along. This was their style. It was time to join them and pick up part of the workload. I knew about flying, systems, procedures, and checklists. I started to figure out how and where to use my background to fit in. It wasn't easy, nor did I expect it to be.

By the time the Gemini program got rolling, I knew my job much better. I looked forward to stepping into the flight director's shoes and taking charge.

6

GEMINI — THE TWINS

The astrologers loved the Gemini project. Gemini was one of the twelve constellations of the zodiac, the sign of the Twins, and one controlled by Mercury — a perfect label for a spacecraft flying a two-man crew. This program would be the training ground for the lunar landing. To reach the Moon, we needed to develop new skills in mission planning, in the rendezvous and docking of two spaceships, in performing on-orbit maneuvers. Computers had to be abruptly yanked out of laboratories and made operational. Our mission duration had to virtually double with each flight until we reached fourteen days, the longest possible lunar mission duration.

Then there was the matter of pride. We were tired of being second best in space. We were reaching for the brass ring, an American manned space record. With two spacecraft, the manned Gemini and the unmanned Agena rendezvous target, we doubled our risk and the burden of responsibility. We now had two guys with fishbowls on their heads, sharing a cramped cabin and flying higher and farther than anyone before them. They were also flying untested state-of-the-art systems.

The Gemini spacecraft looked like the Mercury capsule, but if you peeled off the skin, you could see a profound difference. An onboard computer provided the capabilities for precision navigation and maneuvers. Fuel cells and cryogenics allowed longer mission duration, bipropellant rocket engines were more efficient, and the propellant fuels were storable.

The Gemini technologies were new and alien, the engineering so complex it bordered on pure science. My engineering knowledge had expanded only in a practical sense after my college years. With the exception of a brief experience with a computer on board the supersonic F-100 Super Sabre, my knowledge of digital systems was nonexistent. I was a dinosaur stumbling forward into a technical revolution.

Tec Roberts's digitalized control center was taking shape at the new Manned Spacecraft Center in Houston. Many of us were quite nervous about, even suspicious of, computers, but they were inevitable. Automation allowed us to stay ahead of the escalating risks of spaceflight. With radically improved ground data systems and a deeper knowledge of the Gemini spacecraft systems, we gained greater control from the ground and enhanced our capability to support the astronauts. This was a whole new ballgame.

The tracking stations were now using digital systems and 2.4 kilobit/second high-speed data transmissions. The Mercury requirement for multiple voice communications was relaxed to once per orbit for Gemini, allowing the thirteen manned Mercury sites to be reduced to six, including the two tracking ships. The Mercury veterans gave us a foundation to build on. Now, with only six sites to staff, the control team skills were the highest in our brief history. We reinforced our existing teams with the first generation of college graduates who had grown up in space. Young engineers schooled in the new technologies were matched with the Mercury veterans, and jointly they marched to the edge of knowledge, technology, and experience.

During the Gemini years, the Kranz family continued to grow. We were back to girls. Our daughter Brigid was born on February 15, 1964. We considered her and Mark our "twins" because they were born eleven months apart. That was a pretty tight formation.

After Gordon Cooper splashed down in May of 1963, nearly two years passed between the last Mercury and the first manned Gemini launch — twenty-two months and one week. We needed every precious minute.

The Mercury debriefings indicated that in the unforgiving and fast-moving world of space, the personal abilities of controllers and the quality of data feeding into the console were key to the controllers' performance. "Learn by doing" kicked into high gear in Project Gemini. Flight controllers from my branch deployed to the contractor facilities, returning with bundles of the drawings used to manufacture and test the spacecraft and boosters. We studied the manufacturing and test data and then prepared schematics and performance plots on each of the spacecraft systems. This data was then used in our classroom studies, also taught by the controllers. When the schematics and training were completed, the controllers turned to the flight procedures, then to the mission rules.

Only after we thoroughly understood the design and operation of the spacecraft, did controllers focus on the Mission Control Center, designing our displays and laying out our consoles. During the years preceding our first Gemini mission, we lived as a team, accumulating vital data, preparing the mission plan, and teaching each other. We were ready to start only when we trusted our data and trusted each other.

The relationships that developed between the controllers and their families on Flight Controller Alley were both personal and professional, and it was not long before they turned to the social and the athletic. One evening while standing on the corner talking, Dutch casually stated, "How about coming out to Ellington with me tonight? I'm thinking about starting to play judo again. They have a pretty good coach who is forming a team." This invitation started me on a decade-long love for the sport of judo, the "gentle art." When I was issued the invitation, I did not know that Dutch and John Llewellyn had been playing the sport for years and were brown and black belts, respectively. I was the new guy on the block. Our primary instructor was the Houston Marine recruiting sergeant, and his team was topnotch in the Armed Forces Judo Association, in which we competed for promotion.

We looked forward to moving into our new permanent offices at NASA near Clear Lake, outside Houston. Inside the old Stahl-Myers plant where we had been, the constant hassle to make more room for the incoming rookie controllers had us all but sitting in each other's laps.

I compounded the floor space problem there when I brought in a cockpit trainer for the controllers. Electrical cords stretched across the floor and under the drafting boards and desks. People treaded lightly near the mockup, walking around it like a swamp. But I wanted the control team to know the cockpit as well as the crew did. I spent hours in the plywood and cardboard replica until it became as familiar to me as my console in the Mission Control Center. The controllers practiced blindfold drills, sitting in the cockpit, reaching out and touching each switch or running through each of the checklists until they had the intuitive feel for the crew's every procedure. That was comparable to what we did when preparing to fly jets. We would have a blindfolded cockpit drill, with an instructor leaning over your shoulder. This enabled you to find any switch, at any time, under any flight environment. That was exactly the kind of proficiency and self-confidence I wanted my controllers to develop.

Shortly after noon, on November 22, 1963, Maureen Bowen, Mel Brooks's secretary, burst into my office, her face drained and white. She said, "Kennedy has been shot in Dallas. Connie just heard it on the radio!" The shock spread through the building. We hung on to each radio report. Someone found a television set and we congregated around the drafting tables. Tears were coursing down the faces of Kennedy's moonstruck recruits. John Kennedy had inspired us with his vision. One by one, we left work to grieve in private. The flag was at half-staff in our hearts.

The vice president, Lyndon Johnson, who as the Senate majority leader had been instrumental in passing the Space Act, was sworn in as the thirty-sixth President on the plane that carried Kennedy's body back to Washington. We now looked to him for leadership. We were confident that he would carry out JFK's commitment. But none of us will ever forget what it was like to live through that incredibly sad weekend when America came to a stop, stunned by this tragedy. At Mission Control and throughout NASA, in our hearts we resolved to honor John Kennedy's memory by meeting the challenge he had set for us.

Midway through the interval between Mercury and Gemini, Hodge, Lunney, and I were named as flight directors for the Gemini and Apollo missions. The short intervals between the planned missions (less than two months), high flight rate (up to six missions per year), and long flight duration (up to fourteen days) demanded around-the-clock operations in Mission Control. The announcement in August of 1964 was a hell of a nice present for my thirty-first birthday. I was no longer flying Kraft's wing; I now had my own team to lead. To differentiate the teams for training and mission support, the flight directors chose identifying colors. Kraft chose red for his team, Hodge blue, and I selected white for mine. At a later date, Lunney chose black as his team color.

In Mercury, the decision process seemed to be a shared responsibility between Kraft and Walt Williams. At the conclusion of Mercury, Kraft knew that the flight director must have the authority to make the final decision and more autonomy within his team. The operations director's role had to be advisory rather than supervisory. Kraft's job description, when he was Mercury flight director, had included words such as "directs, controls, monitors, and approves." But Chris had gotten tired of having someone sitting above him. For Gemini, he began rewriting the definition. The mission rules are the flight director's bible, and in that document he inserted this new description:

> The flight director may, after analysis of a flight, take any action necessary for mission success.

This was about as clear as a job description will ever get. So the flight director no longer shared authority. He was now the guy in charge.

We had plenty of other job descriptions to lay out. For example, the Titan II booster for Gemini was equipped with a primary and secondary guidance system, plus the ability to switch between the systems in flight. Since the switchover cues were determined by the flight trajectory, a guidance position was established in the trajectory team.

Responsibility for the Gemini spacecraft's systems was split between two engineers. One was accountable for the guidance, navigation, control, and propulsion systems; he was assigned the communications call sign GNC. Guidance systems provide the computations or corrective actions to achieve a set of orbital conditions, or to reach a target. Navigation systems determine the position, velocity, and orientation of the spacecraft. Control systems use guidance information to compute and provide pointing, steering, and engine start and stop commands. The second engineer was given the electrical, environmental, and communications systems, call sign EECOM. With longer and more complicated missions, we established support staff rooms to assist the front-room controllers. Midway in the Gemini program, we further expanded the team to accommodate the Lockheed Agena rocket that would be used as the rendezvous target.

January 1965

Gemini 2 was the first mission requiring controller support. (Gemini 1 was an unmanned Gemini launch on a Titan II rocket on April 8, 1964. The Cape MCC flight support was limited to an informal evaluation of the Titan rocket systems and launch trajectory support by the booster and trajectory controllers. The dummy spacecraft was not recovered.) The game plan for Gemini 2 took Kraft and me to the Cape for what we hoped would be penultimate Cape deployment. The construction and much of the testing of the Houston Mission Control Center (MCC-H) was completed. John Hodge headed the checkout team in Houston that would monitor Gemini 2 from the new Mission Control Center. After two successful missions in a monitoring mode the mission control functions would be transferred from the Cape to Houston.

Kraft and I were sharing a two-bedroom efficiency apartment to stretch the per diem. The arrangement was convenient, since the living area provided enough room for us to spread out all the material we had to work on to study the endless details involved in a launch. This mission was to be a simple lob shot, much like Alan Shepard's, but without a pilot. The goal in Gemini 2 was to check out the Titan's propulsion and guidance system, and its ability to steer along the planned launch trajectory. This would place the Gemini on a path to test the heat shield on reentry. An automatic sequencer controlled the entire mission. The only role of the control team was to issue backup commands if the sequencer failed. Two ships were deployed downrange to monitor the eighteen-minute flight.

Ed Fendell was a rookie flight controller assigned to lead one of the ship's teams. In Gander, Newfoundland, he was the voice on the radio guiding pilots to safe landings in blizzards and later, at the Cape, he worked in range control. A hard-driving perfectionist, Ed was perfect for Gemini, and he inherited responsibility for a computer-driven remote site system that was marginally ready to do its assigned job.

To prepare for my role as a flight director, I essentially became Kraft's understudy. The pre-mission responsibilities were vast. The flight director developed the mission game plan; ascertained that each test objective was scheduled; verified the network readiness; made sure that his control team was fully trained — and came up with a contingency plan in case things went wrong.

Throughout the pre-mission period, the flight director oversaw the mission control, network, and team readiness, and provided the flight operations Go NoGo at the reviews preceding the start of the launch countdown.

Each evening, Chris and I would meet in the apartment, and I would listen to him run through the long list of open items for the first unmanned launch. As troubles mushroomed in the thruster and seat tests, he became obsessed with ensuring that the control team was capable of detecting and responding to any problems in flight. Kraft gave me the authority to do anything needed to get on top of our job. I told him I had been marching my controllers in this direction for a long time. This was my first experience in the Go NoGo world in which Kraft lived.

At launch minus six days, Kraft got sicker than hell. He couldn't even leave the apartment. I knew it was bad when he told me to cover the final simulation, pad tests, and pre-launch briefings. I wondered what he would do if he did not recover for the launch. Each day as I returned, I waited for his word, half expecting him to tell me to step into his seat. While I prayed for his swift recovery, I have to admit that I also liked the idea of filling his shoes for the first time. Kraft told me to go ahead with the briefings. Nervous as a cat in a rocking chair factory, I conducted the final readiness reviews. It finally dawned on me that I might actually have to do the launch, and I looked around for a security blanket, a way to mentally power down when I needed. A copy of *Sports Illustrated*, with a cover photo of an awesomely attractive young woman in a one-piece black-and-red swimsuit, was on the table. I cut out the photo and slipped it inside the plastic cover sheet of my mission book. Whatever happened, I would have her image with me at the console.

Kraft recovered through sheer willpower. As I watched him get dressed for launch day, impeccable in a crisply starched shirt and tie, I was relieved. With Chris firmly ensconced in the flight director's seat, my job on Gemini 2 was simply to make sure the control teams were go, and the data flow met the pre-launch requirements. During the brief test flight the mission controllers would send ground commands to back up the automatic sequencer in the Gemini. These commands would separate the Gemini from the Titan rocket, fire the separation rockets, initiate the turnaround maneuver for reentry, etc. I used two stopwatches; one started at liftoff, and another started at Titan rocket shutdown to cue the controllers to send their commands. On my call, the controllers issued the commands to back up the onboard sequencer.

The press was allowed inside the control room. Launch coverage was provided by fixed cameras and lights on both sides of the room, with a roving camera coupled by an umbilical to a recorder. We weren't going live with this one but we would be live on later launches. It seemed excessive coverage for a simple lob shot, but it was a launch that kicked off the next round in the race to the Moon.

As the countdown progressed through the last few seconds before launch, the lights turned on, the room momentarily bathed in brilliant white, while the cameras whirred. I had a fleeting urge to call out, "Lights, camera, action!" Then everything in Mission Control turned black as the Titan lifted off. It was so dark I could not read my stopwatches. We had been plunged into a power failure because of the overload caused by the TV lighting. The only illumination in the room came from the small buttons on the Western Electric intercom sets, which were provided with a battery backup.

Working blind, I listened to the reports from the launch pad. Unable to do anything, the controllers in the dark monitored reports coming in from the ships downrange. When Hodge didn't hear my backup command calls, he reported mission status to Kraft from his console at the new control center in Houston. The teams were prepared for anything — except a total blackout.

Gemini 2 was in reentry and the mission virtually over before we were able to restore electrical power. The Titan rocket and Gemini spacecraft performed flawlessly and Hodge's team in Houston tracked the entire mission. Kraft's debriefing was short and curt. "Find out what happened," he barked, "and fix it so it never happens again." We soon determined that when the press powered up their lights and cameras the surge momentarily overloaded the circuit breakers and cut the power to the entire control center at the Cape. Houston made sure there would be no such snafus in the future. Critical systems were reassigned between two separate electrical circuits powered by three different electrical sources. And the press was required to provide its own power.

The mission was declared a success and the team returned to Houston, enduring the cheerful put-downs of the Houston controllers who had participated in the mission. In debriefing, Hodge jokingly told Kraft he should carry a flashlight with him for Gemini 3, the final mission to be controlled from the Cape. Kraft didn't laugh.

March 1965

Almost overnight, it seemed we were back at the Cape in the final days of preparation for the first manned Gemini mission. This three-orbit flight test involved a large number of maneuvers to check out the propulsion and guidance systems and the new onboard computer, the first ever used in space. Gus Grissom had been selected as the pilot of America's first two-man spacecraft with John Young as his co-pilot. Bright and exuberant, Young was a Navy pilot from the second class of nine astronauts. We now had three astronaut groups, including a new breed of scientist-astronaut, competing for a handful of flights.

Since the spacecraft would do three orbits, the sites at Carnarvon, Australia, and Hawaii were critical to the preparation for deorbit and entry. Dan Hunter, who was leading my operations section, anchored the team at Carnarvon. During the final week of testing, the network achieved readiness and I gave the Go to Kraft for the remote site teams.

Many of the early astronauts, in particular Slayton and Shepard, doubted that CapComs who weren't astronauts could stand up to the pressure of time-critical decisions and communications. They believed that only an astronaut should get the job as the remote site CapCom. I disagreed. My remote site teams had matured. In my view, astronauts assigned to remote sites were observers given the job of assisting the controllers if that became necessary. In any event, five days before the launch Slayton deployed astronauts Charles (Pete) Conrad and Neil Armstrong to Carnarvon and Hawaii, respectively, continuing the tradition of stationing astronauts at critical positions on the ground track. Hunter and Conrad were both men of strong conviction. But given orders, they would salute their leader and then execute with few questions asked.

Three days before launch, the mission readiness review was concluded, and the time had arrived to begin the countdown. After a great meal at Ramon's Supper Club, Kraft and I returned to the apartment we were still sharing. We went over the next day's schedule and then retired for the night. It seemed like we had barely dropped off to sleep when I heard a loud pounding on the door in the hall. Someone shouted, "Chris, we got a problem!" It sounded like Slayton. I rapidly pulled on a shirt and pants, wondering what the hell had happened. By the time I emerged, Kraft and Slayton were in a heated argument. Deke was exclaiming, "Dammit, Chris, get your guy under control!" Kraft then went nose-to-nose with Slayton. I felt that within seconds the dispute would escalate beyond shouting. Then, magically, both realized it was time to deescalate but not back down. Like two junkyard dogs, they circled. Slowly, I realized that Hunter and Conrad had tangled at Carnarvon over who was in charge of the site during the mission.

Conrad had quoted Hunter as saying, "Kranz put me in charge, and if you give me any more trouble, I want your ass out of the control room." Conrad, a Navy carrier pilot and new astronaut, was not about to take that from anyone. Besides, Slayton had told Conrad that *he* was in charge. Kraft finally got Slayton calmed down. It was around 3:00 A.M. by then. Chris agreed he would write out some guidance for the teams at Carnarvon and Hawaii in the morning.

In preparation for a mission, the tracking stations worked on the same schedule as the controllers, so Hunter was on site when I called on the conference loop the next morning. I briefly mentioned Deke's outburst and asked him what the devil was going on out there. Hunter said, "Conrad arrived and proceeded to take over. Then the maintenance and operations staff and the site manager came to me and wanted to know who to take their orders from. I told them Conrad wouldn't know an acquisition aid if it fell on him. If Carnarvon wants to support the mission, they damned well better take their orders from me." (The acquisition aid is a piece of equipment used to lock on the capsule signals and point the site antennas.)

"The site staff," he concluded, "is still not sure who is in charge and they want a Teletype directive to cover their ass." I told Hunter I would get on it, terminated the voice conference, turned and briefed Kraft on Hunter's side of the story. With two days until launch, Kraft was not about to get into a hassle with Slayton over roles and responsibilities. Conrad was an unknown quantity to me and I thought we could paper over the differences until after the mission, then resolve it properly. This turned out to be a serious mistake. Working with Kraft, I drafted a message that clearly reiterated the job description of the remote site CapCom job, assigning Hunter the overall site responsibility. When Slayton arrived, Kraft handed him the report and the argument in the apartment started all over again, only this time before a large audience in Mission Control.

"Dammit, Chris," Deke snapped, "if we are not going to put my astronauts in charge, it was a waste to send them." Kraft cut him off: "Deke, I don't have time to argue. We will put Hunter in charge of the site operations and Conrad in charge during real time."

The deal was cut. Kraft marked up the changes and I had the instructions Teletyped to the Carnarvon and Hawaii CapComs. The next day, launch minus one, the entire network was called up for a final review of the mission rules and procedures. At the end of the call, Kraft began polling the sites for any open issues. When Hunter came on the line, he said, "I've got Conrad here and I'd like to understand the Teletype you sent yesterday." I passed Kraft the message and he briefly summarized the content. The whole world, at least our part of it, was listening as Hunter continued. "This message does not resolve anything. When I get back, I am going to frame it and hang it on the wall in my crapper." Controllers around the world listened, stunned. Kraft was speechless, and Hunter knew he had said too much. At the limits of his patience, a furious Chris snarled, "You've got your orders, young man!"

Following the launch-minus-one-day briefing with the tracking stations, we adjourned to the beach house that had been provided for the astronauts by *Life* magazine, who had exclusive rights to publish their personal stories, approved reluctantly by NASA. Grissom and Young, totally unaware of the Carnarvon flap, had invited Kraft, the controllers, and astronauts Cooper, Shepard, and Gene Cernan for a brief get-together prior to launch. I was preoccupied with how I was going to save Hunter's career when he returned to Houston. Since Slayton was not in the MCC for the L-1 briefing, he was unaware of the friction he had helped unleash. Now he walked in and greeted Kraft with a cheery "How's it going, Chris?"

Kraft, still fuming from his discussion with Hunter, didn't respond. Instead, he made it clear he had no interest in talking to Slayton and walked over to Grissom. The customary mixing between the astronauts and controllers was missing. Controllers were in a group on one side of the room, the astronauts on the other, hovering around Grissom and Young. It was like a wedding in which the bride's side and the groom's side were strangers to one another. You could almost hear the usher: "Friend of the bride? Friend of the groom?"

Approaching nightfall, the controllers, especially John Llewellyn, had had enough to drink, and, as we were getting ready to leave, their feelings surfaced. Llewellyn responded to some remark from Shepard. By the time I got there, the two were going at it, Llewellyn yelling, "You better hope that Hunter covers Conrad's ass. If he doesn't, you can kiss Carnarvon goodbye for this mission."

I grabbed John, moving him toward the door with Shepard on our heels. Llewellyn, forever the Marine, then commented on Shepard's Navy background and again, to his face, said, "I got more Purple Hearts than you'll ever see in your lifetime, you SOB." I corralled Llewellyn again and hustled him outside, where a very concerned Mission Control team jammed him into a car and drove him back to the motel. This was no way to run a mission, and I hoped and prayed that cooler heads would prevail the next morning when we prepared to launch. We needed Llewellyn, and we needed a united team — controllers and astronauts — at every site, in the control center, and in the spacecraft.

Living as we did in an environment that combined the temperament of a football training camp and the confinement of a submarine, with ego and pride all around, as well as relentless pressure, I sometimes wondered why an occasional bloody brawl didn't break out.

March 23, 1965, Gemini-Titan 3

The first manned flight in a program evokes many emotions in me. Seated at the console next to Kraft, I was about to enter a new age, not unlike the leap from the Wright brothers' Flyer to the fighter aircraft of the 1930s, bypassing two decades of normal development. With Gemini, we were stepping directly into the future. The incident involving Hunter and Conrad was far from my mind as the countdown progressed smoothly to liftoff. A brief hold just before launch was the only glitch and then the *Molly Brown* was on its way. Just as every fighter pilot gets to name his aircraft, Grissom and Young assigned that name to their capsule. (Since

Grissom had lost his first spacecraft, *Liberty Bell 7*, when the hatch blew off at splashdown, *Molly Brown*, the "unsinkable" heroine of a Broadway musical who survived the Titanic disaster, seemed a logical choice of name for his second spacecraft.)

This astronaut combination proved to be a splendid one. Both had only the single desire to fly and were pure joy to work with. A three-orbit mission has to go by the numbers. There is limited time to experiment, troubleshoot, or innovate. Grissom and Young's first orbit was devoted to abbreviated checks of the spacecraft, and the second included a brief series of experiments. In the third orbit, the propulsion and control systems were tested, gradually lowering the orbit so the deorbit maneuver could be completed with the reaction control jets, if the retro rockets failed. Fortunately, the deorbit was normal.

Molly Brown's flight was flawless, with only minor technical problems to be worked out. The press and medics blew one item out of proportion. Just prior to launch, Schirra slipped Young a corned beef sandwich he had purchased at Wolfie's restaurant. To our amazement, some were offended, up to and including members of Congress. Schirra had thought his catering service was a practical alternative to the bite-size freeze-dried chicken and beef pot roast provided by the medics.

The mission debriefing was held in the auditorium, attended by the staff of the program office, engineering, and the astronauts. After a meticulous walk-through of the pre-launch and flight periods on an event-by-event basis, we listed the open issues for the subsequent mission. At the conclusion of the debriefing, I asked the flight controllers to stay and when all visitors departed, we secured the doors. The room was silent as I again climbed the stairs to the stage. They knew what was coming. "Discipline. If you remember only one thing from this debriefing, I want you to remember one word . . . discipline! Controllers require judgment, cool heads, and they must lead their team. Leadership, judgment, and a cool head were *not* evident at Carnarvon. I understand how conflicts can raise tempers, but I also expect control. We are the last line of defense for our crews. We let a small incident escalate into a major flap that involved the entire operation. It was a distraction that we should have put behind us. We lost track of our objective."

I took a breath and continued, "Our mission will always come first. Nothing must get between our mission and us . . . *nothing!* Discipline is the mark of a great controller. That is all!" The auditorium was silent as I stomped off the stage and left the room.

The military has long used the command and control principle and now it was formalized in Mission Control. The Hunter episode finally defined the role of Kraft's organization. Provoked by the incident, Kraft sat down with Slayton and cut a new deal that gave the crew control of the spacecraft and gave the ground command of the mission. The overall mission responsibility now rested clearly with Chris Kraft, his flight directors, and the remote site CapComs.

Slayton sent his astronauts to the remote sites as observers for one final mission. Hunter transferred to Goddard Space Flight Center and performed as the Madrid tracking station manager during Apollo. Hunter had fought the battle and lost, but he helped to win the war for Flight Control.

7

WHITE FLIGHT

As we celebrated the success of Grissom and Young on Gemini 3, the Russians were also celebrating. We would soon learn that five days earlier, Lieutenant Colonel Aleksei Leonov had become the world's first space walker, venturing outside the cabin and stepping into the void. On his return Leonov delivered a speech from the top of Lenin's tomb, flanked by the Kremlin leadership. Leonov predicted that "The time is drawing close when people will pass over from orbital flights around the Earth to interplanetary flights, and will go to the Moon, Mars, and Venus."

The United States had yet to set a manned space flight record. Every member in Flight Control was aware of our opportunity in the Gemini program to set records for rendezvous, docking, duration, and extra-vehicular operations. We were confident that our turn was coming.

April 1965

The primary objective of Gemini 4 was to obtain long-term flight experience. We were going to four days in space, which would be longer than all of our flights combined in Mercury and Gemini to that point. The Apollo lunar missions were being planned to last from nine to fourteen days, and we had to prove our capacity for missions of that length. From a Mission Control standpoint, this would be our first full-blown test of the new space center and the new technologies, as well as the first mission to operate on a three-shift basis.

We had three teams operating twenty-four hours a day; the shifts were generally eight hours long, with a one-hour handover at each end. We saw no reason to juggle them much more than that. The crew in the Gemini capsule was awake about sixteen hours and working in flight testing for about twelve hours.

When the astronauts were awake they were in the "execute" phase, so termed because they were executing the flight plan. This was the core of their workday. Kraft would normally cover most of that period with his Red Team. My team's eight-hour shift covered the systems shift, which started with the Gemini crew preparing for sleep. During this shift we would put the crew to sleep, then look at any glitches in the spacecraft. We would check how much "consumables" had been used — oxygen, water, and so forth — and develop "workarounds" to cope with problems. We used this information to assess mission progress. We had to keep a very careful count of the rate of use of consumables in order to achieve the mission's desired duration. The third shift — Hodge's Blue Team — was the planning shift. They took the data we had gathered and devised a daily flight plan, gave the crew a wake-up call, and briefed them on the activities they would be carrying out during their workday.

The Gemini 4 mission, with astronauts Jim McDivitt and Ed White, was my "coming out" as a flight director, the first operation totally controlled from the new Mission Control Center in Houston. I liked the systems shift; it was an engineer's dream. My White Team sorted out the problems and developed fixes so that the mission could proceed as planned. During the mission, no one called me Kranz; I now answered to the name of "Flight."

One week after the Gemini 3 mission, I completed the summary of the flight controller debriefing and carried the report to Kraft's office. While giving Chris a brief verbal summary, I noticed he was not paying attention. He turned to me and said, "I hope you're ready as a flight director for Gemini 4 because I have a job I want you to do for me." He walked across the room and closed the door. Chris wanted to know how things were going as I prepared for my first shift as a flight director. Then he turned to a new subject.

"We had a damned good mission," he said. "The spacecraft did well and now it's time to stretch a bit more. We're going for four days, you know that. What you don't know is that we're going to try to do an EVA [extravehicular activity]. Since January, Ed White has been in training for a possible space walk. We have scheduled altitude chamber tests with the space suit, chest pack, and umbilical to be run in a couple of weeks. We've just about caught up with the Russians — now we can set our own records."

Chris never wasted words; his instructions were invariably clear and concise. "I want you to work with the engineering team in the Crew Systems Division. We are going to do an EVA if we can get the equipment ready in time. I want you to write the rules and put together the data package we will need to carry out the mission." I was reminded of the day when Kraft first sent me down to the Cape, simply telling me, "Here's a job that needs to be done and I trust you. Do it!"

Kraft started pacing. "This is risky," he went on, "but I think it is worth the shot at getting a space walk on McDivitt's mission."

So I began leading two lives in flight control. Daily, I worked at running the Flight Control Branch and preparing as a flight director to lead my team on Gemini 4. I left work at 5:00 P.M., went home for dinner, then returned to work on the EVA plan. Each night I worked with the task force's spectacular engineers, sitting in on briefings and studying the space suit's operations, and then went back to my office to work into the night writing the rules for this high-profile mission.

As I was preparing for flight director duties and developing what I dubbed Plan X for the EVA, Marta had to take on all the work and planning for our move to Dickinson, Texas, ten miles south and about a twenty-minute drive from Mission Control Center. Our family had outgrown our house in Flight Controller Alley.

The EVA task force was the most powerfully creative effort I had witnessed to date in the space program. Astronauts, technicians, doctors, and engineers in a huddle — one minute discussing wrapping the twenty-five-foot umbilical in a figure 8 layout in its bag, another minute reviewing movies of Leonov's space walk (obtained from Russian TV), while commenting on his body position and the mechanism used to tether him to the ship. The task force had to meet several deadlines and, as more top-level NASA management got involved, it was tough to keep a lid on what was going on in order to prevent leaks to the press. Our objective was to have the hardware qualified and planning for the EVA in place by launch minus fourteen days.

I was uneasy. I wanted a full briefing on the EVA systems and procedures for the teams prior to deploying to the remote sites. The orders came down: "No briefings." But Kraft agreed I could use my secretary, Sue Erwin, to type the materials for the flight controllers' data pack, and use Ed Fendell, assigned to Carnarvon, to prepare a data pack for the EVA Go NoGo sites. Sue was a tough-minded, tough-talking rodeo barrel race rider who had broken a leg when her horse fell on her, but had never stopped competing. Absolutely intolerant of foul language (of which more than a little was heard in the hallowed halls of MCC) and a woman with a radiant smile, she was the key to sanity in the wild world of flight control during the weeks prior to a mission.

After I had worked many extra hours with the task force for a month, the tight-knit flight controller community started to suspect something was up. The final equipment qualification for the EVA would not be completed before we held the mission deployment briefings. After the mission briefings on May 10, the remote site CapComs were called into my office and handed a double-sealed envelope. Only when given specific direction by me would they open the package. If no direction was given, the packages were to be returned unopened. The CapComs picked up the envelopes, about half an inch thick. Inside the envelope was another package labeled neatly in one-inch letters, PLAN X. The cover sheet for the data pack read:

> The mission rules and flight plan are to be used with the data you already have; however, these rules cover flight plan activities you have not heretofore considered; i.e., booster rendezvous and extravehicular activity.

This brief note was a hell of a way for the controllers to learn of America's first EVA. The remote site teams would be on their own, but at the MCC we had the talent of the engineers who developed the EVA equipment at our fingertips. The second element of Plan X involved an attempted rendezvous with the Titan booster to obtain experience in flying formation (close station-keeping with another object in space).

With McDivitt flying in close enough, White could use his nitrogen-powered zip gun to propel him toward the Titan booster stage. The zip gun was a T-shaped EVA maneuvering device. It had a pistol grip and two small thrusters at the ends and used nitrogen as the propellant gas. I thought this was pretty sporty for the first American EVA.

One of my primary responsibilities was team building. I remembered a key method we used in the Air Force to weld rugged individualists into a cohesive working group. A squadron insignia is used to give a group of fighter pilots a unique identity — it binds them together as a team. While I was flying the Super Sabre at Myrtle Beach, Marta and I had volunteered to paint the squadron insignia on the ready room doors. Marta made scarves with our insignia for several of the pilots. She understood the need for team cohesion in a high-risk business. I discussed my concerns about the White Team's training with Marta one evening. As we talked, she said, "Gene, white is your team color — why don't I make you a white vest to wear when you are on console? You can use it as your team insignia." Somewhat skeptical, I told her to give it a shot. I would make up my mind later on whether to wear it.

The press kit for Gemini 4 was released on May 21, fourteen days before liftoff. In it was a single page describing an EVA from an earlier plan. People immediately started to speculate on the possibility that there would be an EVA on this mission. We had only one final week of training when the word came down from headquarters: "We are Go for EVA." I transmitted a message to all sites directing the CapComs to open the Plan X package, and I set up an all-sites conference call the following morning.

Around the globe, the plan was read, no doubt, with murmurs of disbelief that we would try for our first EVA without benefit of any training. The flight surgeons, in particular, were incensed because they had virtually no data on crew performance during the suited altitude chamber tests. Today we would not dare take this kind of risk — but at this point we were determined to at least catch up with the Russians, if not pass them. CapCom Ed Fendell's team would give the Go for the EVA on the second orbit at Carnarvon, Australia, and then Ed White would step outside the spacecraft at sunrise over Hawaii. The EVA would be concluded over the Cape. The remote site teams burned the midnight oil to learn enough about an EVA to ask the right questions at the scheduled briefing the next day.

Crew systems engineers briefed the remote site and the control teams on the overall plan and details of the qualification testing. Then I briefed them on the EVA Go NoGo mission rules. Kraft wrapped it up with a brief pep talk. I could feel the confidence grow in the control teams worldwide. Kraft's ability to use just the right words for the occasion was a rare gift. He was able to convince everyone who worked for him that no matter how steep the odds or how great the risks, we would succeed. Just the sound of his voice on the conference loop would give a young, inexperienced controller at some desolate site the confidence essential for doing his job. All of us who grew up in Flight Control learned how to use our own versions of this confidence-building technique that Kraft taught by example.

An electric shock passed through John Aaron when he learned of his role during the briefing. John, the son of an Oklahoma rancher father and minister mother, had trained in college to be a teacher of physics and mathematics, but on the advice of a friend applied to the space program at graduation. Now, just one year out of Southwest Oklahoma State College, John was Kraft's Red Team engineer in charge of the Gemini life support, electrical, and communications systems during the first American EVA. To this day John remains the most respected engineer ever to work in Mission Control. He was a superb mentor for younger, less experienced engineers. He would eventually become NASA's Space Station manager and then the station's chief engineer.

June 3, 1965, Gemini-Titan 4

The maturing technical prowess of the launch team gave us a nearly perfect countdown, with only a minor glitch in lowering the erector that placed the rocket in a vertical position for launch. After a brief hold, the count resumed and Kraft gave the "Go for launch." Gemini 4 had a large number of objectives. During a four-day mission, we had scheduled the booster rendezvous, the EVA, an array of in-flight maneuvers, and eleven scientific experiments. Each Gemini mission would explore some of the many unknowns of spaceflight and test the new technologies needed for the eventual Moon landing. With less than two years remaining before the first scheduled Apollo manned launch, we needed to race through our adolescence and grow up fast.

All three flight directors were in the MCC for launch. Hodge had run the count from booster fueling through crew wake-up, with Kraft picking up when the crew squeezed into the Gemini spacecraft. Shortly after liftoff, the Titan went through a brief period of POGO, a violent chugging that, if sustained, can cause the rocket to break up. During POGO, we had heard Jim McDivitt's reports in staccato, then it smoothed out as the Titan raced toward engine cutoff. The new control center was humming, the computers cranking out the Go NoGo recommendations to the flight dynamics team, the controllers smoothly reporting to Kraft.

Five and one-half minutes after liftoff, McDivitt reported booster cutoff. He waited until the booster thrust had decayed and 20 seconds later fired capsule-separation pyrotechnics and then maneuvered away from the booster with the Gemini thrusters. After he turned around, he saw the booster about the length of a football field away. McDivitt braked the Gemini and briefly fired thrusters to close the gap. Minutes later, perplexed, he reported the booster was now moving away and down from the Gemini capsule. In Mission Control, Llewellyn stood up at the RETRO console and started talking over the intercom to Kraft. Gesturing with his hands, like a cab driver in Rome, he was trying to explain what was causing problems for McDivitt.

As the reports came in, and with night rapidly descending, McDivitt again thrust toward the slowly tumbling booster. The range increased to 2,000 feet, then appeared to decrease and then increase again during the night. Coming into daylight, the booster was now over three miles away. McDivitt checked his fuel quantity, conferred with Kraft, and, observing the cutoff limits, terminated the attempt to rendezvous with the booster.

As I sat in the MCC, I was baffled by the problems. Unwittingly, Jim had kicked open the door to the mysteries of orbital mechanics, and I had a new respect for Lunney and Llewellyn, who quickly mastered the mysteries of trajectory control. After the mission we reconstructed McDivitt's maneuvers. Following separation he was ahead of the booster in orbit and as he thrust toward the booster he was performing a retrograde (slowing down) maneuver. As the spacecraft slowed down it went into a lower orbit. To balance the force of gravity, a spacecraft in a lower orbit (no matter how slightly lower) must travel faster than one in a higher orbit. Thus, by slowing down, McDivitt was descending and going faster, pulling away from the booster's orbit.

In the trajectory world you have to reset your mental gyroscope; orbital mechanics are counterintuitive, particularly for someone used to flying an aircraft, which follows an entirely different set of rules. I realized then that orbital mechanics was something else I needed to learn a lot more about — particularly if I wanted to be ready for the far more complex orbital maneuvers involved in a rendezvous!

While McDivitt was chasing the booster, Ed White did the best he could to complete the preparation for the EVA, which was scheduled to start over Hawaii at the end of the second orbit. McDivitt looked at the timeline. Since the EVA would expose the Gemini and its occupants to a vacuum, with nothing between them and instant death but the thin fabric of their space suits, he decided to postpone the EVA to the following orbit to make sure all checklist items were completed.

John Aaron, seated with the crew systems engineers at the console, welcomed the delay. It bought them some badly needed time. They had been having trouble keeping in sync with White as he proceeded with his checklist. Using the delay to the third orbit, the astronauts had ticked off the items on the checklist and were now ready for the EVA. Fendell, at Carnarvon, made the call. "Gemini 4, your spacecraft's looking great. You are Go to depress [depressurize] the spacecraft."

The oxygen pressure in the spacecraft was gradually reduced from 5.2 pounds per square inch — one third of what we experience on Earth

— to three psi. With the pressure suits inflated, White and McDivitt carefully checked out their suit systems, making sure the lock rings and seals held the suit's air pressure. They then gradually reduced the cockpit pressure to zero. If the suit sprang a leak or they blew off a glove while they were still inside the capsule and the hatch was still closed, they would have had to respond instantly and repressurize. Once they opened the hatch and started the EVA, they faced a different set of problems. With his suit inflated, it was damn near impossible for a Gemini astronaut to get back into his seat and get enough leverage to close the hatch through which he had exited for the EVA. If a major leak occurred in the suit with the hatch open the astronaut wouldn't survive.

Minutes later, after a brief report from the Hawaii CapCom, John Aaron turned to Kraft, gave a thumbs-up, and quietly said, "We're Go, Flight." With the Go from Houston, Ed White emerged over Hawaii to his third sunrise in space, decked out in the suit made by highly skilled seamstresses at David Clark, Inc., in Worcester, Massachusetts.

The world listened mesmerized to the interchange between White, McDivitt, and Gus Grissom, the CapCom at Mission Control during the stateside pass. This was one of those magical moments, like Alan Shepard's launch and John Glenn's reentry, that are forever embedded in my memory. Pride, patriotism, and American know-how triumphed that day. We were now neck and neck with the Russians and in the next few minutes would eclipse

Leonov's space walk record. I was happy and proud, almost giddy, but then reminded myself that it was time to get back to business.

There was no real plan for a sequence and evaluation of the extra-vehicular activity, so White improvised as he went along. He used his thruster gun to maneuver around the Gemini, from the adapter section to the nose, trailed by his umbilical, reporting on what he could see in space and on Earth. He floated and tumbled for twenty minutes, and clearly was in no hurry to return to the spacecraft.

As White passed across Florida, Kraft consulted the timeline and told Grissom it was time to terminate the EVA. Grissom briefly stalled Kraft to give White some more time, then Chris repeated, "I said to get him in!" The final crew comments were vivid. McDivitt: "Come on, let's get back in here before it gets dark." White replied, "It's the saddest moment of my life." McDivitt, sternly: "Well, you're going to find it sadder when we have to come down with this whole thing." White, finally yielding to his commander's orders, reluctantly responded, "I'm coming."

The early lessons were always quick in coming. Throughout the EVA, the hatch seal had been exposed to deep space, and with the cold, the seal lost its flexibility. It was a real struggle to get the hatch closed and latched. Once it was closed and the cockpit repressurized, the team decided not to open it again to jettison the EVA equipment. Another rule in Mission Control was "Don't press your luck."

With the hatch closed, it was time for me to occupy the flight director's chair for the first time. Kraft handed me the logbook as if it were a baton in a race. With a broad smile, he gave me a nod. "Young man, it's yours," he said, then left on a high for the post-EVA press conference.

Mission Control was mine. My White Team members completed their handovers, and Kraft's Red Team left for a well-deserved celebration. Marta's resplendent white vest hung behind the console. No one had seen me bring it into the MCC. What the hell, I made up my mind to put it on. I felt like a matador donning his suit as I put on the vest.

Dutch Von Ehrenfried, seated to my left, was the first to notice. Rolling his eyes, he buried his head in his arms, then scooted his chair over to mine and said, "If you're not careful, they're going to haul you away, then I'll be in charge." He rolled back to his console, made a remark on the intercom, and then I noticed the control room TV turn and zoom in on me. Moments later my picture at the console, resplendent in a white vest, was on television throughout the center, as well as the press area. One by one, the controllers reported to me, "Nice vest, Flight!" The next day, a photo of me wearing the vest made newspapers across the country.

Now that I had decked myself out in Marta's vest, it was time to clean up the open items, resolve the "funnies," or anomalies and glitches, get the spacecraft configured for drifting flight, and put the crew to sleep. It was time to go to work as a flight director.

Throughout the eight hours of my shift the capsule's orbit, combined with the Earth's rotation, would take the spacecraft off range and into sparse tracking coverage. At the end of the shift, only the *Rose Knot* Victor tracking ship would "see" (that is, be able to contact) Gemini 4, once every ninety minutes. (Our official call signs or designators for the two tracking ships were RKV — *Rose Knot* Victor — and CSQ — *Coastal Sentry* Quebec.) This period in the mission, with the infrequent contacts, the spacecraft in drifting flight, and the crew asleep, provided an opportunity for detailed analysis of the spacecraft systems and orbital trajectory.

Mission Control borrowed technology from many eras. Most of the systems the controllers used were cutting-edge, only months removed from the laboratories. One notable exception was borrowed from turn-of-the-century technology. A pneumatic tube system provided the means to transmit messages of all types between the three floors and the work areas used by controllers and computer operators. The P-tube carriers were aluminum cylinders, twelve inches in length and three and one half inches in diameter. A spring-loaded hinged door allowed messages to be placed in the tube.

The flight dynamics team was a principal user of the P-tube system, exchanging hundreds of messages daily within the complex. During a particularly hectic shift, when Llewellyn and his console partners fell further and further behind in unloading and returning the P-tube carriers, the empty canisters lay scattered on the floor about the consoles. Surveying the litter of canisters Llewellyn, a former Marine, stood up, stretched, and in a voice for all to hear declared: "I think I am back in the trenches again with my fire control team, surrounded by empty 105 howitzer canisters."

Inspired by this colorful analogy, "The Trench" was subsequently adopted as a nickname by the flight dynamics team, who used it in their reports and media interviews. Within weeks "The Trench" matchbooks were circulating within the control teams' ranks — and each subsequent mission contributed fresh additions to their lore and legend. The stage was set for competition among the mission control specialties. The Trench had thrown down the initial gauntlet.

My White Team quickly settled down to the business at hand. One of the black arts in the Trench, and one of the most critical, was orbit trajectory propagation. As the spacecraft circled the globe, numerous forces worked on the orbit, twisting and shifting it. It is necessary to forecast the spacecraft position hours and often days ahead for flight and maneuver planning. Now flying our longest mission and with only two Gemini missions remaining before the first attempted spacecraft rendezvous, Ed Pavelka, the FIDO, plotted the data, directing the computer controller to periodically input precise changes to the atmospheric drag constants. He worked with the same precision he used when tuning the engine on his hot rod.

As my shift progressed, Gary Coen, my GNC

(guidance, navigation, and control) engineer, calculated how much propellant had been used during the Titan rendezvous. Then he turned to forecasting the usage throughout the remainder of the mission. All engineers were reviewing the telemetry for the entire mission, looking for anything, even the slightest deviations from the expected. As they worked, they smoked, and soon the usual pall of blue smoke hung in the air over the consoles. Stale cigarette butts, cold coffee, and day-old pizza made up the scent of Mission Control.

Mission Control was windowless. No clock referenced us to local time in Houston. Greenwich Mean Time, the local time in England, synchronized us to our stations around the world. Then as now, every action, both in the spacecraft and on the ground, worked to time that started ticking after liftoff — mission elapsed time (MET). Our bodies were the only laggards, responding to the need for food and rest on a schedule corresponding to a sun we couldn't see.

Besides the team in the control center, I had Skinny Lewis stationed on the *Coastal Sentry* Quebec midway between Hawaii and Japan, and the team on *Rose Knot* Victor also stationed in the Pacific, 1,200 miles southwest of Lima, Peru. The ships on their lonely vigil were the only source of information about the spacecraft's status for the MCC throughout the last half of my shift.

Tec Roberts's new control center was working to perfection. After the Teletype of Mercury, the new Mission Control Center was a giant leap, a window to the New World of space. The Gemini telemetry list had doubled from Mercury. With 225 measurements, I could now review, during a remote site pass, one or two samples of the data seen by the controllers at the site. Trend monitoring of the spacecraft systems was becoming a reality due to the computer and communications breakthroughs generated by the space program.

The eight hours passed swiftly and my relief was palpable as I started shift handover to Hodge. In many ways, my first shift was like flying an aircraft for the first time. It was great to get back on the ground in one piece. My first shift as a flight director was over and, thank God, it was uneventful. I took off my vest and hung it on the rack at the side of the room. It was a good start.

Each flight director had another ordeal to endure. After a ten-hour shift (including handovers), he was expected to spend at least an hour at a press conference, feeding news to the hundreds of reporters covering the mission. I depended on the public affairs officer (PAO) just as I relied on my controllers. He would help me through press conferences (they sometimes felt like interrogation sessions) and make sure that the right word got out — and that the flight director's tail was covered. The PAO really earned his pay when things went wrong. He was our first line of defense — and fortunately we had PAOs who were very good, and unflappable when things got a bit dicey during a mission.

They would prep us on how to answer questions likely to come our way, show the flight director how to put his best foot forward — and keep the other one out of his mouth! I never quite got used to seeing my face on TV and in newspaper photos — but Marta assured me that I was the handsomest flight director she had ever married.

Returning to the MCC, I went to the back-room surgeon to receive my whiskey ration. To assist in getting a good sleep in the center's bunk room, the military surgeons prescribed a double shot before bedtime. I poured the whiskey into my coffee and I sat silently next to Hodge for the next hour, watching as he led his team and periodically chuffed on his pipe. Pouring booze into a cup of black coffee was dumb, but winding down after a shift was a problem for virtually everybody, and this was the way we did it. There were some nondrinkers in our ranks. They either took pills handed out by the surgeon — or counted sheep.

Gemini 4 helped create a media misapprehension that I was a Marine. Jim Maloney, a reporter for the *Houston Post*, a morning newspaper, always covered my late night press conferences. Since the Gemini 4 mission was the first flown from Houston and the first with three flight directors, he wrote an article on Kraft, Hodge, and myself. Adding some color he described me as "an ex–fighter pilot who you would trust with your life. Stocky, crew-cut and blond, Kranz is a bloodthirsty model for a Marine Corps recruiting poster." The next evening after the press conference I corrected him,

"Jim, you got it wrong in your article. I'm Air Force, not a Marine." He corrected me, saying, "I didn't say you were a Marine. I said you looked like a poster boy for the Marines! What the hell do you want me to say?" Over the years this clipping was picked up by the national and international media. No matter how many times I corrected it, the image persisted, even to the present day.

The mission continued through the shift rotations and I slept at the MCC. With five kids, Marta felt my best chance for rest was at Mission Control. She was quietly pleased that I had worn her vest and told me that she would make me a new one for each succeeding mission. Later on, if my team did especially well, I would ask her to make me a splashy one for the landing shift. It became one of the ways I signaled to my team that they had done well.

My relationship with Kraft had subtly changed. As in judo, there is a relationship between master and student that develops as the student grows in skills. Then one day the master steps aside and a relationship based on a newfound respect begins to grow. Kraft was beginning the process of teaching his brood to be on their own. The flight of Gemini 4 ended in a nearly flawless fashion, a smash hit with Congress, the American public, and the media.

The mission debriefing indicated we had indeed done well. The team structure was successful and the shakedown cruise of the new control center was a roaring success. We would need every millimeter of technology as each successive mission increased in complexity, duration, and risk. The computers and television displays gave the controllers instant access to hundreds of Gemini measurements. The trajectory data was instantly available to users, and the pneumatic tube eliminated the need for runners — and the distraction caused as they raced around the control room. But most of all, we liked the new Mission Control Center because we no longer had to travel to the Cape and live in motel rooms for day after day.

Summer 1965

The Gemini 5 mission was scheduled to last eight days, twice as long as the previous one. It would break the Russian record for endurance in space. Both inside and outside NASA, doctors had expressed doubt that man could adjust to life in zero gravity. Some went so far as to predict that exposure for a long period would probably be fatal, but the astronauts continued to confound the physiologists and the doomsayers. The concerns of the medical communities had been increased by a string of reports coming from the Russians at medical conferences, citing problems in adaptation, crew performance, and post-mission recovery.

The key to the long duration of Gemini and Apollo flights was replacing batteries with fuel cells for the orbital phase of the missions. Fuel cells represented the leading edge of the science of electrochemistry. They produced electricity from oxygen and hydrogen and, in the process, generated heat and pure water. The oxygen and hydrogen fuels were stored on the Gemini spacecraft in separate thermoslike insulated spheres at temperatures as cold as minus 400 degrees Fahrenheit.

The fuel cell technology was so new there were no textbooks and little engineering data. These devices were a good deal more complex than batteries, and when we took this technology into space for the first time I suspected that we would have problems. I began a crash course on fuel cells and cryogenic technology. I was fortunate to have John Aaron as my EECOM. Aaron tutored me relentlessly on "his" electrical systems. Frustrated at trying to absorb the arcane details of the cryogenic systems, I asked, "John, if I opened up one of these tanks, what would I see?" Thinking a few moments, he looked up, his eyes glinted, and he said, "Gene, it would be like trying to look through a super-dense ice fog."

The principal unknown member on my team was a new astronaut, Edwin (Buzz) Aldrin. Most of the CapComs in the new groups approached their job with an attitude that said, "I'm here, how can I help you?" They learned the people, positions, and prerogatives through the process of training with the teams. Buzz was different, more assured and opinionated from the start. The good news on Aldrin was his pedigree as an Air Force veteran and an F-86 pilot. Four years later I would give him the Go to land on the Moon, but for now he was just a rookie CapCom.

Returning to my office from the final day of Gemini 5 simulations I was surprised to find several of the controllers standing and talking in my office. As I turned to speak to them I saw a spectacular American flag standing in the corner. The brilliant red, white, and blue contrasted with the gunmetal gray desk. Gold fringe and tassels hung from the eagle at the top of the dark oak staff. I was speechless. I had tried to requisition a flag from NASA many times, but all my requests were ignored. The flags were reserved for the top NASA brass. As I admired the flag, Don Bray, a young controller with the talent of an Army supply sergeant, stepped forward. "Flight, this flag was requisitioned for Mission Control. You're gonna have to carry it over when we fly. Between missions I think it should stay in your office." I carried the flag to the MCC for each of the Gemini missions, returning it to my office after splashdown. At the beginning of Apollo we finally acquired a permanent flag for the MCC. The original flag remained in my office and at retirement it became one of my most treasured gifts.

The crew for Gemini 5 was one of my favorites. The commander was Gordo Cooper, my welcome wagon driver from day one at the Cape, the friendliest of the original astronauts, and one who had shown great poise on Mercury 9 when his electrical system went to hell. Pete Conrad, his partner, had the most engaging personality in the second group of astronauts. Notwithstanding his flap with Hunter at Carnarvon, I liked Conrad's toothy grin, intensity, and frank and total openness. You had no doubt that he would do what he said.

August 21, 1965, Gemini-Titan 5

The Kraft, Kranz, and Hodge mission team cycle on Gemini 5 was virtually the same as on Gemini 4. At the first site after reaching orbit we became concerned about an unexplained drop in the fuel cell oxygen pressure. Since this was the first experience with fuel cells and cryogenics in orbit, there was no frame of reference for Kraft's Red Team EECOM.

During the first pass over Carnarvon, Cooper reported that the oxygen pressure was dropping rapidly. Troubleshooting, the crew found that the circuit breaker for the heater that warmed up the spacecraft's liquid oxygen supply for the fuel cells had tripped. EECOM, listening to the crew's report, recommended resetting the circuit breaker and cycling the heater switch. The crew concurred and acted on his recommendations as they coasted across the Pacific.

By the time the spacecraft arrived over Hawaii, the pressure was well below the point at which the fuel cells were expected to keep running — and continuing to drop. The crew advised Hawaii that they had abandoned the rendezvous test and had powered down most of the spacecraft equipment. The operation of the fuel cells required a very precise pressure balance between the oxygen and the hydrogen and the pressure of the water produced in the cell.

We had no data to guide us when the cells were operating at the current level. Additional equipment was turned off as the spacecraft continued on a path across the United States.

It was customary for the second shift of controllers to be present for launch. This gave the launch team some extra controller resources if problems occurred. It also gave the second shift a good start at working and understanding the problems they would be given at handover. John Aaron from my shift had been sitting next to Kraft's Red Team EECOM since launch. He walked over to Kraft's console with the plot of the pressure data, his face reflecting the grim news. The rate of pressure loss, however, was starting to decrease. Faced with a possible fuel cell failure, but needing to buy more time to evaluate the situation, Kraft elected to continue for a few more orbits to see what would happen. The only good reentry options would occur for the mid-Pacific landing areas on orbits four, five, and six; then we would have to shoot the gap of poor coverage, a no-man's-land. Kraft advised recovery to deploy aircraft to the orbit six landing area. Within the hour, six aircraft, a destroyer, and an oil tanker had received orders to proceed to the landing area southeast of Hawaii.

As Kraft continued to weigh the options, my control team was reporting to its consoles wondering whether we would get a shot at the problem. Meanwhile, Kraft was masterfully playing the options and assessing the alternatives. When the fluctuating oxygen pressure finally stabilized, and the fuel cells were still operating, I knew Kraft was going to go for a full day. He queried Cooper: "Gordo, I think the oxygen has bottomed out. We've got thirteen hours on the batteries. I think we should go for it." Cooper's response was immediate. "I was hoping you would say that. Let's give it a go."

I was mentally going through my handover questions for Kraft when he abruptly stood up and started to put his headset away. Surprised, I glanced at his log. At shift handover the controller coming on shift would check the log entries of the controller he was relieving in order to see if there were any outstanding items or problems. The log was a summary of the status of the mission. There was no set plan to follow — and I suspected he was ready to leave for the press conference. So I asked him, "Chris, what do you want to do?" His reply was crisp and curt. "You're the flight director, it's your shift. Make up your own mind." Kraft had given me the job to shoot the gap!

The spacecraft flies sixteen orbits per day. During a mission, Go NoGo assessments are made of the spacecraft and crew to determine whether it is safe to continue to the next point. These decisions are made by factoring in the spacecraft mission status and the locations of the naval recovery forces. The decisions are for orbits one, three, six, and sixteen, and multiples thereafter. The major daily Go NoGo is the one that shoots the gap of poor ground station coverage between orbits six and sixteen.

This event in my life as a controller stands out

as the moment I finally came of age in Mission Control. Chris was right. I was the flight director, it was my shift, my decision, and I had better get going. Like my first solo with Mr. Coleman, it was time to spread my wings and give it a shot. I was damn happy to have Aaron on my shift. He was born to be a systems engineer; he could tell me in plain language what the status was of any element of the spacecraft.

In Mission Control, there is no such thing as a first team. Every team must have the leadership and technical knowledge to sustain the effort during a crisis. The White Team controllers stepped up to the plate and took their turn at bat.

The Trench worked with the recovery forces to select the best landing areas as we shot the gap. Aldrin, in between calls to the crew, worked with the flight planners to develop alternatives to the rendezvous. The systems team led by Aaron refined the power-down procedures in case we had to use the batteries. The team was on top of the job, and its response to the problem was electrifying. Within a few hours the engineers at McDonnell were running a laboratory test of a fuel cell in St. Louis to replicate the problem we were having in flight.

At John Aaron's urging, I decided to run a load test on the fuel cell. As the team talked, we developed a plan to perform a power-up followed by an oxygen purge to the cell. We selected a time for the test so that if the cell failed, we would have enough battery power to get back the next morning to the landing area patrolled by an aircraft carrier. We developed the procedure, talked to the crew through the power-up, and held our breath. All eyes were on the data from the Hawaii site as the power-up and purge progressed. A purge shoots a stream of pure oxygen into the fuel cell to flush out any moisture and impurities in the cell. The spacecraft was again powered down while we evaluated the data.

The fuel cell sustained the load and with a few more data points we were convinced the pressure would gradually increase. We handed our data to Hodge at shift change and recommended a plan to power up incrementally over the next three days.

The press conference was Aaron's show. When I got the press question, "What would it look like if we could open up an oxygen tank?" I handed the briefing over to him. I was proud of my team and Aaron in particular. Sitting next to me was a fresh college graduate teaching the world all about fuel cells.

We limped through the flight of Gemini 5 a day at a time. By day four, we had full power even though the oxygen tank pressure had only risen to 140 psi rather than the design operating pressure of 875–900 psi.

Before liftoff a clock was set in the MCC to read 119:06. At launch the clock started counting down. At the beginning of Kraft's sixth shift, as the clock approached zero, the retrofire officer (RETRO) counted down the seconds over Kraft's voice loop. At zero, Kraft lit up a cigar and proudly announced, "America has just set a new space record!" The Gemini 5 crew had eclipsed the Russian manned flight duration, and it was suggested to Conrad that he should perform a victory roll. Conrad wisely demurred, "I ain't got the fuel, sorry!"

The next two days became really limited. The seesaw effect of powering up and down took a toll, and then the Gemini thrusters started to fail, further limiting the flight plan. With the crew in drifting flight and many experiments canceled, McDivitt, now a CapCom, decided it was time to pipe some Al Hirt jazz up to the crew for wake-up. Another MCC tradition was born.

During my final shift, I cut the mission short by one orbit. Tropical Storm Betsy had turned into a hurricane, and the recovery forces were moved off the storm track, north and west to the vicinity of Grand Turk Island in the Atlantic. The final shift allowed me to pull a "gotcha" on my friend and judo partner, Llewellyn. The Public Affairs Office prepared a news update for the in-flight crew and this one included a report that Scott Carpenter, now also an aquanaut, was descending to the Sealab off the California coast for a thirty-day undersea mission.

Llewellyn had never forgotten Carpenter's remarks implying that Llewellyn did not know where Scotty landed on Mercury 7. John is intense, emotional, and trusting, perfect characteristics for an easy set-up. The defining moment that brings every mission to an end is John's stentorian second-by-second countdown to retrofire. This is an event looked forward to by the entire team, and one John relishes like a fine cigar. I called FIDO, RETRO, and CapCom over to my console. I told them that we wanted to get Llewellyn to believe that Carpenter would perform the countdown to the Gemini 5 retrofire from the Sealab. The team readily agreed to give it a shot and see if we could pull a gotcha on John.

At shift change time, Llewellyn sauntered into the room, exchanging banter with the various controllers until he got to FIDO. When FIDO did not respond, Llewellyn tried to pump him and RETRO to find out what was wrong. RETRO reported, "It's that damn test with Sealab. It doesn't make sense." Llewellyn, concerned, responded, "What test?" RETRO advised him to talk to me about this mysterious test.

It was tough to keep a straight face as a troubled and deeply serious Llewellyn approached the console. He said, "RETRO said to talk to you about the test." I told him to sit down, then I solemnly said, "John, I know how you feel about Carpenter, and you're not gonna like what we have got to do." His frown deepened as I continued. "We have orders to rig communications to Sealab, so that Carpenter can make the retrofire countdown. I don't like it and have argued all night with Kraft, but he says, 'Do it!' "

Llewellyn looked as if he had been poleaxed, first puzzled but then furious. At that moment, rookie astronaut Dave Scott walked by and asked, "John, what's the matter?" Scott now listened to Llewellyn's explanation of the Sealab command control test. Scott became even more furious than Llewellyn, loudly stating this was some kind of a "half-assed" decision and then complaining that he had doubts about whether he wanted to be an astronaut if such crazy decisions were going to be made in the future.

The team let Llewellyn and Scott dangle on the hook for a few minutes more before telling them it was all a joke. The gotchas in Mission Control were usually irreverent and often silly, a way of sticking a pin in the bubble of someone's ego. They relieved the tension, poked fun, or just let one of the chiefs know that you will take their orders and respect their rank, but you won't run scared. But I knew that I was in for a tough bout the next time I stepped on the judo mat with Llewellyn.

With the successful Gemini 5 mission behind us, it was time to go for a rendezvous and docking. The Russians, after their fast start, had been unable to rendezvous two spacecraft. We suspected they were hampered by inadequate computer and guidance capabilities. On two previous dual spacecraft missions, in 1962 and 1963, their spacecraft came within three miles of each other. Close does not count. A rendezvous means achieving a stable position that allows docking. It was now time for America to try for rendezvous.

The rendezvous target was an Air Force Agena upper-stage rocket. In early 1965, I made a fortunate series of personnel selections and formed a new systems branch combining Mel Brooks and Jim Hannigan, giving them responsibility for the Agena target and the lunar landing module. Brooks's innovativeness and can-do attitude were perfectly balanced by Hannigan's conservative do-it-by-the-numbers approach. Together, they built a great systems team for Flight Control.

The Agena was normally used to place super-secret military satellites in orbit. It was operated by an onboard programmer that issued coded instructions to operate the systems. The Agena was modified by NASA to provide a restartable engine, a docking adapter, a status light panel, and electrical connectors that provided a limited display and control capability for the crew. Many of my Agena controllers had cut their teeth at Lockheed and were rock-solid confident in their spacecraft. Other members of the MCC team were less impressed, disparagingly referring to the Agena as a restartable cigar.

I had a hell of a scare with the Agena rendezvous target rocket during a command test. Five days before launch, I was supporting the Agena readiness demonstration. In the middle of the evening, with everything quiet at the MCC, I heard the Atlas Agena test conductor call out on the loop, "Who the hell is transmitting engine start commands?" He continued, "Houston, have you been sending commands?" Just as I was about to respond, "Negative," I heard the voice of my command technician state calmly, "Flight, the command system has failed, we have been in continuous command transmission. We have belched out every command in our inventory!" We were lucky that day. If we had been in orbit and our command system failed, the commands would have been transmitted to the

Agena, possibly starting the rocket engine, changing control system modes, and turning the telemetry system off. The unplanned commands would have wreaked havoc on the mission. We amended our procedures, but we were still uneasy about the next step. Loading new software into new computers and using it for the first time was like playing Russian roulette. It demanded and got a lot of respect.

October 25, 1965, Gemini-Titan 6

Brooks's team was sharp as a scalpel as the test conductor pressed the button and the Atlas/Agena rose majestically from launch pad 14. To the north, at nearby launch complex 19, astronauts Wally Schirra and Tom Stafford sat atop the Titan II rocket in their couches and listened as the Gemini test conductor updated them on the Atlas/Agena launch progress. To set up the conditions for a rendezvous, we first launched the Atlas/Agena. When the Agena passed over the Cape at the end of its first orbit, we would launch the Gemini spacecraft from another pad into a slightly lower orbit to begin the catch-up phase of the rendezvous. Launch of the target spacecraft was normal. The Agena separated from the Atlas booster and the Agena engine ignited. Brooks's data flickered briefly, then stopped updating. FIDO reported tracking was lost. Then came a report that Range Safety was tracking multiple pieces of debris falling into the ocean.

Meanwhile, the Gemini countdown had continued on Pad 19, but the reports going to the crew became progressively worse. The reports from Canary Islands and Carnarvon were negative. At launch minus fifty minutes, we scrubbed the Gemini countdown and gave the crew the bad news. The Agena was destroyed, pieces scattered in the Atlantic. They had no target for their planned rendezvous.

I stayed with Brooks and his Agena team until there was no hope. Brooks was defeated, utterly spent. Many thought that the Agena had lived up to its less than sterling reputation. I was sad for my team. It was time for a few beers at the Singing Wheel before we started to regroup.

This watering hole was a two-story, barnlike building, the place we went when we needed some R&R. It was located a mile west of the center on state Highway 3 and hosted the Gemini-Apollo generation of flight controllers. The floor in the barroom tilted toward the wall to a degree that made it impossible to lean back in a chair without falling over. The Singing Wheel sold Lone Star beer by the pitcher. Nelson Bland, the owner, knew all the controllers and ran remarkably accurate tabs for all of them. A wall-to-wall mirror stretched behind the ponderous and scarred bartop. The tables in the back room were covered with checkered oilcloth. John Llewellyn's wife, Olga, occasionally tended bar. After judo sessions, Llewellyn, Dutch, and I, wearing our sweaty gi outfits (white cotton trousers and a heavy, kimonolike jacket), would drive over for a beer or two after calling it a night.

It was, for all of us, a place of refuge where we could celebrate on the good days — and lick our wounds on the bad ones. Today was one of the bad ones.

8

THE SPIRIT OF 76

November 1965

The rendezvous in space continued to elude us. The Russians had tried twice and failed — but we were impressed by their dual launch capability. We couldn't even get our target rocket into orbit. Accomplishing this became the highest priority for Gemini and the American space program. Until now, a rendezvous in space was something only mathematicians really believed was possible. They worked out elegant equations and said, "If you launch it at this time, and go this fast, in this direction, you eventually are going to catch up to a target. If you perform the maneuvers properly, the two spacecraft will end up side by side."

Proving that this theory would work became not only a goal, but an obsession. If we accomplished a rendezvous, we would validate the software that controlled the Gemini spacecraft as well as the crew's fallback manual backups. There was no time to waste; we needed to dramatically improve our learning curve in order to be ready for the far more complex and sophisticated rendezvous and docking procedures necessary for a lunar landing. Within hours of the launch failure of our Agena target and the consequent scrubbing of Schirra and Stafford's Gemini flight, we were discussing an alternative mission. A proposal from McDonnell's senior management seemed to offer the most promising option. Walter Burke, the McDonnell vice-president and general manager for space and missiles, and his deputy, John Yardley, suggested we take a page from the Russians' script by launching two Gemini spacecraft in rapid succession from the same launch pad. He proposed using the Gemini 7 spacecraft, flying the subsequent long duration mission, as the rendezvous target for Gemini 6.

Frank Borman, the commander of the next Gemini mission, overheard the discussion and became an immediate convert. The proposal got a cold reception from the Air Force and NASA Cape management, so Yardley, a close friend of Kraft's, took the proposal to the MSC director, Robert Gilruth. (The respect for Gilruth was so great that virtually everyone in the program addressed him as "Dr. Gilruth." Only those very close to him, like Kraft, ever called him "Bob.") Within twenty hours of the Agena failure, Yardley and Burke convinced Gilruth to give the dual launch concept to his staff. In short order, the NASA Gemini program manager, Chuck Mathews, and Kraft agreed to check it out with their people. In the early afternoon, Hodge called a division staff meeting, apprised us of the dual launch plan, and gave us an hour and a half to see if we could pull it off. I was short on staff since most of my controllers were returning from the remote sites. After a brief meeting with the remnants of my branch, we concluded that it could be done and that the concept was not fatally flawed. I passed the word to Hodge, who passed it to Gilruth: "Flight Control didn't see anything we couldn't do, or anything we couldn't work around."

Later in the afternoon Gilruth had talked with Dr. George Mueller, the agency's associate administrator, in Washington. After a day of intense discussions and, only forty-eight hours after the Agena failure, a press conference was held at the Texas White House. Bill Moyers, President Johnson's press secretary, announced the planned rendezvous of two manned spacecraft. The mission was assigned the designator Gemini 76, combining two Gemini missions into a single mission by using the long-duration Gemini 7 spacecraft as the target for Gemini 6. Within hours of their return from the remote sites, our guys were at their desks writing the data plans, procedures, and site confidence tests.

The mission concept was simple. Borman and Lovell would be launched first (before Gemini 6) on their fourteen-day Gemini 7 mission. Immediately after launch of the Gemini 7 spacecraft two things were scheduled to take place. First, everybody at the launch site would carefully comb the entire area looking for any debris that might have fallen off the booster as well as checking for any damage to the pad. Then the Titan carrying the Gemini 6 spacecraft crewed by Wally Schirra and Tom Stafford (the previous launch of Gemini 6 had been scrubbed after the Agena failure) would be erected on the same pad used to launch Gemini 7 and the pre-launch checkout could begin. If all went well, we would be ready to launch Gemini 6 seven days after Gemini 7.

The day after the press conference Flight Control was in high gear. My branch went about its work with the kind of cheerful exuberance one experiences all too rarely in life. It was like watching Patton's Third Army break off their offensive, perform a pivotal maneuver, turn, and march 100 miles in the dead of winter to relieve Bastogne. In forty-eight hours, we had redeployed and were back on the attack. The launch was scheduled for early December of 1965.

In order to move ahead to more complex missions while Hodge and I were building up the Flight Control Division, the Mission Planning and Analysis Division was expanding to develop mission concepts, design the trajectories, and write the software for the MCC computers. When I joined the Space Task Group in 1960, the Mission Analysis Branch was the largest organization in Chuck Mathews's division. The branch eventually grew to a division led by a perfectly balanced pair of leaders, John Mayer and Bill Tindall. They were an unlikely pair and, except for the challenge of space, probably would never have met. Mayer was short, dark-haired, with a nose for finding answers to questions that appeared to have none. He had an air of aloofness until he got to know you. With his sharp features and horn-rimmed glasses, he could pass as an accountant for the IRS rather than a space pioneer. Tindall, the deputy, with the easy manner of a farm boy, was tall, blond, and youthful in spirit and manner. He was gregarious, short-tempered but quick to recover from an outburst. Whatever needed to be done at the cutting edge, these guys could do it. Although the scientists and engineers at the Massachusetts Institute of Technology spearheaded the development of the Apollo guidance and navigation systems and software, many of the technical studies and prototype software were designed by Mayer and Tindall's division, experience that would come in very handy when we had to come up with alternative ways to achieve our objectives.

This unlikely pair pioneered trajectory design in Project Mercury. They presided over three sections of engineers and mathematicians. The highly talented and resourceful women of the computing unit, Mary Shep Burton, Cathy Osgood, and Shirley Hunt, started out in Mercury with mechanical calculators, manually plotting the results of their measurements and calculations on graph paper. But in Gemini, with key-punched card decks and computers, they started planning every aspect of the launches, the rendezvous, and reentry. They provided us with options that just months before we did not know existed. We had no choice but to believe in the data and methodology they came up with, so our trust in their work was absolute. They designed the mission, then loaded their software in the computers in the spacecraft and in the MCC. Their work had to be perfect — and it was, thanks to increasing computer capacity, speed, and availability.

The rendezvous on the coming Gemini 76 mission was a trajectory show. The Trench and the flight designers from Mission Planning and Analysis were the orchestrators. Mission Control and the launch team followed their lead to the last note, improvising the music only when things fell apart. During a mission, Mayer and Tindall's division operated a stand-alone computer in the ACR — the Auxiliary Computer Room. Throughout the mission their computer ran in parallel with that of the mission team. If we crashed and couldn't generate the data, they fed the answers into MCC. The MCC frequently had to load new software into new computers, and the ACR was our only backup if we got into trouble.

Mission deployment for Gemini 76 started November 21. Several of the new remote controllers were Air Force officers assigned to Flight Control to prepare for the Air Force man in space program. Ed Fendell was deploying to Hawaii with Bill (Big Shoes) Bucholz, an Air Force captain assigned to my branch. Bucholz was a blond, broad-shouldered Missourian who croaked when he talked. Both of us graduated from Parks College in 1954 and entered flight training the same year. His family had grown rapidly to eight children, and the only affordable transportation for the entire bunch was an old large black Cadillac hearse that seated twelve.

The hearse often served another purpose. After the mission deployment briefings the remote site teams often partied in downtown Houston. When they wanted to come home through the downtown traffic the controllers would form up their cars in a line behind Bill's hearse and turn on their lights mimicking a funeral procession. The ploy worked every time.

The afternoon before the final Gemini 6 simulation, while drinking beer with his Hawaii team, Fendell had been approached by the site-training chief, who proposed that Fendell fake a heart attack during the training run to see if Bucholz and the backup flight surgeon, Dr. Warren Prescott, were capable of taking charge of the site team. Fendell returned to the Hawaii site, conferred briefly with the training boss in Houston, then put the plan into action.

The final network simulation was a full-blown mission dress rehearsal, involving all sites and teams in the MCC. During the second simulated Hawaii pass, as the Gemini 6 spacecraft was closing preparatory to the rendezvous, Fendell turned to the Hawaii surgeon and said he did not feel well. Moments later, he stood up from his chair, grasping his chest, emitted a groan, and then crumpled to the floor.

The team momentarily forgot the simulation as the flight surgeon ministered to Fendell in the course of their own little simulation. Struggling into the CapCom's chair amid the tangle of headset cords, Bucholz stepped on Fendell's chest, punched in his headset, and croaked on the voice loop, "Chris, Fendell just had a heart attack." Kraft, momentarily startled but wise to the tricks of simulation, called the simulation supervisor and asked. "SimSup, is this some of your doing?" Since he had not originated the plan, SimSup responded, "Not mine, Flight!" Kraft then punched up the loop. "Hawaii, keep me apprised. Have the surgeon give Fendell's status to the MCC surgeon." Although deeply concerned about his CapCom, Kraft knew that the clock was ticking to launch and, with the whole tracking network up and operating, not even a heart attack could be allowed to interfere with the mission preparation.

Bucholz passed the test. With Fendell on the floor at his feet, he took control of the site's part of the simulation, provided support to Gemini 7, then reconfigured to support Gemini 6. For the next hour and a half, despite everyone in the loop worrying about Fendell, the simulation went forward flawlessly. As the third Hawaii pass approached, Fendell rolled over, got to his feet, and placed his headset back on. The startled Hawaii team was speechless, then relieved, then concerned how they were going to break the news to Kraft. Kraft was angry, but Fendell, with an ear-to-ear grin, was delighted that his new protégé, Bucholz, had come through.

Bucholz, a pilot, got the critical assignments on the old merchant ships for the remaining Gemini missions. At the end of Gemini, he returned to combat duty flying C-123 transports to relieve the pressure on the Special Forces based in Vietnam. Many young heroes passed through Flight Control in the 1960s. The Air

Force transfers were some of the best, adding seasoned backup to the recent college grads in the evolving Brotherhood.

As rapid as the technology developments were in the entire program, the human factor was still key to our success in Gemini and Apollo — and in integrating the contractors for both into our data and operational loops. From the first Mercury mission through Gemini, the personal, gut-level knowledge each controller brought to his console from liftoff through completion of mission was the key to success — and to survival when things went to hell. Years later, it was the human factor that would save us when technology could not.

It wasn't always easy to get contractors to play the game our way. North American Aviation, for example, was one of the new Apollo contractors, and they strongly resisted the transfer of design engineering data to Flight Control. They grew up building fighter airplanes that were always delivered with their own write-ups of flight manuals and procedures. Control of this function was even more important to them in building a spacecraft. They didn't believe that flight controllers had sufficient knowledge to put together the manuals and procedural documentation. I fought a long battle (as I had with contractors in Mercury) to get two North American engineers assigned to Flight Control to set up a data pipeline linking our offices at the MSC with the North American factories. In any complex and high-risk program like Gemini or Apollo, there is always an understandable reluctance to share the intimate design details and detailed test data. The contractor's design teams often doubted that the flight controllers were technically capable of understanding and correctly using the data. The team building and trust between designer and flight controller demanded sharing the information openly.

December 4, 1965, Gemini 7

Given the magnitude of the change in mission content and direction, it was remarkable the way the launch, flight, and contractor teams collaborated. There are times when an organization orchestrates events so perfectly that the members perform in perfect harmony. It is part of team chemistry, where communication becomes virtually intuitive, with teams marching to a cadence, the tempo increasing hourly and the members never missing a beat.

The cadence continued unbroken and at 1:30 P.M. Central Standard Time on December 4, 1965, the Gemini double launch mission began.

The Gemini 7 flight was a saga of human endurance and spirit. Borman and Lovell were cooped up in a spacecraft smaller than the front seat of a Volkswagen Bug. The ejection seats and instrument panels limited their range of motion. The seats were canted 12 degrees outboard and 8 degrees forward. A console, with the pistol grip attitude controller much like a gearshift in a modern-day sports car, was between the seats. The crew would be virtually immobile for fourteen days. Try to imagine yourself stuffed into a confined space like this for fourteen hours, much less fourteen days.

The crew wore new lightweight suits designed for use only inside the spacecraft. The most recognizable aspect of the suit was the soft hood that replaced the traditional hard helmet. According to the manual, in an emergency, the crew could don the suit in fifteen to twenty minutes. In fact, after much effort, it turned out to be more like an hour.

Food was as limited as the rest of the crew systems, consisting of simple rehydratable meals ("add water and ignore the taste") in a squeeze bag with a feeder spout. The bite-sized foods were dry and tended to crumble. The fourteen days in the spacecraft were like a primitive campout, minus the ability to shower, stand, stretch, or take a walk. Through every day of Gemini 7, the controllers' hearts were with the crew in the spacecraft, and we worked hard to cheer them toward their fourteen-day flight goal.

As soon as the Gemini 7 Titan had cleared Pad 19, the launch turnaround for Gemini 6 started. Both stages of the Gemini Titan arrived at Pad 19 within two hours of the previous launch. The race was now on.

Kraft, Hodge, and I were following the team rotation pattern we had established on the two previous missions. During Kraft's shift, Borman and Lovell flew formation with the Titan upper stage, and then methodically started on the flight's medical experiments. Kraft's handover was smooth, and soon I was up and working my third mission as flight director. The technology of space was sprinting forward, especially in communications. The ships on my shift were in their familiar locations in the northwest Pacific near Japan and in the southeast Pacific off the coast of Chile. For the first time, we used a satellite communications relay from Chuck Lewis's team on the *Coastal Sentry*.

I had a new controller working on the White Team for Gemini 76, who would become a key player on many of my Apollo teams. Gerry Griffin was an experienced Lockheed Agena engineer and in the military flew as a "scope dope" (radar and weapons officer) in the supersonic McDonnell F-101 Voodoo interceptor. Griffin followed in the footsteps of Aldrich as a Gemini GNC (guidance, navigation, and control engineer).

My shift broke down into three distinct activities. The trajectory team worked to pinpoint Borman and Lovell's orbit to support the rendezvous targeting. Griffin and my EECOM split their time between spacecraft support and updating the telemetry, command, and display data for the subsequent Gemini 6 launch. The challenge to the MCC procedures team was to integrate the pad test and controller training schedules for the coming Gemini 6 launch into the team shifting and daily operation of the Mission Control Center as we continued the support of the flight of Gemini 7.

After Kraft's Go NoGo on day three, Lovell removed his suit. We had planned to get both astronauts out of their suits but NASA management got involved, and the decree came down that one crewman would be suited at all times, and both would be suited for rendezvous and reentry. By day four it was obvious that Lovell was a hell of a lot more comfortable and was sleeping better than Borman. The message was clear — the cockpit was cramped, the suit was hot, and it was again time to challenge headquarters' decision. By compiling medical telemetry and data from both men we were able to show the marked difference in things like blood pressure, pulse rate, and quality of sleep between Frank Borman, sweating and uncomfortable in his suit, and Lovell. We even deliberately raised the issue at press conferences, but NASA's top management remained adamant. Kraft finally brokered a sort of compromise — Borman and Lovell would take turns wearing the suit.

We also had to follow NASA's directive to let the media — and through it, the whole world — listen to virtually all communications between the spacecraft and the ground. We had to make some exceptions to give us privacy in certain communications involving things like mission risk discussions or direct conversations between flight surgeons and crew members, so we developed a code word. If the MCC or the crew wanted a private conference either side could request or schedule a "UHF-6" test, which we gradually wove into the daily flight plan, hoping that the media would pay little attention to the "test."

Early in the mission the UHF-6 passes went well, and we were pleased that we could conduct needed mission communications in private. When the UHF-6 was requested, the surgeon at the MCC went to the back room with the communications technician and all lines to the controllers and outside the center were disabled. After the UHF-6, the surgeon briefed the flight directors privately. We thought we had pulled it off and got a bit cocky. We were wrong. Reporters are a sharp and nosy lot (after all, that's why they're reporters). They started to become suspicious and pressed Kraft to explain the UHF-6 business. After repeated queries, Kraft finally said, "Ask Kranz at his post-shift conference. He's the one with the details of the test." At my press conference I tried to bury it in a highly technical discussion of communications and antenna patterns, but I knew the jig was up when a reporter finally asked, "Is UHF-6 a code name for a private medical conference?"

My answer was simply, "Yes." With my response, the press corps applauded, and the atmosphere became friendly once again. They had caught the NASA flight directors fair and square, and for a few minutes had us jumping through the hoops. Virtually every event in my early career taught me some very painful but useful lessons.

We debriefed the UHF-6 episode with the mission team much like any other mission event. We determined never again to try a fast one. Personally, I took the attitude that a press conference was like a high-stakes poker game. I loved playing the game with the press, always telling the truth but showing my full hand only when asked the right question.

Apart from such public relations glitches, things were going well and certain individuals began to stand out as superlative performers. Griffin was a great systems guy. He was an Aggie (a graduate of Texas A&M, which enjoyed a fierce rivalry in football — and everything else — with the University of Texas) and, thanks to his experience as a back-seater in early jet interceptors, had an ability to make good snap decisions. He also had an uncanny ability to grasp complex issues. Next to Lunney, I considered him one of the quickest controllers to recognize problems and initiate corrective action. We were lucky that Kennedy's challenge had inspired people like Lunney and Griffin and so many other sharp men and women to rally to the greatest challenge of our country in a turbulent era. They entered Flight Control as rookies and within a year, if they survived, they had the MCC version of a master's degree in real-time operations. By their second year they had a Ph.D. in flight control.

During the periods of my shift when we were shooting the gap and the communications with the Gemini spacecraft were infrequent, I would drill the remote site teams with a series of hypothetical mission situations. My objective was to get the teams supporting my shift up to speed and on the edge, responding to each other crisply, precisely, correctly, and convincingly. Learning to make the seconds count, I would continue the brain twisters for hours, interrupting them only when the Gemini passed over a tracking site in real time. I did not know it then, but these early drills would pay off handsomely on Gemini 8 when the remote sites would be critically important.

In the midst of all this, the turnaround process on Pad 19 was going with remarkable smoothness. The Air Force 6555th test wing, Martin, the Titan contractor, and McDonnell were pulling off a miracle. All three teams of controllers were pulling double shifts, supporting the ongoing mission and preparing for the coming launch with Schirra and Stafford. After one of the extended shifts, John Llewellyn decided to go home to get some rest. When John did not arrive for shift change, the previous shift's RETRO called his house to find out whether he was still at home.

Llewellyn awakened, realized he had overslept, and charged out of the house into his Triumph TR3, tearing up the road en route to the MCC. Arriving in the parking lot, he circled, looking for a parking space, his frustration mounting by the second. Spotting no spaces he made the only decision possible for a Marine, driving his car up the walk, circling across the grass and then up the steps, stopping at the entrance. Clipping his badge on, he emerged to a surprised group of controllers and security guards, moments later striding into the control room and, with a grunt, putting on his headset and starting the handover.

Outside the building, the security forces mustered around John's car, calling in additional support. Lunney and Hodge were fed up with

Llewellyn's antics, so when they pulled his car pass after the mission John appealed to me as his judo partner to intercede on his behalf. I talked to Hodge, but in his clipped British accent he said, "Gene, Llewellyn has got to learn a lesson. Having him walk on site will maybe make a dent in his thick skull."

The story doesn't end here, however. With no car pass and faced with a mile-long walk from the front gate, John came up with an alternative not covered by the regulations. The first day of his suspension, Llewellyn pulled his horse trailer into the parking lot at the Nassau Bay Hotel across from the NASA main gate. Mounting the horse with his leather briefcase, then showing his badge prominently to the surprised guard, Llewellyn galloped through the gate to Mission Control. For the remainder of the week we knew John was in the office or on console when we saw a horse hitched to the bicycle stand. Llewellyn's legend grew once again.

The simulations before the mission had shown that supporting two manned spacecraft simultaneously with a system designed to cope with only one Gemini spacecraft was a CapCom's nightmare. The job was marginally doable. The slightest procedural glitch would crash the site computer. For each tracking site pass, the team had to load the computer at the site with telemetry and command programs for the first spacecraft the site acquired and then, after the guys at the site got a few minutes of data, they dumped the site computer program and reloaded the computer for the second spacecraft in the pass.

As the rendezvous drew closer, the intervals between the two spacecraft narrowed from minutes to seconds. During this tense interval the site CapCom would be directing the computer switching, relaying instructions to both spacecraft, and reporting and responding to MCC in Houston. Never a dull moment.

The remote site drills on rapidly loading the computer programs had brought the site maintenance and operations teams to their peak. You could feel the gritty determination of the operators. The voice exchanges with the remote sites crackled as I handed over the shift to Hodge and his Blue Team for the start of the Gemini 6 countdown early in the morning of December 12, 1965. Hodge, puffing at his pipe, bid me good night as I beat a hasty retreat to the MCC sleeping quarters. Public Affairs had wisely canceled my post-shift press conference. The press would get its story from Kraft's shift in the morning.

December 12, 1965, Gemini 6

The countdown proceeded without a glitch. The Trench computed the precise liftoff time for the rendezvous and passed it on to Kraft, who passed it to the test conductor on Pad 19. Schirra and Stafford were given the liftoff time, the countdown clocks were synchronized, and the liftoff was scheduled for 8:54 A.M. CST, Sunday, December 12. The controllers leaned forward, alternately scanning displays, reporting the final countdown events, and worrying during the final seconds to liftoff.

The chief worrier at this time was the booster engineer, Charlie Harlan. His console was on the left side of the Trench next to the RETRO. He was located so that if the console communications failed, he could yell out the Titan rocket data to those in the Trench, who needed his data. Seated next to Charlie was astronaut C. C. Williams, observing a small plot board displaying the Titan rocket's fuel and oxidizer tank pressures. Between Harlan and Williams was the red abort toggle switch. The flight director normally executed the abort command, but the booster response to problems was measured in seconds and so booster engineers act on their own during launch. Their decision time was two to four seconds. Booster's two main nightmares were calling for an abort when it wasn't really necessary or ejecting the crew too late — the parachutes wouldn't inflate or the crew would be swallowed in the boiling, explosive, toxic propellants that we called the BFRC, the big fucking red cloud.

Among many of the major differences between the Mercury capsule and the far more sophisticated Gemini spacecraft was a shift from the escape tower system to individual ejection seats modeled on those used in high-performance jet aircraft. To make the Gemini crew fully aware of conditions that would require ejection in the vicinity of the launch pad, we trained a two-man MCC team — an astronaut and one of my controllers. They monitored the booster for low tank pressures, engine parameters, and pad fallback conditions. The astronauts in the capsule had only meters indicating tank pressures and lights indicating thrust level.

Sitting on the launch pad, the Gemini astronauts would eject horizontally. Either crewman could initiate the ejection sequence by pulling a handle between his legs, which would jettison the hatches and fire both of the catapult rockets for both seats. If the Titan rocket engines shut down, or developed insufficient thrust after ignition, the controllers had only seconds to issue the abort command. The decision process allowed no delay and no error if the crew was to eject in time to avoid the fireball in a pad fallback. The ejection seats were the last resort and were irreversible.

Harlan, Williams, and the crew lived by a few simple ground rules. Booster must see two independent confirmations of a problem before deciding on an abort action, then when the abort command was transmitted, it had to be followed by a voice "Abort" before the crew would take action.

The abort command from the MCC illuminated a red abort light in front of both astronauts. If the crew saw the red light and received a voice-call "Abort," they were to eject from the Gemini. At liftoff, to avoid a pad fallback, the overall response had to be within three seconds. Three seconds seemed a lifetime to the booster engineers. Harlan made the abort calls for the Titan engines, Williams for the fuel tanks. Both fingered their mike buttons nervously as the countdown clock approached zero for Gemini 6.

The crew also waited through the final seconds. The eyes of Schirra and Stafford were no doubt glued to the clock, engine lights, and tank pressure meters. In the launch sequence, at T=0, the firing command issues engine start commands. The engine lights in the cockpit will blink on briefly, then go out as the engine thrust builds. After two seconds, at greater than 77 percent thrust, the hold-down bolts fire, releasing the booster from the pad. The hold-downs are mechanical attachments that restrain the rocket until thrust is sufficient for liftoff. When the rocket moves one and a half inches, the electrical umbilicals release, which starts the clocks in the spacecraft and on the ground. At this point the mission is committed and liftoff has occurred.

At 8:54 A.M. Central Standard Time, the engines roared to life as steam billowed from the flame bucket. Harlan saw a blip indicating thrust buildup, and the first motion command triggered the clocks in the MCC to start counting up. Schirra and Stafford in the spacecraft felt the initial rumble of engine ignition, the thrust lights blinked, and the Gemini clock started, but then it was strangely quiet. Like a lightning bolt the same thought had to have flickered through the minds of the crew, Harlan, and the launch test conductor. Had liftoff occurred? Were we in a pad fallback? Was 300,000 pounds of rocket, spacecraft, and crew crumpling back to Earth?

Within the seconds allowed for this case, three separate minds came to the single correct conclusion. Harlan called over the voice loop to Kraft, "No liftoff . . . no liftoff!" In the spacecraft

Schirra and Stafford were icemen. They held fire, calmly reporting the cockpit indication as the Martin test conductor initiated the kill recovery procedures.

A launch kill is the most critical single event the operations team faces in the seconds before a launch. The few seconds between engine ignition and the hold-down release is the kill period. With the rocket engines running, the launch system computer rapidly scans the final performance checks. If all is well, the hold-down bolts are released, but if there is a fault, the computer commands an engine shutdown. A complex sequence of commands closes the valves, engages relays, and returns control of the space system to the blockhouse.

During a launch kill everything has to go perfectly. The safing functions and events are critical to fractions of a second, commanding engine shutdown and locking out the hold-down release. The crew, controllers, and launch team must be super-cool, at the highest state of readiness, and all decisions must be perfect.

We not only dodged one bullet that day, we dodged two. The Titan kill occurred at 1.2 seconds because an umbilical released prematurely. Reviewing the data, we found that the engine thrust was already starting to decrease before the umbilical dropped. Engine inspection found a dust cover had not been removed during the engine assembly months before. That day we measured up to the challenge, but we were also lucky.

The turnaround was short and efficient, and

much of the redundant testing was deleted. Having only five days of mission lifetime remaining for Borman and Lovell provided the needed incentive. It was going to be a horse race to get Gemini 6 turned around and rendezvous accomplished before we had to bring Borman and Lovell home.

December 15, 1965, Gemini 6 – Second Launch Attempt

The Cape test team pulled off another miracle, recycling for a successful launch with Schirra and Stafford three days after their kill.

The launch was almost an anticlimax considering all we had had to contend with before Gemini 6 actually lifted off the pad. The rendezvous plan launched Gemini 6 into an orbit below the target, and since the craft in the smaller orbit travels faster, it would catch up with Borman and Lovell's spacecraft. Ground radar tracking was used by the Trench to compute a series of maneuvers to align the orbits of the two spacecraft and set up the catch-up condition. When the Gemini 6 passed in the smaller orbit below Gemini 7, a maneuver was performed to bring Gemini 6 to a position where the crew could initiate the final braking maneuver.

Six hours after liftoff, on the fourth revolution over the Hawaii site, and following nine maneuvers, Schirra smoothly braked to a standoff position on Gemini 7. Sensing history in the making, Jerry Bostick, the rendezvous FIDO, wanted an American flag for each of the mission controllers to celebrate the world's first rendezvous. Unable to find several hundred flags in the stores, he had sent his secretary on the rounds of funeral homes in the Clear Lake area, collecting the flags they mounted on the fenders of cars for military funerals. As the two spacecraft closed together, Bostick started walking between the consoles passing out small flags to each controller. As Schirra closed to within feet of Borman and Lovell, Kraft gave the command and the flags were raised over each of the consoles. I mentally savored the moment of America's triumph like a fine wine.

With the rendezvous complete, Kraft handed over to my team, and for the next four and a half hours the two spacecraft continued their aerial ballet like two friends celebrating a reunion, only this time in space. Just prior to sleep, Schirra performed a pair of maneuvers to establish a standoff position that separated the spacecraft by ten miles.

My shift was brisk and there was little time to celebrate. My team needed to get the final planning together to bring Stafford and Schirra home the next day, and when this was done, review the status of Borman and Lovell's spacecraft to make sure it could stay in space for the planned duration of the mission. At the conclusion of Kraft's press conference, reporters offered a champagne toast to Kraft's team and America's new space record. I watched the celebration on the TV at my console.

At the conclusion of my shift's press conference, Martin Caidin, one of the great pioneers in aviation writing, Jim Maloney, and others in the press corps stood up and passed out champagne glasses just as I had seen them do for Kraft's conference. Caidin passed me one wrapped in a red, white, and blue ribbon, then filled my glass and then the others at my press table. He poured from a bottle wrapped in a towel until the glasses were brimming, the liquid straw-colored and bubbly. It looked good, and as they offered me a toast, my ego soared. As I drank deeply the taste was familiar, but it sure as hell was not champagne. Caidin then unwrapped the bottle and set it in front of me. It was Canada Dry ginger ale. I had been had, a press gotcha on the White Team. There was no alternative but to laugh with them. It was that way with the media on Gemini; they were a great bunch of talented and dedicated professionals. (My mother came down for the lunar landing years later. Meeting ABC correspondent Jules Bergman was the highlight of her trip. She talked about it for years afterward. I mean, Gene Kranz was a guy she had known since he had been born — but Jules was a star!) Gemini 76 — the biggest and riskiest one so far — had worked. We had calculated the risks and, in space and on the ground, won our bet. It was one hell of a great day.

Borman and Lovell continued their heroic mission. Borman finally got out of his suit, two of the three fuel cells ceased operations, two thrusters failed, and we were down to 4 percent of the orbital fuel when Gemini 7 came home. Their mission was longer than any of the planned Apollo missions and would hold the

U.S. duration record for the next eight years. It was a great triumph.

All too soon, it was another time of change. Glynn Lunney launched the first Apollo Saturn developed by the Marshall team from Mission Control Center's second-floor control room as Hodge and I were preparing for Gemini 8. Kraft turned the last five missions over to his students and began preparation for Apollo flight director duties. Hodge planned to leave for Apollo after Gemini 8, and I would follow him after Gemini 9. We would join Kraft for the first manned Apollo mission. Flight directors Glynn Lunney and Cliff Charlesworth would close out the final three Gemini missions.

March 16, 1966, Gemini 8

Due to staffing limitations, Hodge and I elected to support Gemini 8 on a two-shift basis. This was the dumbest staffing decision we ever made. With the planning, training, mission reviews, and the press conferences, by the time we were ready to fly we were flat-out exhausted. The two-shift arrangement, however, fitted in with the Agena team's staffing. They had only two teams of controllers.

The astronauts for this mission were Dave Scott and Neil Armstrong. Scott, a former Air Force pilot, later flew on Apollo 9 and Apollo 15 as well, racking up more than 546 cumulative hours in spaceflight and more than twenty hours doing EVAs. Armstrong had done it all. Neil was a decorated Navy combat pilot in Korea. Then as a civilian, he spent seven years as a test pilot at Edwards AFB and was one of the few who had flown the X-15 to the fringe of space. He was the first civilian pilot hired into the astronaut corps. Neil would spend more than 205 cumulative hours in space, and would be the first man on the Moon. He had worked with Buzz Aldrin as a CapCom on my Gemini 5 team.

With the successful rendezvous on Gemini 76, it fell to the Gemini 8 crew, supported by Hodge's and my teams, to capture the next big objective: the physical docking of two spacecraft. If all went well, we would attempt our second space walk during orbit.

At last the Agena target performed as advertised, rising from Pad 14 and then five minutes later separating from the Atlas and reigniting its engine to maneuver into a 180-mile circular orbit. At cutoff, Brooks was all smiles when the Agena responded to his commands. Hodge's attention now turned to the astronauts, Armstrong and Scott, and the Gemini on Pad 19.

Since I was spending most of my time planning for Gemini 9, I had little experience with this crew. Hodge had debated the mission rules, run the simulations, and briefed the pilots. My association with the crew was limited to the handful of reentry simulations to get my team pulled together for the mission.

For Gemini 8 we had an experienced Gemini team, a novice Agena team that had never seen a spacecraft in orbit, a crew that would be docking with an Agena for the first time, and MCC and remote site computer systems running brand-new software. Adding to the level of concern was the fact that the Agena had failed on its previous mission — many at MCC considered it the potentially fatal weak link.

The Gemini 8 launch was textbook. Hodge's team smoothly guided the crew through the rendezvous maneuvers to the handover point where the crew took it in on their own. The mission was progressing smoothly, and I arrived for my twelve-hour shift just as the crew and the MCC were comparing data for final rendezvous maneuvers. The Agena was performing flawlessly. During the docking it would be positioned perpendicular to the direction of orbital travel.

To maneuver the unmanned Agena from the ground we had to send time-tagged commands known as stored program commands (SPC) to its onboard computer. The commands were prepared at the MCC and sent to the remote sites. When the sites transmitted the commands to the Agena there was a complicated error-checking routine on board the Agena to make sure the commands were correctly inserted into the Agena computer's memory, which the ground controllers could read using the site computer and could automatically compare with the commands transmitted from the site. This was a new computer program, used for the first time on Gemini 8. As a backup the site recorded the commands as they left the antenna and checked these automatically with the intended command load. We called it an echo check. The favored technique was always to compare the commands actually in the Agena's memory. If you couldn't automatically compare the data, the Agena controller could perform a manual data comparison using the recorded telemetry of the Agena memory. This process, however, often took several hours. To avoid glitching the mission timeline, the controllers would normally give their Go based on the echo check if the auto memory comparison failed.

At MCC in the second row of consoles, Brooks was struggling to complete the SPC load to configure the Agena for the after-docking maneuver. This maneuver used the Agena control system to align the docked spacecraft with the direction of orbital travel. Brooks, after manually assembling the command sequence and adding to the time tags for each event, made a final eyeball check of the load and sent it to the *Rose Knot* Victor where it awaited the arrival of the Gemini 8 spacecraft.

At acquisition Keith Kundel at *Rose Knot* Victor uplinked the command load to the Agena while Armstrong and Scott were in the braking maneuver. The load was accepted by the Agena, but the Agena memory dump and automatic data comparison could not be performed. Perplexed, Chuck Gruby, the Agena systems engineer, reran the comparison routine to no avail, then began the laborious process of printing out the memory data and manually checking every one of the thousands of data bits. Gruby called Brooks and advised him that although they did not get the auto compare, all of the commands were uplinked and, from the echo check, he was certain that the Agena command load was okay. It would take Gruby several hours to do the manual data comparison.

Over the next half orbit, Armstrong slowly closed on the Agena, finally maintaining a station-keeping position a few feet from the docking collar. The *Rose Knot* CapCom, Keith Kundel, looked both spacecraft over, got the nod from his systems controllers, and gave Armstrong the "Go for docking!" Neil had been patiently waiting, standing off a few feet, the nose of the Gemini aligned with the Agena docking adapter. With the Go, he closed on the target, moving a few inches closer each second.

Six hours and thirty-four minutes after liftoff, another American record entered the books as Armstrong reported to the *Rose Knot*: "We're docked, no noticeable oscillations, very smooth." In Houston, a brief cheer rose from the team, then we all settled down to listen to the remainder of the pass.

Kundel and his team were busy, rapidly assessing the status of both spacecraft. As the docked Gemini-Agena approached LOS, Kundel gave Armstrong a "looking good. . . . The planned maneuver load has been uplinked." Chuck Gruby, the Agena controller, remained at the console reviewing the pass record and continued the bit-by-bit comparison of the command load. He advised Kundel once again that he was sure that the load got in properly. On board Gemini 8, the crew began their post-docking checks. Brooks advised Hodge that the *Rose Knot* Victor did not get a "compare" on the load but that he was sure there was no problem in the load or the Agena. Hodge's CapCom, Jim Lovell, noted the load compare problem as something that Armstrong and Scott should be aware of.

After a military coup in Zanzibar in January 1964, President Johnson had ordered that the site be closed, and much of the air-ground communications equipment was relocated to Tananarive, at the center of the island of Madagascar in the Indian Ocean. With the new communications installation, the MCC could communicate remotely through the site transmitters to the Gemini. (We did not get telemetry from this site; only voice communication.) The Tananarive pass was very short, the spacecraft passing low on the horizon at the very fringes of radio coverage. Toward the end of the pass Lovell advised Armstrong and Scott, "The command maneuver load was uplinked but we were unable to verify . . ."

On board Gemini 8 there was a burst of static, and the crew never heard the end of CapCom Lovell's sentence, ". . . but we are sure that it all got up okay!" Approaching the *Coastal Sentry*, Scott, in the Gemini spacecraft, commanded the Agena to execute the maneuver.

The command sequence began, the Agena jets firing to maneuver the docked spacecraft 90 degrees in the direction of orbit travel. Dave Scott clocked the turn and was pressing on through the flight plan checklist when he looked up. The docked spacecraft had rolled 30 degrees off the horizon, according to his attitude indicator. A check by Armstrong also showed a roll. Using the Gemini thrusters, Neil maneuvered back to the correct attitude. When he released the hand controller, the docked spacecraft resumed its motion. The astronauts conferred briefly, both believing the problem was a stuck-on Agena thruster. Dave disabled the Agena attitude control.

(Unknown to Armstrong and Scott, inside the Gemini spacecraft an electrical short was triggering a twenty-three-pound roll thruster to fire. None of us knew at the time there was a problem. At first it was intermittent, and then it came on continuously. With no ground station in view, the crew was on their own.)

For a few moments the problem seemed corrected, then a roll rate developed, causing the spacecraft to spin erratically like a small centrifuge. Armstrong took control, fighting the gyrations with the Gemini's thrusters. As their orbital fuel levels plummeted, Scott cycled the Agena controls once again. Nothing seemed to be working. To isolate the problem they quickly decided to separate from the Agena. Prior to jettison, Scott, with a test pilot's remarkable presence of mind, re-enabled ground command control of the Agena. This would give the ground team the ability to troubleshoot the Agena later in the mission.

When Scott hit the emergency release Armstrong fired the thrusters in one long burst, pulling the Gemini away from the Agena. At undocking the Gemini spacecraft shed almost half its mass. Now, as a much lighter spacecraft, the effect of the continuously firing Gemini thrusters virtually doubled.

The Gemini was now rolling and tumbling. Using every test pilot skill, Armstrong and Scott fought for survival as the spacecraft completed a turn every second. As the orbit propellant dropped below 25 percent, there could be no doubt that the problem was in one of the Gemini thrusters.

The *Coastal Sentry* control team, stationed east of Okinawa, heard the docking report over the air from the *Rose Knot* and had received the Teletype message that all was Go at loss of signal. At the planned acquisition time the *Coastal Sentry* ship's technicians reported difficulty in locking up the Gemini telemetry. They had the impression that the Gemini was spinning. Finally, with the telemetry lockup, Jim Fucci, the horrified *Coastal Sentry* CapCom, reported to the MCC, "The crew has undocked . . . they are rolling 360 degrees per second . . . orbital fuel is down to 20 percent."

All we could do in the MCC was hold on and pray that *Coastal Sentry*'s team could somehow give the crew some help. Scott reported, "We have serious problems, we're tumbling and have separated." Then Armstrong continued, "We're rolling up, we can't turn anything off." Armstrong then activated the reentry jets, and killed power to the orbit thrusters.

On the verge of losing consciousness from the accelerating spin, the crew regained control, using the fuel normally reserved for the reentry phase. *Coastal Sentry* CapCom Jim Fucci related: "They've activated their entry fuel system, they are firing the entry jets, the spinning is slowing, rates are coming down, it looks like they are starting to get control."

Fucci continued reporting: "They've just about used up one entry fuel system, they're getting it just about stable now!" With the Gemini craft stabilized, Armstrong started flicking the switches in the overhead panel to determine which Gemini jet had stuck and nearly killed them.

As the Gemini left the *Coastal Sentry* the mission was a mess. Gemini 8 was now entering the period where the only site coverage was provided by the two ships. Within three hours, we would see the crew only once every ninety minutes, which was not good. Although the docking objective had been satisfied, the rest of the mission was a bust. We were out of orbit fuel and half of the fuel in one of the two reentry fuel systems had been used up. The reentry fuel was of course intended for reentry. It was time to come home. No mission rules covered the exact predicament we were in. To get to the next day's landing area, Hodge would have to limp through sixteen hours with no fuel and limited tracking station coverage. His alternative was to call it a day and bring the mission down in the West Pacific prior to entering the period of limited site coverage. Hodge moved quickly to terminate the mission and bring the spacecraft down in the last suitable landing area covered by the recovery forces. He turned and told Kraft he was bringing them home.

My team had come into the MCC for the docking and we were critically aware of the mission status. Hodge, drained and groggy after eleven hours on the console, told me, "Your team is trained for entry. I want you to bring Armstrong and Scott home." I stood up, put on my white vest, and over the voice loop told my controllers to get on the console and start the handover.

Within minutes, my team had messages circling the globe, activating the recovery forces. The West Pacific recovery forces were alerted that the planned target point was 500 miles east of Okinawa. Within minutes, four para-rescue aircraft with life rafts and parachutists, as well as an amphibian (seaplane), began the race to the landing point. The destroyer USS *Mason*, 160 miles from the landing point, turned east and put all four boilers on line.

The White Team handover in the midst of the crisis was seamless. Our job was simply to get the crew home in the West Pacific landing area. My assets were the three remaining site passes and the MCC team. We had a single shot to do it right. A wave-off would send the crew off the tracking network critically low on fuel, shooting the gap for another thirteen hours before reaching the next suitable landing area. It was like changing quarterbacks in the middle of a handoff. In simulations we had practiced this kind of switchover, just as a football team rehearses plays.

The Trench, the EECOM, and the GNC pulled out the checklists and smoothly prepared the briefings at the next two ship sites. The Trench read the targeting orders to the computing complex, then relayed the resultant landing data to Recovery and then prepared the detailed deorbit data for the crew. Forty minutes after Armstrong's undocking, my team had completed the deorbit planning and had passed the reentry data to the two ships.

Fucci, on the *Coastal Sentry*, in crisply measured tones read the long string of the deorbit data, recovery call signs, pickup times, and last-minute reminders to the crew. This was our last go-for-broke pass, and the remote site performance was superb. We would hear no more until the recovery forces made contact. Gemini would be on its own, deorbiting out of station contact over Central Africa for the West Pacific landing site.

A rescue aircraft was overhead as Gemini 8 splashed down, and fifteen minutes later, a para-jumper landed with rafts and flotation gear within swimming distance of the spacecraft. The USS *Mason* arrived after the crew had spent three hours bobbing in the heavy seas in their spacecraft.

Gemini 8's flight lasted ten hours and forty-one minutes, but the flight of the Agena, thanks to Dave Scott, continued under a carefully crafted plan for the next two days. Brooks performed ten maneuvers with the Agena, expending all maneuver fuel and electrical power and vindicating the record of the Agena in the process. Over 5,000 commands later, the Agena died, a valiant first effort for Brooks's team. They were ready.

My debriefing was short and pointed: "We got our crew home safely and the control teams did a damn fine job under real-time pressure. I know this is going to sound like Monday morning quarterbacking, but the lessons from this mission are how we screwed up in planning and training.

"The crew reacted as they were trained, and they reacted wrong because we trained them wrong. We failed to realize that when two spacecraft are docked they must be considered as one spacecraft, one integrated power system, one integrated control system, and a single structure."

I continued: "The next thing is that many of us did not trust the Agena. Only Brooks's team thought it was a great piece of hardware. If we don't trust a spacecraft, we have to fix it. We were lucky, too damned lucky, and we must never forget this mission's lessons."

Treating docked spacecraft as a single system was one of the most important lessons to come from Gemini. It had a profound effect on our future success as flight controllers. The lesson learned on Gemini 8 would be invaluable on Apollo 13.

9

THE ANGRY ALLIGATOR

Spring 1966

John Hodge departed the ranks of Gemini flight directors to prepare for the unmanned Apollo Saturn rocket flight testing. Glynn Lunney and Cliff Charlesworth joined me as flight directors for Gemini 9. Both men had grown up in the trajectory world, but there the similarity ended.

Glynn was an early entrant to the Space Task Group. He was smart as a whip, boyish and trim, the youngest of the flight directors, free-wheeling, with a tendency to get ahead of his team in moving to a solution. Cliff was the oldest, a civilian who had previously worked in the Navy and in Army ordnance labs. He had a laid-back personal style that earned him the nickname "The Mississippi Gambler." I regarded Glynn Lunney as my friendly but intense competitor; he ran the Trench, and I ran the Flight Control Branch. Our branches were two of the seven in Hodge's Flight Control Division that staffed the Mission Control Center. Glynn and I both wore two hats, as flight directors in the MCC and as branch chiefs in Hodge's organization.

I had the good fortune to have grown up under several outstanding leaders who had given me a lot of hands-on experience with people and technically complex missions. I didn't have Glynn's innate talent, so I surrounded myself with smart people and relied on them to work with me as a team to get the job done. My credo: always hire people who are smarter and better than you are and learn with them.

My team respected me because I did the dirty work and never pulled rank. I assigned the work responsibilities and once I set up the plan I stuck to it. I took the risks to let people stretch and grow and I took the heat, and deserved it, especially when I decided there would be no holidays, no vacations. Which meant that I got to answer the calls from agitated wives. I was blessed that Marta was not among them. She understood the pressures at NASA and, in turn, I tried not to bring them home. In truth, there was no room for them. Our sixth child, and fifth daughter, Jean Marie, was born on April 16, 1966, between Gemini 8 and 9. The house we had moved into just a year before was crowded again.

With the increasing frequency and complexity of the flights, we now had to prepare for several missions simultaneously. To integrate the Gemini mission design, training, and planning, Hodge, in his role as division chief, assigned a lead flight director to coordinate the mission strategy of the flight directors, astronauts, and program office. During the mission, the lead provided a strategy overview, resolving any conflicts among the flight directors.

Risk is normally highest during the launch phase. Decisions must be made rapidly and the teamwork must be precise, for the results are irreversible. With Kraft closely monitoring us, Hodge and I worked up to the position of launch flight director by demonstrating our skills at risk judgment and rapid and correct decision making under pressure. Working with Kraft in Mercury and early on in Gemini I had been an understudy for the flight director's part. On occasion during training or testing for these early missions Kraft gave me the helm at Mission Control. Hodge had launched Gemini 8 and now he assigned me to launch Gemini 9. He made me the lead flight director. Kraft, the teacher, was now handing over the baton to the new generation of flight directors.

A Gemini launch took about six minutes. The launch simulation begins with the final five minutes of the countdown, the Mission Control Go NoGo, and then liftoff. Some launch simulations might end only seconds after launch with a call for a crew ejection. Other exercises might be hours in duration if we screwed up and got into orbit when we should have aborted. In those situations we were required to solve the problem we had created.

The launch phase of the mission was the toughest to prepare for. The flight director, the Mission Control team, and the astronauts had to be tuned to perfection for this phase to make the most fundamental decisions — continue the mission or abort. The real-time decisions were made with the entire world listening and watching. So our simulation dress rehearsals had to come as close to reality — and the unpredictable — as possible. We took a quantum leap forward when we got digital computers and systems that worked faster and faster with each new upgrade. They brought us into that virtual reality that made simulation training almost indistinguishable from the real thing, particularly as the missions became more complex. This technology also replicated the atmosphere — the tension and intensity — that prevailed in actual missions.

Technology and training were pushing us to the ultimate standard: failure was not an option. In simulation and in real time controllers knew that if the team made the right decisions, we would accomplish the mission and bring our men home safely. If we were wrong in real time, we would ruin the mission and the crew might be killed.

In the course of an abort training session, for example, eight to ten simulated launches were run in a six- to eight-hour day. Some sessions might be only seconds in duration, demanding instant decisions for an on- or near-the-pad abort. These were perhaps the most intense. I remember one when Kraft, in the middle of a training run, his nerves and reflexes set at a hair-trigger level, unexpectedly threw the abort switch and shouted "Abort! Abort!" After the crew in the capsule simulator got the abort information from Control and responded right on the money, Chris got on the intercom and asked, "Who said 'abort'?" In a somewhat embarrassing postmortem debriefing, it turned out that no controller had called and there was no reason to perform an abort. The only voice Chris had heard was his own. We kidded him about this for years afterward — but we all knew that any one of us, when we were primed and on edge, could have done the same thing. *Not* doing it was an important part of what a simulation was all about. As a Catholic, I found debriefings were almost like confessing my sins to a priest — except that this was done over a microphone, so the whole "congregation" heard my mea culpas, particularly when I had to say what all of us learned to say: "I don't know." Knowing what we didn't know was how we kept people from getting killed.

May 1966

The Gemini 9 mission was a brute, and the two-month turnaround went far too swiftly. Gene Cernan and Tom Stafford were my crew. Cernan was the easiest of the third group of astronauts to know. On the first four Gemini missions, he worked as the booster tanks monitor or a CapCom.

In the eyes of the controllers, Gene was one of us. I knew Cernan in a different way. The Catholic priest at my church, Father Eugene Cargill, was also Gene's chaplain. The padre was invited to the Cape for launches and was a familiar face at the crew's splashdown parties. Father Cargill followed our missions closely and always gave a special blessing at the morning mass before my missions as flight director.

Although I knew Cernan, Stafford, with a wide smile and a perpetual hoarseness to his voice, was a bit harder to get a fix on. He kept his opinions to himself, generally letting Cernan talk. Tom was in the second astronaut class, out of the Air Force, studious and balanced, while Cernan's antics were characteristic of Navy pilots who land airplanes on aircraft carriers. Cernan was my favorite for his carefree and jovial attitude, unabashed patriotism, and close personal relationship with the controllers. He was also a skilled systems guy.

Gemini 9 had virtually everything packed into the first seventy-two hours of the mission — three different rendezvous techniques being tested for Apollo, a docking, and a space walk by Cernan with a jet backpack. The Air Force proposed that Cernan fly the jetpack without being tethered to the Gemini. Slayton and Kraft made their position clear: "He will fly without a tether over our dead bodies." The Air Force lost that argument.

May 17, 1966, Gemini 9

Things did not go well on the Atlas/Agena launch. Twenty seconds before cutoff, we lost control when the Atlas engines swung abruptly to the side, spinning the rocket. For the second time, the Agena target was reduced to junk as bits and pieces crashed into the Atlantic. There was no question the controllers were beginning to feel snake-bit as we passed the word to Stafford and Cernan and scrubbed the Gemini countdown. Before we left Mission Control, we had received orders to develop a backup mission. There was no party that night at the Singing Wheel, but by now our team had developed a resilience that we believed could overcome this and any other difficulty.

The Gemini program office had directed McDonnell Aircraft to develop a backup rendezvous target. The target was called the augmented target docking adapter (ATDA). It was assembled using the nose (aerodynamic) shroud, docking collar, and command system from an Agena and a reentry attitude control package from a Gemini. The whole lash-up was launched on an Atlas rocket. The backup could perform every Agena function except on-orbit maneuvering. The most distinguishing characteristic was the ten-foot aerodynamic nose shroud that opened like a clamshell and was jettisoned after launch phase to expose the docking system. My control team, aware of the backup option, had developed procedures, rules, and plans for the mission. The remote teams were advised to remain at their sites and we began a two-week turnaround for the backup mission. To keep the paperwork straight, the mission was renamed Gemini 9A.

The ATDA launch on June 1 went well, reaching the planned circular orbit at an altitude of 160 nautical miles. I passed the good news to Stafford and Cernan on Pad 19 and continued counting down to the second launch. As the target passed in its orbit over Bermuda, the ATDA controller, Jim Saultz, passed a warning on to me. "Flight, I think we've got some problems with the ATDA. We're using the attitude control fuel like crazy, and I did not see telemetry indications of the nose shroud separation." Five minutes later, after reviewing the target's telemetry data, the Canary Island CapCom advised me, "Flight, we're really hosing out the fuel. I recommend we secure the attitude control jets before we lose it all." My response was brief, "Go ahead, shut it off." In less than twelve minutes of flight we had used one of the two tanks that supplied fuel to the ATDA attitude thrusters.

I turned to Saultz and ordered, "Jim, keep me advised of any further developments. I'm going to follow the Gemini-Titan launch countdown from now on." The mission rules for launch were simple: as long as the ATDA was in an orbit suitable for the three planned rendezvous demonstrations we were Go to launch the Gemini 9. Docking with the ATDA was considered a secondary objective.

The Gemini countdown entered the scheduled hold at launch minus three minutes, while waiting for the precise liftoff time needed to set up the orbital conditions for rendezvous on the third orbit. During the brief hold in the countdown, the ground test equipment computed the exact steering information to guide the Titan into the ATDA orbit. When the countdown resumed, the ground support equipment failed to provide the update to the Titan guidance. The launch was scrubbed for forty-eight hours.

During the two-day turnaround period, we conducted several tests with the ATDA using ground commands to extend and retract the docking mechanism and fire the attitude control jets to kick the shroud loose. We concluded that the target nose shroud was only partially deployed. While we regrouped, the Titan team fixed the electronics box that had failed to send the update.

The crew, the control team, and myself were briefed by the team at the Cape that had installed the shroud. They concluded that a safing pin had not been removed from the band and as a result the sequence had started but the band had not separated, leaving the shroud unopened.

June 3, 1966, Gemini 9A Second Launch Attempt

The countdown was virtually perfect for our third try at getting Stafford and Cernan off the ground. After giving the launch team the word, "Mission Control is Go for launch," I stood up and put on my vest. By now everyone was used to the bit with the vest. But this one was radically different from any I had worn previously. All earlier vests, while different in style and material, were solid white. After my second Gemini 9 launch scrub, Marta made a splashy vest of gold and silver brocade over white satin. She thought I needed a bit of good luck for my third launch try.

When I put this vest on, Kraft made a few wry comments. Looking up and through the glass into the viewing room, I could see people pointing. The vest had made a hell of an impact on the visitors; now I just hoped it brought my team and the crew a bit of luck.

There is no feeling in the world like a launch day. The controllers, launch team, and crew are a single entity bound by a mission, the atmosphere brittle and electric. The clock provides the cadence as we grind through the procedures, events, and tests. In the final minutes prior to launch, I began the flight director's ritual, locking the doors of the control room after the final status check.

For the last sixty seconds the voice calls are all programmed, there is no superfluous chatter, all reports are crisp and formal. We were like sprinters in our blocks, waiting for the starter's gun. Alas, the automatic launch update again failed, so my GUIDO (guidance officer) manually transmitted the commands, ramming them home in the allotted forty-second window.

In the final seconds, it turned eerily quiet in the control room. The controllers scanned their displays, absorbing and assessing hundreds of pieces of data. The only sound was the incessant finger tapping against the consoles or the nervous clicking of ballpoint pens. Relief from the tension comes only when the launch team calls, "Auto sequence start . . . five . . . four . . . three . . . two . . . one and engine start!"

Approaching zero, I felt like I was flying an aircraft for the first time. My adrenaline reached a peak, and then there was icy calm at the moment of commitment. I was ready and I felt great. In a few moments, when the rocket cleared the launch tower, the ballgame was ours. I never had a controller get the shakes during a mission — the nervous types were weeded out or else looked at the job and knew it wasn't the right one for them.

"Flight, liftoff, 13:39:33 [7:39 A.M. CST]. The clocks have started." Recording the liftoff time, I decided this new vest was really lucky. We were finally on our way, the crew and controllers crisply reporting launch events as the Titan accelerated, arching skyward, reaching for its target orbit in space. The work, sweat, and frustration now paid off with a perfect orbital cutoff. After separation, and then another maneuver, Stafford and Cernan were racing toward their rendezvous target. Their maneuvers inexorably closed the distance between the two spacecraft, the crew and ground perfectly executing the procedures, the team harmony and rhythm fluid and dynamic. Approaching the end of the third orbit over Hawaii, Gemini 9 slid smoothly into position, flying formation with the target.

As Stafford maneuvered, Cernan gave a running commentary during the approach, finally confirming our suspicion that the shroud had not separated. From a distance of a few feet away Stafford said, "The clamshell is open wide, the band holding it together is at the front. I believe the rear bolts were fired, and we can see the springs. The band is holding the whole mess together. It looks like an angry alligator."

While the crew remained as observers, I directed my team to send a series of rapid attitude maneuver commands to attempt to shake the shroud loose. The crew saw the target's motion as the thrusters fired, then reported, "No joy. It's not doing any good. You might as well save the fuel." With this new data, and to buy some time to develop other alternatives, we passed the crew maneuver data for the second planned rendezvous of the mission. This maneuver thrusted them away from the Earth and, if perfectly executed, placed them onto a football-type trajectory, returning to the target in exactly one orbit.

My Agena controllers left the console to attend a meeting with engineering to assess any reasonable shroud jettison alternatives. (Just to make life more confusing, we continued to have our controllers use the Agena call sign, despite the fact that on this mission the target vehicle was not an Agena rocket but a target made up of spare parts from Agena and Gemini.) After an hour and a half, they returned, reporting no one saw any way to separate the shroud. With Gemini's orbit moving toward the area of sparse network coverage, I felt it was time to call it a day and get the crew moving toward the sleep period. A maneuver was passed to separate from the target and set up the conditions for the third and final rendezvous on my shift in the morning.

Lunney had been standing by during the second rendezvous, and at handover his Black Team took to the consoles in Mission Control. During my press conference most of the correspondents tried to engage me in speculation about a comment made by Stafford to Cernan that, "We might be able to nudge the target shroud off with the nose of the Gemini."

During my shift I quickly discouraged Stafford from contemplating such a maneuver, and I did not want to give it any more credibility via the media. I advised the press, "Our priorities are to accomplish the three rendezvous, a complex EVA, and a bunch of experiments. Docking, while an objective, is now only frosting on the cake." I paused, then added, "There is a lot of energy still stored in the thrusters in the shroud. When and if that band comes loose, I want Gemini long gone."

After the press conference I went home for supper and some fresh clothing, then returned to the sleeping quarters above the lobby of Mission Control. Prior to hitting the sack, I went to the control room to check in with Lunney. I was surprised to find Cliff Charlesworth in Lunney's chair. Cliff said, "They called Glynn to a meeting in the controller ready room to discuss tomorrow's EVA." I hit my boiling point in a second as I exclaimed, "What EVA?"

Cliff responded, "They want to do an EVA to

release the shroud."

Next to the sleeping quarters was a ready room for the controllers to observe TV mission status and listen to the other controllers and crew communications while relaxing before or after a shift. The lounge is on a floor midway between the two control room floors and I took the stairs two at a time in my haste to get there.

When I opened the door, the ready room was loaded. Present was an array of NASA's top leadership: George Mueller, NASA associate administrator; Chuck Mathews, program manager; Robert Gilruth, center director; as well as Kraft, Slayton, Dr. Charles Berry, and assorted astronauts. Glynn Lunney, the only flight controller present, looked up as I entered, rolled his eyes, and silently raised his hands in exasperation. Standing up, he walked over and said, "They're talking about doing an EVA to release the shroud."

I was livid. The last word I had left was that in no case would we plan to try to release the shroud. Kraft saw my expression, walked over, and said, "Dammit, control yourself and settle down. No decision has been made yet." As I watched, I concluded that astronaut Buzz Aldrin was the principal proponent of the gamble, since he was animatedly discussing possible procedures to release the band. William Schneider, the Gemini 9 headquarters mission director, motioned me to the corner, briefed me on the discussion, then asked me for my opinion. My response was tart: "Bill, there is little to gain, the risk is high, and I don't want to compromise the planned objectives. This is nothing I will support." Kraft, standing, was studying me, wondering what else I was going to say when my turn came.

I had been expecting some wild scheme since we first learned of the shroud problem. While politics was not my long suit, I knew enough to address my comments to the real players, Mueller, Mathews, and Gilruth. They seemed to be the swing votes. I could not yet figure out who was for the EVA besides Aldrin.

It was obvious that the top brass knew little about the shroud mechanism, so for about ten minutes I briefed them on the details. Then I talked about what the ATDA EVA would require Stafford to do — station-keeping with the ATDA while Cernan, free-floating on the umbilical with no handholds or footholds, tried to cut the band or pull the safing pin. Summarizing my argument, I said, "There is a lot of stored energy in the ATDA shroud mechanism. It is cocked and ready to fire and I don't see any way to safely get it loose. When it separates I don't want an astronaut or even the spacecraft in the vicinity. This is only our second EVA and it is already sporty enough. When we return to the target tomorrow morning, we will have completed our primary rendezvous objectives. The planned EVA is long and complex and Cernan should do the one he trained for. When we get this done, we will have done damned well on this mission."

The discussion continued for another half hour until finally Mathews, the Gemini program manager, made his decision: "I think we should do an EVA to see if we can release the shroud so we get our docking objective. I think we can make the risk acceptable." As he summed up the debate, I was surprised that no one in the room challenged his decision. Mathews closed the meeting saying, "Does anyone have any more comments?" Walking out the door, I took my last shot: "This is a stunt . . . a dangerous and unnecessary one and we're going to kill someone."

By the time Lunney and I got back to the control room, Charlesworth's team was on the console and the Black Team had gone home. Lunney turned to me and said, "What are we going to do?" I replied, "Get smart." I don't react well when management starts second-guessing me, but I knew I had to keep a lid on my anger.

I was surprised and frustrated that Kraft did not shoot this crazy idea down. As I walked by his console I fired at him. "Chris, this is my last damned mission. I am through." Kraft looked me straight in the eye, saying, "You got your orders, now do your job." Red-faced, I turned away, thinking, "Screw the role of the flight director. When push comes to shove, the flight director is just another management flunky." Later, Kraft denied he had directed me, but in the heat of battle it sure seemed like direction to me.

As quickly as my anger came, it went. If I had to implement a bad decision, I would make it come out right. My job had not changed. If it looked as if Cernan or Stafford was over their head, I would wave it off during the EVA. In the heat of real time, none of my bosses would be in a position to turn me around.

There was no sleep for me that night. I had to build a plan for the new EVA prior to starting my shift at 4:00 A.M. I had missed the early part of the meeting, so Lunney summarized the details of the proposal by Aldrin. When John Aaron heard about the shroud, he rounded up the crew systems team we worked with on the original EVA. The team and I worked through the night, searching for any hazards to the integrity of the space suit and the umbilical.

The shroud area provided the largest single hazard, littered with razor-sharp edges and items that could snag Cernan's suit or catch the umbilical. Our plan was to have Stafford fly formation at the open jaws of the shroud. Cernan, on the umbilical, would check to see if the band was under tension, then verify that the springs were in the ball socket. From there the procedures became vague. Some believed that if we pulled the safety pin the band would release. Others thought Cernan should try to cut the lanyard with some medical scissors. This was the best we could do to keep Cernan away from the stored energy. Stafford, the commander, would have to use his judgment on when to call it a day if he didn't like the setup.

When Charlesworth briefed me at shift handover, he said that astronaut Dave Scott had been in the Los Angeles area and, prior to returning to Houston, examined a shroud at the Lockheed plant. Later, in the shift over Carnarvon, Scott briefed the teams on his observations. Everyone was pitching in to try to keep Cernan out of trouble during the ad hoc EVA.

After I came back into the control room for my shift the next morning, Charlesworth advised that the crew was completing the third rendezvous in the first twenty hours of Gemini 9.

Approaching the end of our first day in orbit, Charlesworth gave the Go for EVA preparation over Carnarvon, then quickly handed over to my team. Armstrong and Lovell, my CapComs, briefed Stafford on the timeline and the areas of the shroud to avoid. During the air-ground discussion I started picking up vibes that none of the astronauts involved in the planning felt warm about the EVA. I wondered if Deke Slayton had polled his guys and was having second thoughts. Stafford and Cernan listened to the proposed plan and advised, "We'll get together on this at the next site."

Nine minutes later, over Canton Island, Stafford and Cernan gave us their input. Stafford took the lead. "Both of us are pretty bushed, we're low on propellant, and by the time we finish with the prep we're going to be mighty low. I think we should knock it off for a while and consider EVA for tomorrow." Due to the low maneuvering fuel levels we all knew that there was no way to re-rendezvous with the target the next day, and that waving off today was tantamount to saying NoGo to the shroud EVA. I could have kissed the crew. Neil Armstrong turned in his chair and looked for my input. Smiling for the first time in hours, I gave Neil a thumbs-up, followed by a resounding, "Flight concurs."

Before I received any further top-level input, I had the controllers give the Agena a few more jolts to see if we could kick the shroud loose. But the angry alligator was not about to surrender. Twenty-three hours into the mission, the exhausted crew separated for the final time from the docking adapter. Within one revolution Cernan was asleep and Stafford was soon to follow. The crew had made the right call.

Sometimes you need luck. The program dodged a bullet when the crew waved off. They also saved me from eating crow. I was glad that I never had to follow through with my words to Kraft in the heat of the moment that this would be my last mission. The flight director's job was my life but I was still too proud to say, "Dammit, Chris, I was out of line!" I could easily have bailed out of the program, since I had recently received an attractive offer to go back into aircraft flight testing. If two stubborn types like Kraft and me locked horns, Kraft would have let me accept the offer. In any case, both of us were saved by the crew's decision.

EVA skills, like those of rendezvous and docking, were an essential element of working in space. The skills were needed for lunar surface operations and for crew transfer between spacecraft in the event that the crew could not dock or was unable to open the hatches. Cernan's EVA was developed to add to the knowledge base of the engineers and mission planners as they designed the more complex missions to come. It was also intended to test the utility of a jet backpack.

Cernan's EVA the next day was as tough as it ever got in the MCC. Years later in his book he would describe it as "the EVA from hell." We were behind the timeline from the beginning. It seemed that every EVA activity imposed a high workload on Cernan. In the tradition of great test pilots, his incredibly detailed and graphic reporting made us intimately aware of the problems he was encountering. He worked so hard that his helmet started to fog over from perspiration. The rest periods recommended by Dr. Berry gave Cernan a break, but no more. Cernan, partially blinded by the fog, continued aft to the Gemini adapter module, where the backpack was stored. Entering the adapter by raw force, Gene was further confounded by poor lighting, limited restraints, and inadequate footholds. After a brief rest, he resumed the EVA. The oxygen umbilical was a few inches short, making it difficult to don the maneuvering unit. Nothing seemed to be going right.

Before the mission Cernan had walked through the EVA procedures using the Gemini mock-up in what we called 1G training. Cernan had asked me to observe the training so I would be familiar with the sequence and terms he would use. I now could visualize his effort, virtually blind, trying to adapt and invent ways to anchor in a position, any position, that would let him get his work done or give him some respite. As flight director, I was helpless, hostage to a chain of events occurring 150 miles above me half a world away and with only scattered bits of UHF communications. It was like watching a movie in short segments, with twenty- or thirty-minute gaps between the segments. Only Tom Stafford or Gene Cernan could make these Go or NoGo calls.

As I looked around the room, controllers sat silent at the consoles, scanning their displays, hands tightly gripping the handles of their TV monitors, all empathizing with Cernan. Dr. Berry periodically would notch up the anxiety level of the room as he reported pulse and respiration rates. Each report from Cernan heightened the tension as we strained to hear. "My heel is caught on something. It touched the spacecraft and got a torque that won't quit." Every statement was punctuated by labored breathing and an occasional grunt. I found myself thinking, "God, am I glad Stafford had the sense to scrub the shroud EVA."

In the second hour, the reports weren't getting any better. Cernan selected a high oxygen suit flow in an attempt to keep cool and clear the helmet face plate. Then Stafford started to have problems communicating with Cernan. "Can you see out at all, can you read me?" Finally, Cernan conceded. "Tom, I'm going to call it quits. Nothing seems to be working."

The struggle to unstow the maneuvering unit left Gene totally fogged over, sweat now pouring into his eyes and no way to wipe them dry. When Stafford called I quickly concurred in his NoGo of the remainder of the EVA. Frustration rang in Cernan's voice as he said, "Sorry about the maneuvering unit."

After a brief rest, the blinded astronaut started back to the hatch area. The only place on his visor that had cleared was at the tip of his nose. Peering through this small view hole, he secured the adapter area and started forward. Every time the tiny opening fogged over, Gene would rest for a few minutes until the plexiglass in front of his nose cleared, then he continued on. We had a blind crewman outside the spacecraft feeling his way with his hands back to the cockpit. I thought, "God, those guys are like icemen, chock-full of guts!" Two hours and ten minutes after opening the hatch, the crew started to repressurize the cabin. For the second time in the mission, I felt I had been granted a reprieve. We had walked the edge. Cernan was back inside and he had avoided disaster on the EVA.

We were blessed with a dedicated, well-informed, and highly professional press corps in the 1960s. (Unlike so many "reporters" today, they knew the difference between objective reporting of news and hyping things up to entertain the audience — and bump up their ratings.) The press conference was almost as much of an ordeal as the mission. Reporters were as confused as we were, asking, "Ed White's EVA on Gemini 4 came off like clockwork. Why are we now having problems on Cernan's mission?" They asked the tough questions, but they respected us and the work we did as long as we didn't try to mislead them. For the first time at a conference I found it tough to give them specifics. I was as confounded as they were. In retrospect I am amazed that we had such a hard time figuring out what was causing the problem. We were still thinking in earthbound terms. Space has an entirely different set of rules and dynamics. We also had no telemetry to analyze what was happening, only the crew's subjective impressions and some hazy pictures from the onboard TV camera. As a result of this nerve-wracking experience, most of the Gemini flight directors and Kraft developed a decided and long-lasting lack of enthusiasm for EVAs.

Wrapping up the press briefing, I said, "The nature of flight test is working at the boundaries of knowledge, experience, and performance and taking the risks needed to get there. There are times where the things you learn are things that you don't expect. Often the plan you execute is different than the one you started. I believe this is one of those times."

The Gemini 9 mission left me with mixed emotions. The rendezvous objectives had been satisfied, demonstrating the Apollo techniques and options for rendezvous and rescue. The docking objective was a bust, but based on the Gemini 8 experience I believed that any good formation pilot could dock. The EVA was a different story. At the debriefing party, Cernan's words rang in my mind: "Geno [Cernan and I always called each other Geno], EVA is a tough SOB. There is nothing that prepared me well enough to do the job we had planned." No one could reconcile White's success with Cernan's problems. We would just have to press on and get more EVA time.

With mixed feelings I left Gemini to join Kraft and Hodge in preparation for the first manned Apollo mission. I was proud to be named to the manned flight to kick off the Apollo program, but I hated to leave the console with three Gemini missions remaining. The flight director's console was my life, the White Team my squadron mates. Missions were living, walking on the edge, feeling the camaraderie of the Brotherhood. Leaving the console was like walking away from the cockpit of my beloved Sabre. The flight director's ultimate training comes at the console, working real problems, facing the risks, making irrevocable decisions. I had a lifetime of learning ahead as a flight director, and I envied Lunney and Charlesworth the experience they would gain in the final three missions. I volunteered to work the crew sleep shift if they needed me.

Lunney and Charlesworth directed the ever more aggressive missions of Gemini 10 and 11, conducting rendezvous, then docking and riding the Agena to record high altitudes. Their missions measured the space radiation environment and were crammed with significant new scientific objectives. The glow from the successful rendezvous, docking, and science experiments, however, could not compensate for the continuing difficulties encountered in EVA. Mike Collins's experience on Gemini 10 further demonstrated the need for positioning aids, restraints, and realistic planning of space walk activities.

Collins's debriefing was fed into the training and planning for Dick Gordon's EVA on Gemini 11.

I was an observer seated next to Charlesworth as Gordon's EVA went to hell. As Dick moved from the hatch to retrieve a tether on the Agena, he lost his grip and drifted in an arc floating aft to the Gemini adapter. Conrad pulled him back toward the hatch with the oxygen umbilical. Gordon again moved to attach the tether to the umbilical, but it was obvious he was struggling. Charlesworth faced the same dilemma I had faced on Gemini 9.

Dick's struggle while holding himself in position with one hand created a workload heat level beyond the capacity of the suit to cool. Sweat was running into his eyes, stinging and blinding him. With no way to wipe them, he groped back to the hatch. Kraft leaned over his console behind Charlesworth and demanded, "Get Gordon in!" By the time Gemini arrived at Tananarive, Conrad had already cut the EVA off after only thirty-three minutes.

On the final Gemini missions, Charlesworth, Lunney, and I found the limits of the flight director's role. During the EVAs, we could only listen to the crew and watch over the spacecraft systems. Only the commander's view from the cockpit afforded the perspective to make real Go NoGo decisions. But the experience we gained with Gemini stood us well for Apollo EVA planning.

With one mission remaining and the EVA objectives not satisfied, a headquarters review board assessed the results. Their conclusions stated, "NASA has designed overly complex and demanding EVAs, based on analyses, theories, and concepts that are not entirely accurate." The board recommended that we train the crew in a "neutrally buoyant" mode with the crewman suited and ballasted in a water tank. "This mode of training most closely replicates the environment experienced in EVA."

To get ready for Gemini 12, the crew, controllers, and planners all went back to the drawing board. (Only weeks before, Gemini 12 had been a mission without objectives. Now the urgency to conduct a successful EVA had placed it on the critical path in the preparation for Apollo.) Buzz Aldrin, the EVA crewman, proved an apt student in the neutrally buoyant environment and the combination of extensive underwater training and improved tethers and tools unlocked the door to a successful EVA on the final Gemini mission.

Lunney and Charlesworth had been pulling twelve-hour mission shifts since I left, so they called me back for the sleep periods on Gemini 11 and 12. It was great to close out Gemini surrounded by the greatest crews and controllers who ever lived. We were a unique clan.

Gemini developed the tools and technologies we needed to go to the Moon, but even more, Gemini was an essential step for the crews and controllers. The culture of early Gemini operations centered on Kraft and Slayton, strong individuals who stepped up to the risks and with courage knocked them aside.

In the process, they defined the leadership

qualities needed for success in space. Their words were clear, their expectations high. They knew they needed to develop a second generation of leaders. They used Gemini to select and test those individuals who would carry the torch in Apollo.

10

A FIRE ON THE PAD

The 750 members of the initial Space Task Group that Dr. Robert Gilruth led to Houston had grown to almost 14,000 civil servants and contractors by the end of the Gemini program. Now the Manned Spacecraft Center was a technical powerhouse of scientists and engineers with vast responsibilities, charged with implementing the manned spaceflight program. Center responsibilities ranged from program offices charged with directing the design, development, and operations of the Mercury, Gemini, and Apollo space programs as well as the manned medical and lunar sciences and related experiments. The center also had lead responsibilities to integrate the design and operations activities of the Marshall Space Flight Center, in Huntsville, Alabama, in developing the Saturn boosters and the Kennedy Space Center in developing the launch facilities.

An Engineering Directorate supported the program offices in the design, development, and program integration of the space systems; a Science and Applications Directorate supported the lunar and space sciences; and a Medical Research and Operations Directorate supported space life science investigations.

The operational responsibilities of the center were assigned to two directors, Chris Kraft and Deke Slayton. Kraft's organization was composed of four divisions responsible for trajectory design, MCC design and operations, spacecraft landing and recovery, and flight control. John Hodge led the Flight Control Division and I was his deputy. Slayton's responsibility included an astronaut office, aircraft operations office, and a large division supporting flight planning, crew procedures, and simulator operations. The MSC team had been working on the Apollo program since the initial NASA-industry planning session in July 1960.

January 27, 1967, Apollo 1

There are not many days in Houston that begin with a shiver. This one did, and it was not a premonition. A cold front was pushing through southeast Texas that Friday morning. A rare freeze was expected as I left the house for work, in the dark. I was ready to roll. The details of the day's agenda were dancing in my head as I pulled out of the driveway.

Command and Service Module (CSM) 012 arrived at Kennedy Space Center from North American's factory at Downey, California, on August 26, 1966. Systems tests were completed in September and were followed by altitude chamber tests to verify spacecraft pressure integrity and validate the system's performance in a vacuum. During the chamber tests, problems with an oxygen regulator and later with the cooling system delayed completion of chamber testing until the Christmas holiday period. The Kennedy launch team carried most of the burden. They were working around the clock except on Christmas and New Year's Day to complete testing so the CSM could be mated to the Saturn booster in early January 1967. The MCC had its own problems. Telemetry and trajectory computers were crashing for undetermined reasons, and astronaut training was falling behind due to simulator problems.

The CSM was moved to the launch pad at complex 34 on January 6 and the mechanical and electrical hookups ("plugs") between the CSM, booster, and launch complex progressed smoothly and were completed January 18. A plugs-in test with the CSM and booster was supported by the MCC and was successfully completed the following day.

The launch day countdown is rehearsed several times to run through procedures, to check facilities, and to train the launch and flight teams. Rehearsals for the Apollo 1 launch day countdown consisted of two tests. The plugs-in test verified the procedures used to check out spacecraft and booster systems using electrical power supplied by the launch complex at the Cape. The plugs-out test verified the procedures for the final three hours from the entry of the crew into the CSM through launch. Ten minutes before the simulated launch the CSM was switched to internal power. (The plugs-out test configuration used a set of batteries mounted outside the CSM. The fuel cells were not active.) The plugs-out test concluded with a simulated launch and a series of orbital tests supported by the MCC.

The Cape is responsible for making sure that the CSM and booster are safe and meet the pre-launch requirements specified in the test procedures. The MCC's responsibility is to ensure that all telemetry, command, and computer functions meet the mission rule criteria and that the MCC is ready to assume control of the mission. The testing sequence is similar to that used in Gemini, although more complex.

A second plugs-in test was conducted starting at 3:00 A.M. CST on January 25 to refine countdown procedures and troubleshoot launch day communications. The test was supported by the Apollo 1 backup crew, Wally Schirra, Walt Cunningham, and Donn Eisele, and my team in the MCC. Due to problems with the Kennedy Spacecraft Center ground checkout equipment, the planned ten-hour test staggered through the next day, finally concluding around 2:00 A.M. January 26. Since my team did not get much hands-on experience during the test I volunteered to power up the MCC and support the early hours of the following day's plugs-out test.

The plugs-out test was not classified as high-risk. This classification was reserved for tests at the vacuum chamber or those involving propellants, cryogenics, high-pressure systems, or live pyrotechnics.

The sun was not yet up when I arrived at my office that morning. I called the Kennedy Space Center test conductor; his report that there were no more deviations (changes) to the test and checkout procedure was reassuring. [The TCP was a several-inches-thick manual for the test that synchronized every action of all elements of the launch team, booster, MCC, and CSM crew. It was not unusual to have twenty to thirty pages of changes that had to be inserted into the manual in the hours before a test.] I shoved a thick stack of the papers from my desk into a briefcase and walked the short distance to Mission Control, arriving about 7:00 A.M. local time.

When I put on the headset and moved to the flight director's console, John Hatcher, the ground controller, informed me, "Flight, the MCC is Green!" (Colors were used in controllers' verbal status reports. Green was Go, Amber indicated problems, and Red indicated a NoGo.) John continued, "All voice, telemetry, command, and radar interfaces between the MCC and KSC [Kennedy Space Center] were checked. We are now receiving CSM telemetry. The MCC countdown is on schedule and my test team is in place." Hatcher, as usual, had everything under control, so with less than a month to the planned Apollo 1 launch, I turned to my action list, nailing each item with a to-do date.

My office mail included a note from Kraft, indicating that headquarters was conducting a meeting in Washington to lay out the Apollo schedule. Chris knew that a recent report by General Sam Phillips, the Apollo program director, had been critical of Mission Control. Phillips expressed his concern about our ability to support the approaching flight schedule. His report said: "Mission Control has two problems, the software is not written and the computers are not working." Kraft's action memo to me was clear: "Get your damn systems guys to stop the gold-plating. I want you to personally justify every one of their requirements."

I penned a brief "roger" on his memo and left it in the flight director's log for him to review when he reported for countdown support in the afternoon. Then, in the early morning, I brought the MCC on line for the test scheduled to take place early that afternoon. MCC personnel participating in the plugs-out test would arrive an hour before the crew entered the cabin, about 11:00 A.M.

As I listened, the countdown preparations progressed smoothly. The banter between the communications technicians at Houston and the Cape told me that the Apollo team chemistry was developing rapidly. The launch and flight control teams had reaped the benefits of the previous two days of plugs-in testing. There had been a bundle of communications problems, but at this point the voice communications sounded crisp.

Hodge appeared just after 11:00 A.M. Shortly thereafter a mixed team of controllers straggled in from their nearby offices. Our handover was short. “John, the interface tests are complete, communications solid, and the MCC is Go. I have a team in place on our end to troubleshoot any communications problems if they reoccur.”

Settling into his chair, Hodge adjusted his headset and neatly placed his pipe and tobacco pouch below the voice comm panel. In his clipped British accent he remarked, “I sure as hell hope it goes well today. We need to get a break from this blasted testing and get going on our own work.” I nodded as I packed my headset in a pouch and shoved it in a drawer. I didn’t want to get John started on the long litany of open work remaining for Flight Control. Just past noon in Houston, I handed the console log to Hodge and departed for the office area. I had planned to leave early that day.

At the launch pad, the test conductor gave the Go for crew entry into the command module. Gus Grissom, Ed White, and Roger Chaffee had arrived at the sterilized White Room, then ambled across the twenty-foot catwalk to the hull of the capsule. Shortly after entering the spacecraft, Grissom activated the suit circuit oxygen flow, and immediately noted an odor like sour milk. The spacecraft test conductor continued the countdown hold, as technicians performed air sampling of the suit circuit. When the sampling was completed, Hodge ran a brief status check and gave the pad test conductor a Go to resume the count at 1:25 P.M. Houston time.

Chris Kraft had been listening to the MCC and pad voice transmissions in his office. Since the countdown process is often erratic, we normally listened to the countdown over the squawk boxes in our offices until it was time to report for our shift in the control room, all of a five-minute walk away. When the count resumed and the launch team completed the hatch installation, Chris left to report to the MCC for the final hours of the simulated launch countdown.

Chris adjusted his headset and started to track each step of the test. Kraft and his Red Team would launch Apollo 1 and then support the first three orbits before handing over to my team. My White Team was to pick up from Kraft five hours after launch. I had the systems shift, responsible for dealing with any CSM problems and getting the crew to sleep. Hodge then picked up with his Blue Team, developed the next day's flight plan, awakened the crew, and then handed back to Kraft. The three-shift sequence was the same one we successfully used in early Gemini.

Shortly before 3:00 P.M., Kraft polled his team and waited for the test conductor's call-outs for the MCC abort command checks. Hodge, with his handover complete, briefly considered returning to his office, then put down his tweed jacket, shrugged, and ambled toward the coffee pot. If the test dragged on (like the previous day's plugs-in test), he planned to relieve Chris later in the day so that Chris would have the option to go back to his office and do some paperwork or take off for the weekend. As usual during such tests we had a mixture of members from different teams involved so that all controllers got a chance to see the CSM telemetry.

Since everything seemed to be going well, I left the office early to avoid the traffic.

Marta and I had not had many nights out since the birth of our sixth child, Jean Marie, now nine months old. I had promised her an evening at the Athens Bar and Grill, a popular Greek restaurant on the Houston Ship Channel, where on any night the impromptu entertainment might include, for example, a large, sweaty woman slinging a skinny sailor around the dance floor. I also wanted to see what the big deal was about eating food cooked in grape leaves.

Arriving home, I had to hit the deck running and get dressed if we were to beat the evening supper crowd. Marta had the kids all lined up and fed. With the help of the older ones, a single baby-sitter could handle the whole gang and, since it was a Friday night, the kids didn't have any homework.

At the launch complex in Florida, meanwhile, the spacecraft voice lines started glitching. The crew was having trouble communicating with the launch team as well as with each other. It was approaching sunset at the Cape when the countdown was held to permit troubleshooting. Kraft kidded George Page, the test conductor at the Cape, saying that he was keeping score on who called the most holds.

The communications systems used between the crew, launch team, and MCC were incredibly complex. Hundreds of engineers, operators, and technicians were wired to their support teams. In Mission Control there were more than a hundred communications panels, each with forty-eight talk-listen buttons. If you wanted to talk to someone sitting next to you, the voice communications went through dozens of connections. From the MCC to the Cape the communications were carried by numerous telephone lines. When anything broke down, the simplest problem might take hours to troubleshoot and resolve. Doing it during a pad test bordered on the impossible.

The launch team continued trying to work around the communications problems. Attempting to resolve the comm problems, Grissom and Ed White exchanged their suit audio/electrical connectors while Roger Chaffee and the launch team rehearsed the procedures for a dry check of the spacecraft thrusters. At 5:20 P.M., the countdown entered the scheduled hold point at T minus ten minutes where the spacecraft would switch to its internal power. The communications problems had to be fixed before proceeding. Kraft entered the hold in his console log, punched at his voice comm, and said, "Ground Control, see if you can get a handle on the voice problems. The rest of you can take ten."

At 5:31:04 Houston time a brief voice report jolted the launch and flight teams. It was perhaps the defining moment in our race to get to the Moon. After this, nothing would be quite the same, ever again.

"Fire!"

"We've got a fire in the cockpit!"

"We've got a bad fire . . . get us out. We're burning up . . ."

The last sound was a scream, shrill and brief. The elapsed time of the crew report: twelve seconds.

There would be no final agreement on who in the Apollo spacecraft shouted what. But even today, just reading the words on paper is chilling.

There was a gallant but futile effort to rescue the trapped threesome. The pad rescue team as well as crewmen from North American, mechanics and technicians, grabbed fire extinguishers and rushed toward the inferno. At least twice, shock waves and secondary explosions drove them back, knocking many to their knees. Some got close enough to struggle with the hatch. The heat of the hatch burned through their gloves and the smoke sent them staggering, choking and blinded. The call went out for firemen and ambulances. By then it was too late.

I had just finished hurriedly dressing to go out to dinner when I heard a knock on the door. Thinking the baby-sitter was early, I buttoned my shirt as I walked down the stairs. Again I heard the knock, only this time more insistent. I yelled out, "Hold on, I'm coming."

Opening the door, I was surprised to see my neighbor and fellow flight controller, Jim Hannigan. He strode in, agitated and breathless. "Have you heard what happened?" he exclaimed. Bewildered, I raised my hands as Jim walked across the room to turn on the TV. He then blurted out: "They had a fire on the launch pad. They think the crew is dead!"

I had a sudden apocalyptic vision of a gigantic explosion that had taken out the flight crew, the Saturn rocket, and the launch complex. Marta had raced down the stairs as Hannigan's wife, Peggy, visibly upset, walked in the door, crying out, "It's just awful. I can't believe it." Since the details in news bulletins at that point were few, I assumed, from the emotional state of Jim and his wife, that one of the controllers had reached him on the phone. (Kraft had moved quickly to cut off all outgoing calls.)

Confused, I rapidly switched TV channels. There was no new information, only a brief report that an accident had occurred. I grabbed my badge and plastic pocket protector full of pencils. Nodding briefly to Marta, I jumped into our black Plymouth station wagon and tore out the driveway, shooting through traffic lights on the ten-mile drive to the Space Center. I practically dared a cop to get in my way.

I tried to figure out where in the countdown the accident had occurred. Given my awareness of the command problems in Gemini, my mind raced through the current tests. A thousand questions filled my thoughts as I tried to rule out the MCC as a cause of the tragedy on Pad 34. Nothing I knew about the situation made sense. This had been a very low-risk test. I kept telling myself, "The propellant systems are not loaded." I kept thinking about it the way I would analyze an aircraft accident — did some part of the plane fail, was it pilot error, did someone on the ground screw up?

The radio was still reporting only sketchy details of the accident as I swung into my parking slot behind the MCC building. I bolted from the car and raced to the entrance. Getting inside was difficult. With the news of the fire, every controller was reporting to the MCC to find out what happened. Cars were parked haphazardly behind the building.

Kraft had declared a total freeze on operations to protect the data, terminating phone calls and directing the controllers to write down every event, any and every recollection of what they had seen and heard. With any ground or flight accident, it was essential to the investigation to bring everything to a dead stop while memories and data were still fresh and uncontaminated by the inevitable aftershock, confusion, and second-guessing.

At the Cape, they had been able to keep news of the disaster under wraps for about an hour, but leaks were inevitable. Wives of some of the technicians had received tense phone calls from their husbands, saying only that they would be home late. Sensing that something horrific had happened, the wives called the newspapers and radio stations with anxious questions. Reporters began to put pressure on their contacts who worked at the Cape and at MCC.

Security had barred further entry to the MCC without the permission of Kraft or Hodge. I waited for the guard to break through the busy signal on the phone at the flight director's console. Cursing in frustration, I walked around to the rear entrance, bluffing my way past the guard, saying, "The main elevators are locked out. I've got to take the freight elevator to the second floor."

Once I was in the control room there could be no doubt that something catastrophic had happened. All I had to do was look into the eyes of the controllers. They seemed stunned, talking in short snatches, all wondering what the hell happened. I finally reached Hodge. Kraft was standing by the surgeon, listening more than talking. Hodge was unusually quiet, muttering under his breath, "It was gruesome," then lapsing into silence. Clenching his pipe in his teeth, he fought to retain enough composure to stay focused. It wasn't easy for him. It was impossible for the younger controllers. They were milling around, standing, then sitting, too agitated to stay still. They kept playing back the telemetry recordings, looking for clues, desperately clinging to their belief in their data, expecting to find answers.

Kraft hung up the phone after a lengthy discussion with Slayton, then solemnly returned to the flight director's console. "Deke thinks we were damned lucky," he reported, "that we didn't lose a hell of a lot more. There was fire coming from the capsule, molten metal dribbling down the side of the service module."

It was not a good time to be talking about luck, but in times of crisis your defenses kick in. This is especially so among people who have loved to fly. They go on autopilot. Their instincts take over. Nothing could be done for the crew. The important thing now was to find out the how and why — to protect the living and to keep moving forward.

Death had come to the space program in the most unimaginable way during a test — to three men, helpless, not in the air, but in a cockpit just 318 feet above the ground. The fire had flashed through the cabin in seconds. You tried not to think of the horror of it. We all thanked God it had been quick, but how long is quick? How long does it take to suffocate, to burn, to die?

Hodge and I had come from flight testing and knew the risks. Kraft was intimately aware of the dangers from the day he launched Shepard in the first primitive Mercury capsule. We knew there was a high probability that some men would die at some point in the program, but none of us could accept losing our crew on the launch pad. We all had assumed that when a calamity struck us, it would be in flight. Our nightmare was an explosion during launch, or a flying coffin, a faulty craft stuck in endless orbit.

Dutch Von Ehrenfried had been at the guidance console during the crew's last seconds. He was white as a sheet, face drawn, for once speechless and on the edge of tears. The poise I had seen so often on the judo mat and in competition had left him. He was now just a vulnerable young man who had witnessed his friends' deaths.

John Aaron, filling in on the EECOM console, passed the minutes playing back data, seeing the brief electrical current spike, then the rise in cabin pressure and temperature. He pushed himself beyond exhaustion and finally had to be driven home.

In these harrowing hours and the days that followed there was no way to comprehend or accept the loss of Grissom, forty, White, thirty-six, and Chaffee, thirty-one. If there was anything that could be retrieved from this tragedy, it was the evidence — it was right there in front of us on Pad 34. We had a chance to discover the cause of the fire before another spacecraft was put at risk.

The fire did something else. It reminded the American public that men could and would die in our efforts to explore the heavens. It re-created the tension and uncertainty of the early flights of Shepard, Grissom, and Glenn. The Russians worked in secret, but the entire world could watch our flights on television. Success had become almost routine for us . . . until now. The country had gotten complacent. Only many years later would the full count of losses become known: these three Americans plus seven Russians, all brave, good men who ran out of luck, whose technology failed at a crucial moment.

We were torn between feelings of fatalism and defiance. The United States had catapulted men into space sixteen times without a casualty more serious than a stubbed toe — although we had lived through some very scary situations. In our series of ten Gemini trips, Americans had repeatedly broken all records for survival in space, had strolled casually into the void, had navigated their craft through complex maneuvers, tracking down and docking with another spaceship.

With each flight the bar had been raised higher. No one knew how many orbits Apollo 1 would attempt. Grissom, White, and Chaffee would have been blazing yet another path, an open-ended mission, a bold departure from the rigid, limited spaceflights of the past. Theirs was to be essentially an engineering flight, a shakedown for the Apollo systems.

Built by North American, the Command and Service Module was by far the biggest and most sophisticated space vehicle ever designed. We had come so far, so quickly, from Alan Shepard's pioneering fifteen-minute flight. When reporters asked Shepard what he thought about as he sat atop the Redstone rocket, waiting for liftoff, he had replied, "The fact that every part of this ship was built by the low bidder." It was a funny crack, but with an edge.

In marked contrast to the tiny Mercury capsule, Apollo was, in spaceflight terms, practically a luxury liner. It had hammocks for full-length sleeping, hot and cold water, and a primitive galley. The spacecraft was the transportation for the crew to Earth orbit, to lunar orbit, and back home. It consisted of two sections, or modules, the upper one cone-shaped, the lower a cylinder. In the top section, called the command module, the astronauts occupied three cockpit couches looking up at a maze of controls — gauges, dials, switches, lights, and toggles. The service module was essentially an engine room. It housed the fuel, the crew's oxygen, the basic electrical system, and a large rocket with 22,500 pounds of thrust that would supply the propulsion required to enter and leave a lunar orbit. The CSM was thirty-four feet long and weighed about thirty tons when fully fueled.

The Saturn booster rockets were enormous. Towering 223 feet above the launch pad, the two-stage Saturn IB rocket provided 1.6 million pounds of thrust at liftoff and was used for Apollo Earth orbital missions not requiring an LM, or lunar module.* The S-IB was a prototype for the Saturn V and used the same S-IVB upper stage as the more powerful Saturn V.

The Saturn V rocket with 7.7 million pounds' thrust at liftoff was the largest rocket ever developed by the United States. Standing 363 feet tall, it was used for missions that carried both the CSM and LM.

The LM was a buglike, rocket-powered craft that two astronauts would board for the descent to the Moon's surface. The LM, with landing legs folded, was mounted on an adapter at the forward end of the S-IVB, which was in turn enclosed by four tapered conical panels with the CSM perched on top. For lunar missions the S-IVB (the third stage of the rocket) injected the CSM and LM into Earth orbit, and after a checkout period the engine reignited to place both spacecraft into a lunar trajectory.

* LEM versus LM: In the early planning of the Apollo program, the term "LEM" was used, but by the time the program got started, the "excursion" (E) was dropped from the vernacular and it simply became Lunar Module.

The capsule of Apollo 1 was a total loss, charred and blackened both inside and out, its sensitive instruments ruined beyond any useful purpose — except for whatever clues it might surrender. The three bodies had been left strapped in their seats for seven hours while the first anguished experts tried to sort out the causes of a fiery accident that traumatized an entire nation.

One by one the controllers left after securing the records, the logbooks, and the voice and telemetry tapes. Almost by reflex, everyone drifted over to the Singing Wheel, the controllers' watering hole.

As the word of the Apollo 1 fire spread through the Clear Lake area, Nelson Bland, the owner of the Singing Wheel, cleared the building except for the controllers. Throughout the evening, more drifted in as others left. Worried wives came looking for their husbands, clustering in one of the back rooms. It was like the nights of years earlier, when you lost a squadron pilot and a good friend. All that was lacking were the songs we used to sing back then, our way of saying, in the words of Dylan Thomas, "death shall have no dominion." This night, however, was one of limited and subdued conversation. We mourned our crew and the loss of whatever naïveté we had left.

We had known setbacks before. We had lived through some bad days, but we had never taken a knockout punch like this one. I wished there were some way to get in a judo match. I just wanted to feel some physical pain. The beer was not helping anything.

When we returned to our homes that night, we were changed in ways none of us could describe.

The next day was no different. The controllers wandered between the offices and the control center, their minds now moving to the question, "What's next?" Kraft was nowhere to be seen. I guessed that he was probably working with Bob Gilruth, Deke Slayton, and Joe Shea, the Apollo program manager, to put together an investigative team. The day stretched on forever as dribs and drabs of data filtered into the offices. There were many rumors, few facts.

As I was sitting in my office, a picture of an antique biplane hanging in a tree caught my eye. I had carried it with me to keep me on track since my time in flight test at Holloman Air Force Base. A caption below the picture read:

> Aviation in itself is not inherently dangerous. But to an even greater degree than the sea, it is terribly unforgiving of any carelessness, incapacity, or neglect.

No one understood the risk any better than Gus Grissom. He had been quoted as saying, "If we die we want people to accept it. We hope that if anything happens to us it will not delay the program. The conquest of space is worth the risk of life."

Roger Chaffee had been CapCom for my first mission as flight director. I had worked with Gus and Ed White in Mercury and Gemini, but we hadn't spoken for some time except for brief phone calls or crisp debriefings after a test. Now it was too late. They were gone, and all I had were some foggy memories of three Americans who had died in the race for supremacy in space. A rational feeling or not, I felt that I had personally let down the crew of Apollo 1. But I also knew that I had to put aside these feelings and take the lead in rallying the controllers to get us moving forward again. I had seen Mueller, Low, and Williams get out in front and lead when we had had problems and setbacks. Now it was my turn to set an example.

Monday morning, I told Hodge I was calling a meeting of my branch and my flight control team in the auditorium. Hodge, deeply disturbed by the fire and still searching for his own answers, agreed and expanded the meeting to include the civil servants, Philco controllers, and our spacecraft contractors.

The auditorium in our office area held 250 and was half full when John and I arrived. The controllers were still in the bewildered state they had been plunged into on Friday night — muted and somber, feeling, as I did, that we had failed our crew, but not knowing what to do about it. Hodge spoke first, citing the known facts of the accident, then describing the newly appointed review board and the investigating team headed by the director of the Langley Research Center, Floyd Thompson.

I recognized a few of the Thompson committee members. Frank Borman, the Gemini astronaut, and Max Faget, the director of engineering, were from MSC, and John Williams was from the Cape launch team. When Hodge completed his briefing, I still did not know what I wanted to say as he motioned me to the microphone. Emotionally I had come out of the shock, and my feeling now was one of pure anger. Anger that we in Flight Control in some way had let the crew down.

I climbed the four steps to the stage, looking at all those faces of people I knew so well. I wanted them to get beyond shock, then say, as St. Peter did in one of his epistles, "Let us get good and angry — and then let us make no mistakes." Yes, we had experienced a terrible tragedy and a devastating setback, but this was not the end. The testing and the program would go on and we were the ones who would carry it forward. It was up to us to make sure that the Apollo 1 crew had not died in vain.

I started talking about my feelings, and the words finally poured out. I didn't quite know where they came from, but I spoke slowly, deliberately, and with conviction. "Spaceflight will never tolerate carelessness, incapacity, and neglect. Somewhere, somehow, we screwed up. It could have been in design, build, or test. Whatever it was, we should have caught it.

"We were too gung ho about the schedule and we locked out all of the problems we saw each day in our work. Every element of the program was in trouble and so were we. The simulators were not working, Mission Control was behind in virtually every area, and the flight and test procedures changed daily. Nothing we did had any shelf life. Not one of us stood up and said, 'Dammit, stop!'

"I don't know what Thompson's committee will find as the cause, but I know what I find. We are the cause! We were not ready! We did not do our job! We were rolling the dice, hoping that things would come together by launch day, when in our hearts we knew it would take a miracle. We were pushing the schedule and betting that the Cape would slip before we did."

My remarks were received with silence, no movement, no shifting in the seats. The controllers, each and every one, knew what I meant. I was just putting their thoughts into words.

"From this day forward, Flight Control will be known by two words: 'Tough and Competent.' *Tough* means we are forever accountable for what we do or what we fail to do. We will never again compromise our responsibilities. Every time we walk into Mission Control we will know what we stand for.

"*Competent* means we will never take anything for granted. We will never be found short in our knowledge and in our skills. Mission Control will be perfect.

"When you leave this meeting today you will go to your office and the first thing you will do there is to write 'Tough and Competent' on your blackboards. It will *never* be erased. Each day when you enter the room these words will remind you of the price paid by Grissom, White, and Chaffee. These words are the price of admission to the ranks of Mission Control."

The specific cause of the Apollo 1 fire was never identified, but the conditions that led to the fire were clear. We had a sealed cabin, pressurized with oxygen. There were extensive combustibles in the cabin, including a lot of explosively flammable Velcro. The wiring and plumbing systems were vulnerable to damage and, in retrospect, we made the wrong hatch design tradeoffs. It is easy to see all of this in 20/20 hindsight. Like so much in technology, there was a necessary tradeoff. The hatch was a two-piece design. The exterior opened outward while the interior pressure hatch opened inward. It was a brute, heavy and awkward. Given the design, a rapid escape from the spacecraft was impossible. But the NASA and North American designers hadn't been as worried about escape contingencies as they were about the possibility of a hatch popping open into the vacuum of space or another inadvertent opening during a water landing. The premature opening of Gus Grissom's Mercury hatch and the loss of his capsule was a lesson not easily forgotten.

A fire on the ground was considered such a remote possibility that the cabin contained no extinguisher. Even if there had been one, it probably would not have worked quickly enough in a time frame of a few seconds. Today's Halon gas full-flood system might have worked. The fire involved a pure-oxygen cabin atmosphere, flammable materials, and an ignition spark from somewhere in the spacecraft. Before we could fly again, we had to eliminate one or more of these elements in the interior of the spacecraft. Everyone — designers, launch team, MCC, and even the crew — had not given enough thought to what an oxygen-rich atmosphere could do, particularly in a cabin stuffed with flammable material.

I worked with the controllers, assembling the data for the Thompson committee. After putting the data together, I listened for the last time to the final minutes of the plugs-out countdown. We took all the tapes and other records — everything from MCC and the Cape — and shipped them up to the investigating committee.

As we fought back from the tragedy, *Tough* and *Competent* joined with *Discipline* and *Morale* in defining the culture of the controllers. These words became our rallying cry. The controllers gave me a T-shirt with the words stenciled across the chest. I was proud of their gift and proud to wear it. The ultimate success of Apollo was made possible by the sacrifices of Grissom, White, and Chaffee. The accident profoundly affected everyone in the program. There was an unspoken promise on everyone's part to the three astronauts that their deaths would not be in vain.

At the time of the accident, every element of the program was in trouble. The command and lunar modules were behind schedule, the software was late, and the systems were often failing during testing. The Saturn had had problems also. The second-stage (S-II) rocket was an engineering and production nightmare. After a second S-II explosion, in ground testing, there were some contractor changes at the production and test facilities. There were recriminations, but no excuses.

Engineers were having difficulty moving the leading-edge technologies from the laboratories to the production line. At North American and in the U.S. Congress, the report written by General Sam Phillips before the fire raised questions about competence, quality, and workmanship by the manufacturer. If they sneezed, we caught the flu. Every spacecraft design change triggered more changes in Mission Control and in the simulators. The traffic piled up and engineers found they were making changes on changes.

By late spring, however, the program emerged from the chaos of the fire. Momentum began to build again.

In March 1967, the mission designations were changed. After the Apollo 1 fire, there would be no Apollo 2 or 3. Two unmanned Saturn IB flight tests — AS201 and 202 — were not redesignated with a sequence number. The next mission after those two Saturn IB flights was designated Apollo 4, the first flight of the Saturn V. For our own internal purposes, mission types were given letter designations. The controllers preferred this letter sequence since it denoted the broad objectives and was used in lieu of numerical designations.

- First manned Command Service Module (CSM) — C
- Manned test of CSM and Lunar Excursion Module (LEM) — D

• High Earth orbit (up to 4,000 miles) and test of the CSM at lunar reentry speeds — E
• Full lunar dress rehearsal, with CSM and LEM — F
• First lunar landing — G
• Subsequent lunar missions — H1, H2, and H3
• Extended lunar surface missions — J1, J2, J3, and so on.

The general public, of course, knows the main flights in the lunar sequence by number — G was Apollo 11, H2 the all-too-well-named Apollo 13, and so forth.

The summer and fall of 1967 were the busiest times I had ever known. Nothing seemed stable. Change was constant. The two certainties were that Wally Schirra would fly the first manned CSM mission, and the lunar landing goal for 1969 was unchanged. We had two and a half years to pull it off. Everyone went back to the drawing board. The command module would be redesigned at a cost of $75 million, and a safer spacecraft emerged.

Among the changes was a unified hatch that combined the exterior launch protective cover with the pressure hatch. (The launch cover protected the CSM surface from the rocket blast when the escape tower was jettisoned during launch.) The entire hatch mechanism swung out and could be opened by the crew in ten seconds.

11

OUT OF THE ASHES

When we completed a mission, it was like putting pictures into a scrapbook and then turning to a fresh, blank page. Someday we would have the luxury of looking back and remembering all the moments captured in those earlier pages, but the press of events gave us no time to indulge in reflections, to celebrate past accomplishments — or to grieve. For a time we simply could not dare to look back at the Apollo 1 inferno. We could only look forward to the next blank page, the next mission. But there was no way that any of us could escape those thoughts that come unbidden in the dark hours of the night: we would dream about those terrible last seconds. They would be with us forever. We would not leave the sadness behind until we accomplished what Gus Grissom, Ed White, and Roger Chaffee wanted America to do — land on the moon.

Spaceflight forced you to live with risk by focusing on the task at hand. I would compare it to a pilot walking away from an accident, muttering, "Son of a bitch, that was close!" Then, still shaken, he lights up a cigarette, picks up his helmet and parachute, and starts reviewing his actions and identifying what, if anything, he would do differently the next time. After hoisting a few with his squadron mates, he gets ready again to climb into his cockpit home.

At Mission Control, certain things were understood. Every mission must achieve its objectives, and it must be accomplished on schedule if we were to keep John F. Kennedy's pledge to land a man on the moon in this decade.

While we were recovering from the fire, the space scientists sponsored by NASA continued their work to develop a follow-on exploration program for the Moon. I was sent to the University of California, Santa Cruz campus, in August 1967 to brief a group of government, individual, and university scientists on the Mission Control Center's mission responsibilities and on the techniques we used to develop mission rules.

Preparing for the briefing in the campus library, I realized how narrow my world had become because of the intensity and isolation of my work over the last seven years. I had never been on a West Coast campus. What I saw was beyond my belief, the TV headlines coming alive. It was my first live encounter with the hippie generation. Their songs and chanted slogans dimly penetrated the library as we worked. When I left I was glad to get back to a world I understood. But would these young people comprehend the meaning of all we had been trying to accomplish for so many years — the greatest use of economic and technological power in history for peaceful purposes? The Vietnam War was only one challenge facing (and, unfortunately, dividing) our country. Countless American lives were going to be lost before that long war was brought to an end. I honored those who served; I could not sympathize with those who did not honor members of their own generation, young men who were far removed from college campuses and demonstrations, who had no choice but to fight and be killed or maimed. I returned from that campus in California wondering what the young people I saw there would make of the legacy we were trying to pass on to them — and to the rest of mankind.

November 9, 1967, Apollo 4

There was little fanfare the day NASA recovered from the shock of the Apollo 1 event and resumed the space race. Arthur Hill of the *Houston Chronicle* reported from the Cape on the launch of the unmanned Apollo 4, the first flight test of the Saturn V, the world's mightiest rocket. It was the only machine powerful enough to launch the two Apollo spacecraft, the CSM and LM, into Earth orbit and then hurl them toward the Moon.

"The powerful engines shook the press stands," Hill's story began, "rattling light fixtures and bouncing tables up and down. It was an awesome sight as brilliant yellow fire engulfed the launch pad at liftoff." This time the fire was with us. We sent Saturn into space on the most immense pillar of flame ever seen at the Cape.

In Mission Control, all of us felt elated as America resumed its voyage to the Moon. The

The Mercury, Gemini, and Apollo spacecraft, showing their relative sizes, and the Mercury-Atlas, Gemini-Titan, and Apollo-Saturn V boosters (boosters to scale with each other but not with spacecraft).

Saturn performed perfectly, blending new and old propulsion technologies in each of its three rocket stages, then as the mission ended, the command module was hurtled earthward at seven miles per second to test the heat shield during reentry.

The Apollo 4 test, more than any other, demonstrated George Mueller's fearless "all-up" approach to testing. It showed that we had the right guy filling the job as NASA's boss of manned space flight. "All-up" meant that every element of the space system was on board and operable. There were no "boilerplate" spacecraft. If you were successful, the concept was labeled brilliant, and you could focus your energies on the next step, the next set of unknowns. If you had problems, you found them early and somehow made time to fix them while keeping on schedule. If you failed, a lot of expensive hardware was reduced to junk and the schedule shattered.

I didn't know much about the NASA hierarchy. Our Administrator, Jim Webb, lived in another world, Washington, D.C., from whence came our funding and our mandate. Webb, boss of the whole organization during the years of Mercury, Gemini, and early Apollo, had had a long, distinguished career, including serving as Director of the Bureau of the Budget and undersecretary of state in the Truman administration. A profoundly serious man with a vigorous manner and an ability to deliver a great speech when one was required, he knew every bureaucratic pitfall there was to know and how to navigate around them, inventing new strategies as needed. He was adroit at securing funding from an often reluctant Congress — and at keeping NASA's critics at a safe distance from his people who were doing the work. His style was low-key and effective. He knew how to delegate and give people like George Mueller and George Low the authority they needed to achieve the goals in each mission.

The miracle of the NASA rebirth after the fire

was due to four of the best leaders the program ever had. George Mueller, the boss of manned spaceflight, was a modest man, trained as a research engineer, with a great feel for the complex details of operations. He provided the foundation before, during, and after the calamity, and took the heat from Congress. Above all, he stood up for his people throughout NASA and provided an unwavering direction with his all-up test concept.

In 1966, the year before the Apollo fire, Goddard Space Flight Center advised me that they were not installing consoles for controllers on the two Apollo tracking ships. GSFC, the operator of our communications network, believed that the rapid advancements in communications technology would allow transmitting data and communications by satellite by the time the Apollo missions began. Since I had worked many shifts with the ships in Gemini, I was critically aware of the support they provided in covering key mission events and providing orbital gap coverage. I wanted a controller team aboard the ships for Apollo. I was not willing to risk the crew or mission objectives by making the MCC dependent on "may happen" technology.

I expressed my concerns to Kraft and after a brief discussion he stated, "You're going to have to convince Mueller. He considers himself a communications expert and is the only one that can turn around GSFC's decision." The following day I flew up to Washington to sell my recommendation to Mueller.

This was not the first time I met Mueller. I had

a lot of respect for the way he blocked for his team and took the heat when things went wrong. During a particularly rough press conference after the Gemini 9 Agena failure he sat with seven of us at the press table. Late in the conference a reporter asked, "This is the fourth straight mission where you have had some major problems. When are you going to start kicking some ass and —" That was as far as the reporter got before Mueller tore into him. He described the problems, the actions taken, then concluded with supportive remarks about his team. His vivid response brought a cheer from the other reporters.

Mueller was busier than hell at NASA headquarters, trying to get the Apollo program up to speed. As I sat outside his office I watched grim-faced engineers and project managers carrying the bad news into his office. During the summer of 1966 the Apollo program seemed to be unraveling.

I waited in the secretaries' office as the time for our appointment passed and the afternoon turned into evening. About 8:00 P.M. he came out, apologized, and told me he had reservations for two for supper at the Georgetown Inn, so we would have our meeting there. During the meal, this man who knew more about communications technology than I ever would, listened politely as I briefed him between courses on why we needed controllers on the Apollo tracking ships.

I was impressed by his patience and courtesy, the force of his technical arguments, and his willingness to consider my ideas. To this day I am awed that a man with so much weighing on his mind would spend an entire evening with somebody way down the chain of command. He listened thoughtfully and then told me to go back to Houston; he would make a decision on the following day. Early in the afternoon word came down: my argument had prevailed. GSFC was directed to place controller consoles on the tracking ships.

George Low, the son of an Austrian immigrant, joined NASA's predecessor, NACA, after his Army discharge and worked his way through the government ranks. After the fatal accident Low replaced Joe Shea as the Apollo program manager. He was a master at getting people to work together, creatively channeling their energies and thus building the momentum to achieve the objective.

The flight directors knew Low well from his middle-of-the-night visits to Mission Control during a flight, where he sat silently in the viewing room. Low worked both at MSC and back at NASA headquarters. He had a rare blend of integrity, competence, and humility. You would do whatever he asked you to do, regardless of the odds and regardless of the risk.

Rounding out the NASA management that directly affected us were Sam Phillips and Frank Borman. Phillips, an Air Force lieutenant general, came in from the Minuteman ICBM program. He possessed an uncanny ability to spot problems, define solutions, and keep the complex development processes moving ahead.

Borman, the astronaut who toughed it out on

the fourteen-day Gemini 7 mission, was one of the most respected of the second class of astronauts. Flight controllers saw him as a table-pounding "let's cut out the bitching and get on with it" type of guy. He was the one who finally stood up during the agonizing over the redesign of Apollo 1 and said, "Enough. Let's get on with the job. It's time to fly!"

We moved from disaster to flight in less than a year because we had leaders of this caliber — and because they trusted us.

In June of 1967, as Apollo forged ahead, fate reached out and grabbed me when I was made deputy to Hodge for the Flight Control Division. The division, the home base for the flight directors, controllers, and instructors, had grown to 400 personnel. Virtually every malfunction procedure, schematic, or mission rule used in training, or carried aboard the spacecraft, was produced by this division. The division planned and was now flying an average of six missions each year, a punishing load, and I was glad to give John a hand. I also welcomed the opportunity to step into division management because of the challenge to reach beyond my experience as a flight director and start developing broader organizational skills. I believed I had the capability to do more.

I immediately acquired new respect for Hodge because of his ability to perform as both division chief and flight director. To ease the burden on Kraft and Hodge, the original plan was for Lunney, Charlesworth, and myself to work two missions, skip one, then work two more, alternating as the lead flight director for every third mission.

After the fire, Kraft had his hands full leading the four divisions — Flight Control, Landing and Recovery, Mission Planning and Analysis, and Mission Support — in the Flight Operations Directorate. As a result of his workload, Chris would never again sit in the chair as flight director.

Now that I had moved to Hodge's deputy position, the flight director staffing changed again. Looking at the workload, I decided that I could cover only about half of the missions and I changed the sequence so that I worked only the odd-numbered missions, starting with Apollo 5. Aware of the coming overload, Kraft selected two more flight directors, Pete Frank from Mission Planning and Milt Windler from Recovery. I believed all flight directors should be selected from the ranks of Mission Control and was surprised by the selection of two virtual unknowns. Since they would need time to come up to speed, I successfully lobbied Kraft to add Gerry Griffin, a top-notch Gemini controller, to the list so we could get some immediate help.

Working as Hodge's deputy was one of the most enjoyable times in my life. Initially, I didn't think I would make a good deputy. I am too impatient, love to work with people directly, and like to lead the charge myself. I am used to giving orders, not offering suggestions, and get impatient when I know a team can move faster. In the case of the Flight Control Division in 1968, it turned out that Hodge and I were a perfect fit.

Where I was direct, Hodge was philosophical. Hodge studied the alternatives; I was quick to pick a direction. Our balance of temperaments allowed us to lead the division well. Hodge provided the vision, the long-term strategy, while I concentrated on the tactical. Hodge dealt with finances, I rallied the people. We both worked on the organization and structure.

I liked the way John put his thoughtful comments in the flight directors' logs, the way he characterized his decisions. I also enjoyed him as a person. Hodge was typically English in his approach to work, that is to say, a real gentleman. He got more done without the continual bluster of many of his peers. Above all, he had consideration for others and their opinions, which stood him well with his peers — but not necessarily his bosses.

It was time to let the missions begin. The division was a powerhouse, knee deep in talented leaders and team members. We were indeed *Tough*, *Competent*, and ready for Apollo.

Unmanned missions in every program are forgotten except in NASA's record books, something that annoys controllers, who know how difficult it is to control a virgin spacecraft and booster, and operate with software, all fresh off the production line. The controllers had to do the crew's job without the benefit of their presence. Using ground commands in place of the crew's switches, we performed all the maneuvers and tests called for in the flight plan. Every controller loved the unmanned missions. We were the first to fly each new spacecraft. No man would fly until these missions were successful.

Among the more exacting (and exasperating) tests was that of the Lunar Module — that buglike frail craft that would put two astronauts on the Moon while the command module circled over them in Moon orbit. The LM was a two-stage spacecraft, standing twenty-three feet tall on four rather spindly legs. The lower or descent stage had the propulsion systems and propellant used to get the craft down to the surface of the Moon. Triangular bays supported the batteries, water tanks, and helium used to pressurize the propulsion system. The landing legs supported a "porch" and ladder for the crew's descent to the Moon. When the EVA was completed, the descent stage provided the platform for the ascent (or upper) stage's launch off the Moon.

The ascent stage contained the living quarters, controls, displays, and the attitude control, guidance, navigation, and radar systems used for each maneuver. The brain of the LM was housed in a state-of-the-art computer with 36,864-word fixed and 2,048-word erasable memory. This stage also contained the ascent engine, propellants, batteries, life support, and the communications and data systems.

Directly over the crew's heads was the hatch that provided access, when docked, to the command module through a tunnel. The crewmen stood in the lunar module, looking forward through a small triangular window on each side, with the commander on the left and LM pilot on the right. The external skin of both stages was paper-thin aluminum, the lower stage covered by multiple layers of gold Mylar insulation. You could easily poke a pencil through the side of the spacecraft. Portions of the interior were covered with netting to save weight and catch anything that might fall into nooks and crannies inside the LM. Designed to operate only outside the Earth's atmosphere, the LM looked ungainly, had no heat shield, and was incapable of safely entering the Earth's atmosphere.

I was flight director for Apollo 5, the unmanned shakedown cruise of the LM. The test plan consisted of a series of descent engine maneuvers to simulate a lunar landing, a "fire in the hole" abort, and a sequence of ascent engine maneuvers simulating a rendezvous of the LM with the CSM. The LM ascent engine is buried in a cavity in the top surface of the descent (landing) stage. The fire-in-the-hole test ("Fire in the hole" is a term used in mining when explosives are about to be detonated) involved shutting down the descent rocket, blowing the bolts that attached the ascent and descent stages, switching control and power to the ascent stage, and igniting the ascent rocket while still nestled to the landing stage. All these events occurred in fractions of a second, just as they would in a real aborted landing close to the lunar surface. The fire-in-the-hole abort was the most critical test of the mission and one we had to accomplish successfully prior to a manned mission.

On a personal level, this was the start of my journey to the lunar landing. The mission brought me face-to-face with the team of controllers that would take an American to the lunar surface. I dove into the mission as if it were the last one before the Moon. My Apollo 5 White Team was a curious mixture of youth and experience. Jerry Bostick was breaking in a new FIDO, Dave Reed, while John Llewellyn had an old grizzled World War II bomber pilot, Jim I'Anson, under his tutelage. The contrast between Reed, a city slicker, quick to respond, and I'Anson, a bushy-mustached West Texas rancher with a slow drawl, set the extremes of the team. Jack Craven and Don Puddy were my LM systems controllers.

The mission was a flight controller's dream, consisting of a Saturn launch followed by a continuous string of eight maneuvers spread over five orbits. The entire mission was scheduled for only eight hours. If all went well, the mission would be flown totally under the command of the LM computer. This was the first mission for the new LM team and the most complex unmanned test we would ever fly. The LM's contractor, Grumman, was also new to the space program. Grumman was understandably nervous, and they worked very closely with my team to get through the first flight of their frail but essential contraption. I anticipated, per Murphy's Law, that if anything could go wrong it would. In the months before the first test I put heavy emphasis on making sure that if the LM automatic systems failed, the MCC team could take over and do the job. By the time we approached launch readiness, we had developed several different routes to achieve the primary mission objectives, incorporating eight ground-commanded alternates to the basic mission plan.

Three days before launch I faced a new problem. Jack Craven, my LM control engineer, responsible for the guidance, navigation, attitude, and propulsion systems, had been in a traffic accident. His Volkswagen was demolished, with Craven taking the steering column in the chest. There were no broken bones, but he was beaten up. Just breathing hurt. He was unable to speak beyond a hoarse, raspy croak. With only a single team, I was faced with scrubbing the planned Apollo 5 launch. Since unmanned missions were executed by ground control, the loss of an experienced controller made us terribly vulnerable. Craven, a former Navy "hurricane hunter," was one of the most technically qualified controllers ever to step up to a console. Older than the rest and often a bit cranky, he had come from the recovery division to Flight Control in order to get a piece of the action. What we didn't know at this time was that he was suffering from an increasing hearing loss. Even so, as Apollo progressed, he was given a troubleshooting job, often assigned as second man at the console for critical events. Dwight Coons, my flight surgeon, trained in medicine at the University of Toronto, volunteered to get Craven ready to fly the mission.

A launch countdown is a massive undertaking, like writing the score for a symphony. Putting one together for the first time is an experience not easily forgotten. The Apollo 5 lunar module, launch support equipment, software, and procedures were exercised in an integrated fashion for the first time in a countdown demonstration on Thursday, January 18. The one-day test stretched to almost three days and, without a gap in testing or a day's break, we began the launch countdown. One of the many things NASA operations at MCC had in common with the military was that rest was a scarce commodity. If you are standing watch and then doing ship's work at sea you run on about six hours of sleep in twenty-four; same goes in intense aircraft operations or field deployment in the infantry. You learn to live with fatigue for very long periods — and not let it erode your focus or dull your edge.

At the launch-minus-one day review Dr. Coons reported that Craven could support the mission, but he would be virtually immobile and have difficulty speaking. Bob Carlton drew the job of responding to Craven's grunts and mumbled comments, selecting displays, issuing commands, and communicating to the control team. Bob would later become my LM control systems engineer for the lunar landing.

January 22, 1968, Apollo 5

Late in the afternoon, after a ragged countdown and six hours of delays, I finally gave the call to the test conductor: "MCC is Go for launch." Dr. Coons had done well. Craven was at the console in a stiff-backed chair, headset on, incredibly erect, unable to move head and body. He was a big-time coffee drinker and I knew his body was aching as much from lack of caffeine as it was from Coons's therapy. Seated next to him was Carlton, serving as his voice and hands. The stakes were high, but failure was not on my mind. We had been virtually wedded to the LM and Saturn booster for a week.

The launch was smooth as satin. The LM, once separated from the booster, coasted through the second and into the third orbit, the control team snuggling up to their consoles. The only sound was a periodic hoarse grunt from Craven to Carlton. As the ship off the Australian coast acquired telemetry, the CapCom, Jim Fucci, reported, "Signal strength good . . . mission sequence five cued . . . clocks in sync." After a final check with his controllers, he said, "Flight, we're Go." I acknowledged, listening as Fucci counted down the final seconds to the first test of the LM's engine. The action was about to begin.

Fucci called out the final events: "The computer is in control . . . engine arm . . . plus X jets firing . . . engine start . . . 10 percent."

I instantly thought, Go for it! Then Fucci suddenly called "SHUT DOWN!" The words came almost as an expletive, something that we were not expecting. In seconds, Fucci transmitted commands to secure the system, then more commands to burrow into the guts of the LM computer to find out what had happened.

We were now getting the telemetry in the MCC from Australia. After briefly assessing it, Gary Renick, my guidance officer, came on line. "Flight, we had two alarms, 'DPS DELTA V' and 'FORGET IT!' "

I thought, What the hell are they?

A quick check of the LM showed no apparent problems, then Craven grunted instructions to Carlton. Precisely measuring his words, Carlton said, "The alarm indicates that the thrust did not build up fast enough. The time set in the computer for thrust buildup was too short. We need to change the computer timer."

Renick said, "That makes sense, Flight. The FORGET IT alarm indicates that when the command was given to throttle the engine, nothing happened."

I was proud of my guys. Within minutes of the alarm, they had decoded the problem — caused by an incorrect computer instruction — and were moving toward an answer.

The tracking station coverage was soon going to go to hell. We had only three and a half more hours before the lunar module went beyond the ground network. While the LM crossed the Pacific to the United States, we developed a plan to change the computer timer, delay the mission plan a revolution, and attempt to return to automatic LM computer control.

On the third pass over the States, a problem that had been previously just a nuisance now became critical. Mission Control was having difficulty commanding the spacecraft. The signal strengths were so poor that it took three or four tries for each command. We quickly decided to start a go-for-broke ground command sequence on the fourth and final pass across the States. The control teams struggled to get the target updates and maneuver information into the lunar module computer, punching the commands in manually from the consoles.

Kraft and the Apollo program manager, George Low, joined me at the console as we rapidly discussed the options. Kraft, unfamiliar with the team's jargon, said, "I don't know what you're doing, but keep it up . . . good luck!" Then he returned to his console behind me. Low remained at my console, asking me what I intended to do. I said, "George, I am going to try a Hail Mary pass. I am going to go for the full set of objectives using manual sequences. I will have a backup if we run out of time and tracking stations." Kraft motioned Low to move away from the console and give me room to work. I said a prayer that we could get the ground commands in when needed.

I was at a crossroads. I wanted to accomplish every LM test objective, but I could not risk the loss of the fire-in-the-hole test. To hedge my bet, I set an MCC wall clock counting down to the time of the final tracking station pass where I would settle for the last-ditch plan.

The team's dialogue between positions in the last minutes became so rapid, crisp, and intense that I could hear Craven yelling instructions to Carlton, completely ignoring his pain. Other controllers, their work done, hunched over their consoles, trying to figure out how to help us. I would select the mission sequence, the commands would be executed by Renick, my GUIDO, and the event calls would come from Carlton. The three of us had to be perfectly synchronized. The fact that the schedule for the lunar landing was now hanging in the balance never entered our minds. We were committed to success. We were after the whole enchilada. Time no longer had a meaning; we were locked into orbits, elapsed time, targets, and command sequences. We had to get the engine testing and the fire-in-the-hole objective completed in the next orbit or the mission was a failure.

A new problem was bugging Reed, my flight dynamics officer, as he edged up to my console: "Flight, on the next maneuver, if the engine burns too long we will splash the LM in the Atlantic Ocean. We have to get the engine cutoff in at the right time." I nodded to Reed, silently assuring him that somehow we would get the commands to the LM to stop the burn. If it didn't, we would lose all of our flight test objectives. This was no time to take counsel of our fears.

The ship off the coast of California sent a command that started the descent engine. Renick and Carlton, in perfect sync, called out events and times and snapped out the backup commands. The descent engine shut down after one minute, twenty seconds, coasted, and then restarted. As the engine continued to burn my mind clocked the objectives as each milestone passed by. The call from Carlton — "Fire-in-the-hole, abort stage . . . we are stable" — made me smile and several of the controllers gave a brief cheer. I heard Craven's chuckle over the hubbub in the room as the LM ascent engine burned briefly, then shut down as planned.

Reed reported, "Flight, the engine shutdown

occurred at the right time in orbit. The high-speed tracking indicates we have passed minimum perigee [lowest altitude point]. The altitude is now increasing." You could feel the collective sigh of relief that we had not splashed into the Atlantic. A half orbit later we commanded the final ascent engine maneuver, completing all of the LM objectives for the mission.

As a result of the command problems, we had ruptured a control fuel tank, blown a jet nozzle off the LM, tumbled the gyros, and expended all ascent rocket fuel. But we had satisfied all objectives in our last-ditch maneuver sequence. I was so ecstatic I felt like starting our party in Mission Control. As the LM left our telemetry screens, the spacecraft was heading toward its fiery reentry off the west coast of Panama.

It was past midnight when we finished, but not too late to say thanks to a great bunch — young kids and two salty old-timers. I regretted that the Singing Wheel was closed because of the late hour, denying me the chance to buy my team a few well-earned beers. Although this was just an unmanned mission, the lunar module showed the resilience, the flexibility, the margin we would need to go to the Moon. We had dodged bullets before, but this time we caught one in midair and spit it out. The morning newspapers declared, "APOLLO MISSION A SUCCESS, LUNAR PROGRAM ON TRACK!"

I am poor at committing to memory vast amounts of information, so I developed a series of indexed handbooks that I could refer to instantly at the console. These documents were my bridge to the controllers. I color-coded the books and highlighted the key constraints. By the time a flight was ready to launch, I had spent hundreds of hours with my systems handbook, mostly at night at home, long after Marta and the kids were asleep. I wanted to know the guts of the spacecraft the way I had known about the components of the aircraft I had flown. By acquiring this knowledge in detail I was able to communicate with my systems controllers at a level deeper than the other flight directors. This enabled me to get answers faster and make decisions quickly in real time.

My greatest fear approaching launch day was that I would lose one or more of my books. To assure that they were easy to find, I used pictures of various striking young women from the *Sports Illustrated* swimsuit edition for all my book covers. The controllers knew about my book covers and, if one were missing, this virtually guaranteed a prompt return. I went back and forth daily between the office and Mission Control with my four large distinctively covered volumes under my arms.

No final lunar schedule was yet in place when Kraft called a surprise staff meeting on April 8, 1968. We had flown three unmanned Apollo missions since the fire, all of them little noticed by the public and the media. At the meeting, Kraft reviewed the results of the missions up to Apollo 6, then he started to speculate on a possible alternative sequence if problems kept pushing against the timetable for the lunar landing. Finally he drove right to the core: "The lunar schedule is in trouble. We must understand and fix the problems with the Saturn." (Apollo 4 and 5 had gone very well, but on Apollo 6 we had first-stage rocket thrust oscillations that caused the Saturn to bounce like a pogo stick. Minutes later, two of the second-stage engines shut down, and when we got to orbit we could not restart the S-IVB engine.) Kraft then continued, "The LM is overweight and the software for its computer is not ready."

None of this was news to me and I wondered where he was heading. Kraft then continued, "Each mission in the flight sequence from now on must clearly resolve some flight unknown or add a new capability to our missions. The E mission does not make sense to me. It only goes to a 4,000-mile apogee [highest point of the orbit]. That is not high enough to check out the CSM lunar navigation and verify the navigation and tracking software we will use in the MCC during a lunar mission. If we are going to do the E mission, I don't see why in the hell we don't go to the Moon and test the techniques and software we will use for lunar navigation and tracking."

The chief of mission planning, John Mayer, had been waiting for this opening. Within hours of the meeting, his conceptual flight planners were on their computers. Within a month, Mayer's team had developed a basic plan with a lunar flyby and a lunar orbit alternative. Satisfied with Mayer's planning, Kraft directed the work to continue and to be expanded to involve all of his divisions. Chris, always the master at balancing risk and building options, now had a lunar mission alternate, and, given the opportunity, I was sure he would use it.

12

THE X MISSION

August 1968

I was suddenly the acting division chief for flight control. The Manned Spacecraft Center director, Dr. Robert Gilruth, concerned about the lack of planning for the post-Apollo era, assigned John Hodge to study how the Manned Spacecraft Center should be organized to meet the space programs of the future.

I had my hands full juggling flight director and division chief duties when I received a call to report to Kraft's office for a one-on-one meeting on Friday morning, August 9. I was hoping that the meeting would be short and it was. With no preamble, or even hello, Kraft announced, "George Low wants to fly a mission to the Moon this year. He believes he can have a CSM available in December." The shock on my face must have been evident as Kraft continued, "George wants to drop the E mission from the schedule and then use the Apollo 8 crew for a lunar orbit mission using the CSM from Apollo 9. Your [Apollo 9] mission will be slipped two months to get training for Borman's crew."

Kraft had accomplished much of the mission planning with the studies he had commissioned in April. Now in a gutsy move George Low picked up Kraft's lunar mission plan. Low saw that it provided a way to continue to move forward on the lunar landing schedule and flight-test the lunar navigation and tracking while the LM program resolved its problems. The LM spacecraft deliveries were lagging due to a broad range of developmental problems. I recognized Low's plan as a bold move that would let us get to the Moon by the most direct path and buy us some badly needed schedule time . . . provided it worked.

Kraft asked me to give him a list of the minimum number of people needed to assess the plan. After reflecting a few moments I gave him names of five controllers. Kraft's response was succinct: "We don't need to get the training, booster, or LM people involved yet. Let's keep it to Bostick and Aldrich. [Jerry Bostick and Arnie Aldrich were flight controllers in my division.] I'm flying to Huntsville with Low and Slayton this afternoon to see if they can get a Saturn ready for the mission. We need to get Marshall Center leadership behind the plan." As he motioned me out of the office he concluded, "I will need your assessment by Monday if not earlier. Keep this under your hat." I walked away thinking that Wernher von Braun's Germans and my trajectory team were in for one hell of a surprise.

I had mixed emotions returning to my own office. I am conservative in my planning and had long believed in a thoroughly planned and incremental approach to the lunar goal. Personally, I believed the best track to reach the Moon was the current sequence. Low's plan would heighten the risks, but by moving ahead on several fronts at once, it would buy us time.

On his return Friday night from Alabama, Kraft, with Low, called the flight designer, Bob Ernull, and Jim Stokes, the computer boss, to his office and got right to the point. "I need launch window data for a December lunar mission, and I need it by Monday morning."

With no hesitation, Ernull replied, "I'll need all the computers in Buildings 12 and 30, and I'll need them through the weekend." Kraft turned to Stokes and ordered him to give Bob everything he needed to do the job.

It wasn't going to be easy to carry out Kraft's new marching orders, but Apollo succeeded at critical moments like this because the bosses had no hesitation about assigning crucial tasks to one individual, trusting his judgment, and then getting out of his way. In 1968 computers were still incredibly slow by today's standards. We sometimes needed a run of six to eight hours to come up with a single answer. Computer data entry was time-consuming and the complexity of the data entry often introduced errors in the data input. The long running time and the drain on the memory often resulted in the machine crashing just before it could crank out the answer. With four mainframe computers at his disposal, running around the clock, Ernull was barely able to generate the mission data.

Emerging in the early hours Monday morning, he provided Kraft with options for an Atlantic Ocean splashdown in November, December 1968, and January 1969, and a single Pacific Ocean launch window from December 20 to 27, 1968.

Early Saturday morning, I received a call to attend another meeting with Kraft. His secretary had already called a half dozen new principals, including Bostick and Aldrich. The meeting was again short. Kraft indicated that MSFC was studying Low's plan and that he needed a commitment from his four divisions in two days. Leaving the meeting, Aldrich and Bostick were on top of the world, unable to believe their luck — they were going to lead the planning for the first lunar mission. We were fully aware of the intense workload ahead, and the reshuffling of priorities and the risks that we had to address. Our decision processes might seem unstructured and extemporaneous, but those involved, even the very young, had the requisite experience and were masters at the art of risk judgment. All were acutely aware of the consequences of failure. The teamwork used to respond to changes and problems during a mission is the same used to respond to planning actions before the mission. Our technique assured a rapid, competent, and multidisciplinary response.

Working on weekends had become a habit, at least a half day just to catch up and get ready for the training or testing every Monday. Bostick made a few phone calls when he returned to his office. When you received a call from work on a weekend you dropped what you were doing and just reported in. A half hour later as three controllers walked into his office Bostick began, "A few of us just had a meeting with Kraft. George Low wants to go to the Moon this December. By Monday, Kraft wants to know whether we can do it or not."

A full-scale debate erupted: "Geez, Jerry, we've never been out of Earth orbit before. We don't even know if we can compute a lunar injection maneuver. Christ, we don't even know if the booster guidance can do the job!"

Unruffled, Bostick rolled on, "I want you to get together a small team of the best people you have, give them the job, and turn them loose. I need an answer by Monday."

Bostick then turned to Chuck Deiterich, a young, thin, lanky Texan with a Pancho Villa mustache. "Chuck," he said, "I've tagged you as the lead RETRO for the first lunar mission. You will have my full support." Deiterich, a specialist in reentry trajectories, had never worked a manned mission. He was momentarily overwhelmed by his new role. His other two teammates would be his section and branch bosses, who would be working for him. Such was Flight Control in the final year before the lunar landing. Assignments and opportunities came like a lightning flash. There were no precedents, no guidelines. All of a sudden you were given a job and you just did it — whatever it took.

A lunar trajectory consists of a string of

maneuvers, one to leave Earth orbit, several to adjust the trajectory en route to the Moon, and then two maneuvers to enter lunar orbit. The return to Earth requires another maneuver that is again adjusted during the return. At the time Kraft asked us for a decision we did not have integrated trajectory software; no maneuver in this sequence had yet been fitted end to end with any other.

Kraft was in his element Monday morning as he assembled the pieces of the plan, weighed the alternatives, and sorted out his options. Bostick said, "Chris, if the MCC support team can get the lunar programs into the computers, I don't see any reason why we can't do it." I gave Kraft the Go for my division and finished with the staffing plan for the flight directors and the control teams.

The group then selected the week of December 20 as the launch window. This meant a CSM end-of-mission landing in the Pacific Ocean. This was a major early decision; the Navy could cover only one ocean as the primary landing area. Once the decision was made, we had to live with it for the rest of the Apollo program. The Pacific gave us the promise of a large ocean target and warmer, calmer waters — or so we hoped. The Navy liked the decision and began planning for operations from Pearl Harbor.

To keep this mission clearly separated from the current plans, I designated Apollo 8 as the X mission. Until the mission was approved, we had to keep all mission data for the originally planned E mission. The X mission was now joining the ranks of the Gemini 4 space walk, the Gemini 76 rendezvous, and Mueller's all-up Saturn test concept as examples of the high-risk, high-gain leadership we had in the 1960s. The decision to go to the Moon with Apollo 8 was made before we had ever flown a manned Apollo spacecraft.

The X mission meetings occurred daily, each one pounding another element of the plan into place. By August 16, one week later, the team had expanded and for the first time I felt I was dealing from a full deck. We were continually admonished to keep what we were doing a secret, but it was like hiding an elephant in your bathtub. The constant closed-door huddles, the changing work priorities, and the longer hours gave us away.

A press conference was held on August 19 to announce the official changing of the mission sequence, moving Frank Borman's crew into the December launch slot and formally designating the mission Apollo 8. The announcement described this as a high "Earth orbital" mission, with a lunar option. It took only a few seconds for the press to figure out what the plan really was.

Kraft wanted to use Lunney, Charlesworth, and me for both Apollo 7 and 8. I advised him I intended to stay with the current lead flight director assignments, shifting Charlesworth forward to cover the Apollo 8 mission.

In the fall of 1968, I was like a guy juggling grenades wrapped in barbed wire. I was grappling with my new duties as acting division chief. I discovered there is a hell of a difference between being a deputy and being the boss. Now I had to cope with politics, budgets, job assignments, and direction of an organization of 400 amid the rapidly evolving flight program. I didn't know it at the time while I was working as Hodge's deputy, but Chris and he had had disagreements on a number of policy issues. I believed that Chris thought John was too conservative to be a flight director. Looking back, I see why Hodge let me run Flight Control. I suspect he felt that his days as a division chief were numbered.

My salvation was Hodge's former secretary, Lois Ransdell. Lois adopted me into the office and showed me the ropes. Lois was precise and direct, and had a fiercely protective attitude about "her division." She became my trusted adjutant, the guardian of the office door and my schedule. Years later, she was given an honorary flight director title, selecting the color pink. In the history of Flight Control, only two others, Bill Tindall and John O'Neill, my deputy director during the shuttle era, have been awarded this recognition.

All around us, the tumult of the 1960s continued. The war in Vietnam had intensified. Television brought the casualties into our homes at night, but we did not yet realize we were losing. Campuses across the land were seething as students protested the war and marched for civil rights. Race riots had broken out in major cities in the summer of 1967. Then, after Martin Luther King was shot and killed on April 4, 1968, there were riots in more than a hundred cities. In June Robert F. Kennedy was killed while campaigning for the Democratic nomination for president.

Even the space program was picketed and bomb threats were reported. Everything we carried into the Mission Control Center was inspected. Security guards roamed our parking lots during missions. We practiced bomb threat evacuations from Mission Control, always leaving a small team to hold the fort if we had a crew aloft. These events provided a violent background to our final charge to reach the Moon. Fortunately, the public's support for the lunar program remained high. Apollo was a bright glow of promise in a dark and anxious era.

Apollo 7 would be the first manned Apollo mission, and the shakedown cruise for the redesigned Command and Service Module. Each of the spacecraft systems would be tested in flight, the recorded data analyzed by both controller and engineer during the only flight test to qualify the CSM before actually going to the Moon. Wally Schirra, Walt Cunningham, and Donn Eisele had been assigned as the backups for the ill-fated Apollo 1 two months prior to the pad accident. Now it was their turn to fly after nearly three years of training.

Schirra, the first astronaut to fly all three programs, was the veteran whose cool performance during the Gemini 6 pad shutdown and whose almost fanatical preparation for missions put him high in the ranks of those regarded as "heavy hitters" by the controllers. Wally, ever caustic, never kept his opinions to himself. While preparing for the mission, he let it be known that he was damned unhappy with the inclusion of the TV camera in the spacecraft and the planned "dog-and-pony shows" that would be broadcast from the command module. He considered it an invasion of the privacy of the crew and the sanctuary of the spacecraft. But Kraft was equally adamant that the American public, which was underwriting the program, get an opportunity to see space flight in action through these live video broadcasts. The TV camera won.

October 11, 1968, Apollo 7 – Return to Manned Flight Testing

A launch countdown takes two days. When it starts, there is relief because the tedium of the training period is over. Once launched, the only option is to move forward, facing problems, identifying solutions, forging ahead. A flight control team is an elite force, playing in a sort of Super Bowl with each mission. The real difference, of course, is that we are not playing a game and losing is never an option. If Apollo 7 succeeded, we would be on schedule for the lunar landing. If we failed, the chances were high that there would be no lunar landing before the decade ended.

During the final hours before launch, every engineer in the program has the right to voice any and all concerns he might have by sending last-minute memos and making phone calls to tell the flight team what worries him about some aspect of testing or some unexplained glitch. The maiden launch of a manned spacecraft brings many systems on line for the first time. We were given a lot to worry about from the new North American engineers.

The launch team attempts to give us a perfect system for liftoff. But no matter how hard they try, with thousands of components, 850 crew controls and displays, and 350 telemetry measurements, there is no such animal as a perfect spacecraft. We always have some glitches, some uncertainty. The same can be said for the MCC ground system. We had our share of hiccups.

The greatest focusing mechanism in the space program was the countdown. Clear, crisp, and unequivocal decisions had to be made during the final hours and minutes. As the count progressed, people in each area of the program came forward. After assessing the technical issues, all made their calls. Everyone swallowed some problems, bit their own bullets. Launch day was like a fresh start, a new day, and I loved it. My team started the countdown and checked out the MCC, then handed over to Griffin's team for the CSM and Saturn systems testing. Lunney picked up for the launch, the handoffs between the three teams going flawlessly. Shortly after 10:00 A.M. in Houston, the race to the Moon got the wave from the starting flag.

The launch went smoothly, the Saturn rocket blasting the CSM into a low Earth orbit. To television viewers, as the engines ignited, there appeared to be one heart-stopping moment of hesitation. But because Apollo and its two-stage launch rocket weighed 1.3 million pounds, the launch acceleration was gradual, taking ten seconds to clear the tower.

The late morning liftoff dictated the orbital shift schedule. Lunney's team, with all the crewmen generally awake, worked the day shift. I had the swing shift with Donn Eisele on watch in the spacecraft, and Griffin got the graveyard shift, staying in touch with Wally Schirra and Walt Cunningham. Schirra had set up a "duty watch" on board the command module, so that an astronaut would be awake throughout the entire mission. This plan was counter to the experience we had in Gemini and none of the flight directors thought Wally's "watch" was a good idea. It was tough enough to sleep the first days in space and if someone is awake, rustling around or communicating, it is impossible.

Glynn Lunney had handed my team a clean spacecraft at the beginning of the sixth revolution. The trajectory experts in the Trench worked the maneuver sequence to set up a rendezvous with the Saturn IVB booster on Lunney's morning shift. The CSM was trouble-free, so my principal concern was a report by our weatherman that a low-pressure system was developing off Cuba, 750 miles south and east of Houston. With Houston's proximity to the warm waters of the Gulf of Mexico, a hurricane could make it tough on Mission Control.

A half hour before we were to hand over the controls to Griffin's Gold Team, Schirra said,

"Houston, I have developed a head cold and have taken two aspirin. I've gone through eight or nine Kleenexes with some pretty good blows. I'm thinking about taking a decongestant or antibiotic."

My team surgeon recommended the decongestant only. I took him to the press conference at 2:00 A.M. and was surprised by the large turnout. The doctor turned out to be the star of the show and, with few problems on the spacecraft, Schirra's cold — the first space illness — made the headlines of newspapers across the country and grabbed time on the network telecasts.

At liftoff, three flight tests remained before Apollo would go for the lunar landing. We had a lot to get done. The flight control team tracked each objective and added new ones to exploit each opportunity. With a single shot to qualify a spacecraft, little was left to chance. The Apollo 7 flight plan was incredibly precise, breaking objectives down and literally keeping each minute and second chock-full of activity. The mission objectives are listed in a thick manual that spells out every detail of the required test. The flight plan was designed to cover all of these objectives — and if some weren't accomplished, they would be added to the workload of the next flight. As the flight progressed the test results we received led us to update the flight objectives, add new tests, or modify existing ones. It always has been this way in spaceflight and will continue to be as long as missions are measured in days and weeks. Schirra knew the lunar game plan and understood that we had a lot to get done before we could take the next spacecraft to the Moon.

As the mission continued, Wally's cold was as much a test of the flight control team as was flying the mission. The flight directors were hard pressed to satisfy a cranky Schirra and push ahead to clear the deck for the next mission. There was little that pleased Schirra about what we were doing at MCC, and the discomfort and irritability caused by his cold soon made him pretty testy with Cunningham and Eisele as well. Glynn Lunney, in particular, always seemed to be at the helm when Wally was testy with the ground team. By the midpoint of the mission, I realized how lucky I was to be working the night shift.

The video reports, seven to eleven minutes long, had caught the public's fancy. They were dubbed "The Wally, Walt, and Donn Show" and aired once each morning during the Apollo pass between Corpus Christi, Texas, and Cape Kennedy, the only two ground stations equipped to pick up the transmissions. By the third day, Schirra canceled the daily TV broadcast with a clipped, "No further discussion." We were left with the task of convincing a skeptical press that all was well between the operations team and the crew. Deke Slayton, embarrassed by Schirra's outburst regarding the telecasts, murmured on the voice comm: "Christ, Wally, all you gotta do is flip a switch."

By the fifth day, the headline in the *Houston Chronicle* declared, "Captain Awakes Grumpy."

The press started getting in their licks and the controllers counted the days until we could get a new crew. None of the mission rules discussed dealing with a grumpy commander.

Schirra finally relented on the broadcasts, and at one point the astronauts, trying to make amends, held up crudely lettered signs that read, "Hello from the lovely Apollo room, high atop everything."

With Schirra and Cunningham asleep, my team would listen to Eisele talking in a hushed voice from his astronomy lab on high. He identified the stars and remarked on the vista from his platform as his partners, the "sleeping beauties," rested.

With the lunar mission scheduled less than two months away, we started releasing our backup computers to the mission designers during the day to check out the new trajectory software coming on line. At night, with the Apollo 7 crew asleep, Charlesworth, using the same backup computers, started launch-abort training one floor above us for Apollo 8.

Meanwhile the low-pressure area had turned into a hurricane, crossed Cuba, and entered the Gulf. We were still keeping a close watch, but it appeared the full force would hit the Mississippi and Alabama coastline and not Houston. But I developed a contingency plan for the control center if the storm moved farther west.

Schirra continued to make life difficult and by the seventh day of the mission, both Kraft and Slayton were involved full-time, now arguing with Wally over an unsuited reentry. Schirra had been taking his shots freely at the controllers, but I was amazed when he started zinging Kraft and Slayton.

With a head cold, ear blockage during entry would be annoying at best, and at worst, painful and potentially disabling. If the astronauts reentered without their helmets they could pinch their nose and blow to try and clear the ear blockage. This is the technique used to clear ears when descending in an aircraft. The designers, however, pressed for a suited reentry in case of a sudden loss of cabin pressure. It was one of the classic risk trade-offs we run during a mission, but this time the argument was going public.

While the bosses argued with Schirra on the voice comm, the teams continued grinding away with the planning, chalking off the objectives, patiently explaining each and every "funny" to the crew as we were able to develop answers. Controllers use the term "anomaly" or "funny" to describe something in the CSM or LM systems operations that is not as expected. Every item of this nature is logged and pursued until it is understood, and each is discussed extensively with the crew.

Lunney set a standard for every future flight director, giving real meaning to the word "discipline" in the flight controller's vocabulary. Refusing to rise to the bait of Schirra or the press, he kept the flight directors and teams on track. Two days before reentry, after a series of flight plan updates, the mild-mannered Eisele got into the act and complained about a flight plan maneuver update. "I want to talk to the man or whoever it was," he said, "that thought up that little gem. That one really got us."

When Jack Swigert, the CapCom, one of the fifth class of astronauts selected in 1966 responded, "Okay, Donn," Schirra cut in: "I have had it up here today and, from now on, I am going to be an onboard flight director for these updates. We are not going to accept any new games like adding fifty feet to the velocity for a maneuver, or doing some crazy test we never heard of before." (Prior to a mission the velocity for each maneuver is specified in the flight plan — but as the mission goes forward the Delta V [change in velocity] is updated to trim the orbit for reentry, as well as to set up daylight conditions at landing.)

Lunney's log said it all: "Refer to the crew voice transcript; I can't stand to write it." The handover for the first time indicated his frustration. "I have finally had enough of this crew." In the final days of the mission, the control teams, CapComs, and flight directors, covering for Wally, felt like embarrassed parents of a kid throwing a tantrum.

In retrospect, some of the exchanges seem sophomoric, except that the stakes were high and discipline and teamwork were victims of this feuding. I regretted it and still do, partly because the pettiness that crept into the mission obscured the fact that Apollo 7 was carrying on with the task that was interrupted by the Apollo 1 fire — a task that had been left unfinished for nearly two years, and one we owed to Grissom,

White, and Chaffee.

My tenth and final shift passed peacefully. Griffin knew that Schirra had been counting the hours till his return to earth and was ready to come home. At crew wake-up on his final shift, Griffin as a joke threatened to keep them up another four days to equal the American space flight duration record set during Gemini. The crew, of course, vetoed the idea, and then Griffin handed them over to Lunney to bring them home, after eleven days. Despite all the interpersonal static, Apollo 7 did the job. Only twenty-six discrepancies were detected in flight. Over half were related to the instruments and communications. This was America's second longest manned space flight, and the Command and Service Module checked out beautifully.

I never figured out why Schirra had such a burr under his saddle. Perhaps he just could not deal with the irritation of having something as piddling as a cold invade the trip of a lifetime. In any case, the careers of two younger astronauts suffered. Neither Cunningham nor Eisele flew in space again.

The control team cheered when Lunney later received a medal for the mission from President Johnson at the LBJ Ranch. His performance went well above and beyond the call of duty.

Three years later, when we designed the emblem of the flight control team, we remembered our best days with Schirra. As the central theme of the controller's patch, we used the

Sigma from his Mercury spacecraft, representing the unbreakable link between the crew and ground. We made our peace with the grumpy commander.

13

THE CHRISTMAS STORY

The successful Apollo 7 flight cleared the way for us to land on the Moon in the coming year. A lot of flight and ground testing remained, and I was sure that there would be surprises, but we had developed the momentum required to pull off a miracle. Our greatest worry was that we had to complete three virtually flawless missions and achieve every major test objective before we could shoot for the lunar landing. I didn't think much about the odds, but since every mission would be a first, the odds had to be stacked against our success.

In the late 1960s our simulation technology had progressed to the point where it became virtually impossible to separate the training from the actual missions. The simulations became full dress rehearsals for the missions down to the smallest detail. The simulation tested the crew's and controllers' responses to normal and emergency conditions. It checked out the exact flight plan, mission rules, and procedures that the crew and controllers would use for the flight. The problems thrown at the controllers and crew by the SimSup (simulation supervisor) prepared them for the real crises that might come in any phase of the mission from launch to splashdown. Simulation attempted to make events that could happen in real time — malfunctions in any one of the many spacecraft systems, trajectory problems, or failure in the ground systems — as realistic as possible. With hundreds of possible malfunctions and many time-critical mission events, the training opportunities were limited only by the hours and weeks available to train. We simulated every mission phase under a variety of normal and emergency conditions. By the time the training period for a mission ends, the astronauts and the MCC teams must be thoroughly familiar with the pre-mission plan. They must know what should happen and be capable of making a correct decision to continue the planned mission or execute a mission abort under any set of circumstances.

A lunar mission consists of a series of time-critical maneuvers strung end to end. Two and one half hours after the Saturn liftoff from the Cape the lunar phase of the mission normally begins with the translunar injection (TLI) maneuver. Midway through the second revolution in Earth orbit the Saturn IVB stage is re-ignited, increasing its velocity from 25,500 to 35,500 feet per second. After S-IVB engine cutoff, the CSM separates from the booster rocket. The velocity from the TLI maneuver places the spacecraft into an orbit 250,000 miles high with the Moon at the highest point of the orbit.

The next phase of the mission is called

translunar coast (TLC) and lasts about three days. Small maneuvers are performed during this period to trim the trajectory to pass sixty miles in front of the Moon three days after the TLI maneuver. Fifty-two hours into this period, the CSM leaves Earth's gravitational field and enters the lunar gravitational field. During the TLC phase the Mission Control Center is in continuous communication with the crew.

Three days after liftoff the astronauts perform the lunar orbit injection (LOI) maneuvers with the CSM service (main) propulsion system engine. LOI consists of two maneuvers that place the CSM into a sixty-mile circular orbit around the Moon.

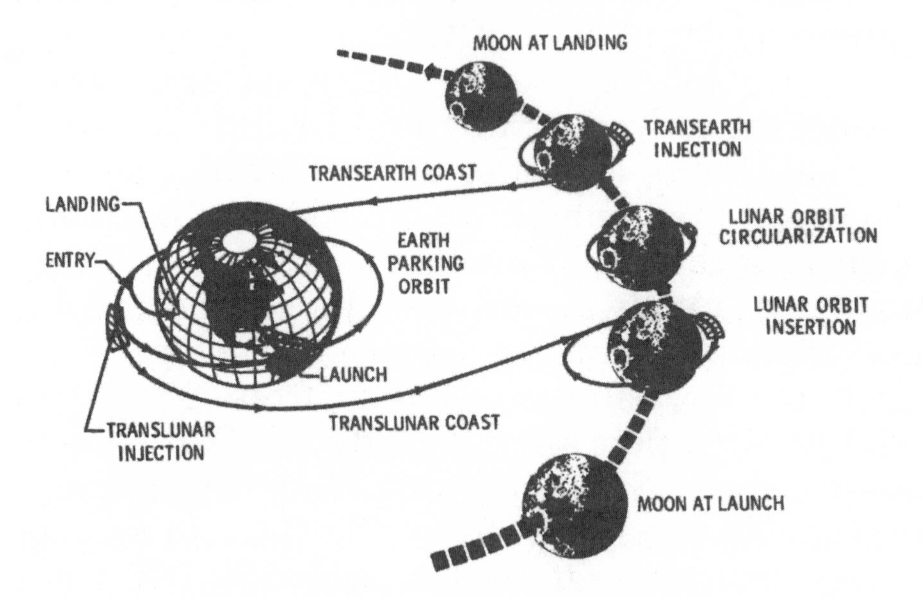

The major phases of a lunar mission (does not include the descent, EVAs and ascent).

After the lunar phase of the mission is completed, the CSM service propulsion system is again used for trans-Earth injection (TEI). The return period is called the trans-Earth coast (TEC) and takes about sixty hours prior to reentry into the Earth's atmosphere and splashdown.

Eight days after Apollo 7 returned to Earth, Charlesworth and his Green Team began the first lunar mission simulations. The postmission assessment gave the command module a solid Go. The next spaceship, in a schedule based almost entirely on gut instinct, would go to the moon in less than sixty days.

As we approached the lunar prize, NASA's future was far from certain. Starting in 1967, Congress had made significant budget cuts in the manned programs and shortly after the Apollo 7 landing they announced there would be no space program beyond the Apollo Application Program, a planned mini–space station that would use Apollo hardware. Responding to the uncertainty about our future direction, MSC director Dr. Gilruth established an advanced programs organization element within the MSC, which reported directly to him. Since John Hodge had been outspoken about NASA's lack of planning for the future, Gilruth selected him to lead the effort. Hodge's new job was to seek out new NASA opportunities in space, develop a rigorous and logical program plan for the future, and establish a more businesslike structure for NASA.

With Hodge moving into his new job I officially became the chief of the Flight Control Division (FCD), the administrative home for the majority of the MCC flight controllers. I reported to Kraft as one of his four division chiefs.

The FCD included the MCC flight directors, assistant flight directors, trajectory controllers (the Trench), booster and spacecraft systems engineers, science and procedures officers, and the simulation instructors (SimSup and his team). Sixteen of the twenty-one controllers normally present in the main control room, as well as SimSup and his team, were provided by the division. The FCD comprised seven branches and two small groups corresponding to the major MCC operations functions and had about 300 personnel. The operations branches were Flight Dynamics (trajectory), CSM and LM systems, Experiments, Mission Simulation, Flight Control, and a Requirements branch that assured the MCC configured correctly for simulations and missions. The flight directors and booster engineers from Marshall were two small groups at my staff level.

The FCD controllers developed the mission strategy, performed premission planning, developed CSM, LM, and experiment schematics and troubleshooting procedures. They wrote the mission rules, supported the design and check-out of the spacecraft and MCC, and performed the integrated crew-controller training. With the exception of the headquarters mission director and mission scientist the remaining

MCC controllers were provided by other organizations at the MSC.

While Charlesworth and Lunney pulled together the teams for the lunar mission, I started preparing with my team for Apollo 9. Mission planning and preparation takes about one year, with the final training starting about three months before launch. The objectives for each mission were vastly different from the preceding mission and now, with the launches spaced at two-month intervals, every flight director and controller was working several missions simultaneously, constantly juggling schedules and priorities. The workload was punishing. Sixty- to seventy-hour workweeks became commonplace.

From the early days of space virtually all of the trajectory data coded in the MCC originated from the Mission Planning and Analysis Division (MPAD). MPAD consisted of several hundred mathematicians and scientists, supported by a large array of high-tech contractors. John Mayer was the boss and Bill Tindall was the deputy. In late 1968, Tindall was reassigned as a staff engineer for George Low. In the restructuring after the fire, Low gave Tindall the task of uniting the entire Apollo team, civil servants and contractors, into a working group to determine how to use the hardware and software most effectively to achieve each mission's objectives. Tindall's genius was his ability to focus on issues and coax diverse people to work together. He combined the friendliness of a puppy with a comic wit. His operational intelligence was brilliant. We formed a particularly strong bond, and our families spent a lot of time together at his beach house. Although our technical backgrounds were very different, we were both emotional about our work, perpetually optimistic, and gave our people unconditional support.

Bill Tindall swung into the Apollo 8 mission with zest, resolving issues from the simplest to the difficult. While we were slugging it out with Schirra on Apollo 7, Tindall was holding daily meetings to work out how we would navigate to the Moon, and how to get into and out of lunar orbit. Allegiance to Tindall did not come easy for the Trench. For a while, Bostick's team believed that Tindall was really doing their job. Bostick's deputy, Phil Shaffer, and Llewellyn complained about these turf issues, while Tindall tried patiently and persistently to gain their support. By the time of Apollo 8, however, the Trench had become Tindall's most zealous group of converts, actively supporting, debating, and testing his plans, carrying into the training his decisions and mission rules.

We were, in a sense, in a race against ourselves, every event and decision converging on the launch date. Tindall was unsinkable. Only a month away from the Apollo 8 launch, he was still arguing with Frank Borman on the best way to navigate the return journey from the Moon.

To the men of the Trench, Apollo 8 was *the* mission; it would be their greatest achievement. Living in the world of pure mathematics, they were the first generation fully at home with computers — incredibly young, dreamers and visionaries who were venturing in their imaginations and theories with the crew into the unknown, working at the very edge of our knowledge and primed to overcome any difficulties that came their way. Their work, coded into computers and plotted in piles of charts and graphs littering their consoles, was the foundation for every computer instruction in the Saturn rocket and aboard the spacecraft. The Trench and the trajectory designers were totally dependent on the millions of lines of code that they wrote in a variety of computer languages such as COBOL and HAL. These computations would hurl the Saturn toward the Moon, and then would swing the CSM into lunar orbit.

Apollo 11 would be the flight for the ages, but Apollo 8 was a very big leap that drew on one's spiritual and moral resolve. For us it would become the second greatest Christmas story ever told. Think about the imagery of a rocket soaring through limitless space, so close to heaven the passengers could reach out and touch the face of God.

After the methodical intensity of the testing, the frequent crisis meetings, the incessant intrusion of the media, and the briefings of "just one more" VIP, the last couple of days before launch always seemed strange. All of a sudden time and motion stopped, as it seems to on a ship caught in the doldrums. I initially welcomed this brief and strange interlude preceding each mission as the final time to catch a breath. Then as the clock kicked over into the last twenty-four hours, the minutes seemed to hang.

This was my first mission as FCD chief. Success belonged to the team; failure was ultimately my responsibility. Even though I was not flying this mission, I went through the same emotional and physical process as my controllers. It was tough to stay away from the control center — and stay out of the way of the guys doing the job — especially during the final hours before Apollo 8. The team understood my anxiety and called me to report, "The count is on schedule and they are in fueling. Why don't you have a beer and get some sleep. We'll call if anything comes up."

The evening before the launch of Apollo 8, a visitor arrived whose presence told you something powerful, something historic, was taking place. He was Charles Lindbergh. He belonged to a more romantic time, when flight was still more an art than a science. His career and his life created a kind of vapor trail that stretched across the years. Lindbergh was with us, as he should have been, when Americans reached for the Moon, so long the object of man's curiosity and dreams. Perhaps more than anyone in the history of flight, he had inspired human beings to explore the skies above them. Commandingly tall, his hair gray and his manner both reserved and modest, he was an honored guest at the invitation of Wally Schirra at a very private party given for the astronauts and a few of their friends.

The plane Lindbergh flew from New York to

Paris in 1927 was powered by a single engine. Lindbergh had sailed through uncharted skies, "hacking it out," as Wally put it, "with the most primitive of technical equipment." No radio, no radar, a windshield a bird could break. Lindbergh's presence was a kind of laying on of hands. I felt that he had handed the stick and rudder over to the astronauts.

December 21, 1968, Apollo 8

The Green Team started arriving at Mission Control two hours after midnight. Cliff Charlesworth was at the flight director's console, backed up by a group of Trench controllers barely out of college. To a great extent, this was their show. For the first time, man would leave the Earth's gravity and be captured by the gravity of another heavenly body. The Trench would provide the guidance and navigation. Working closely throughout the early morning hours with John Mayer's mission planners, they fine-tuned their equipment, their techniques, and themselves. Mission Control is a big, big space, but there is no room in it for ego, only for flawless teamwork.

Sitting in the control center and surrounding buildings were a bunch of very nervous designers, engineers, and computer programmers. All of their work since Kennedy's speech in 1961 was about to be tested. Every assumption, trade-off, and decision they made in creating the system was about to be put on the line. They were threading the needle, shooting a spacecraft from a rotating Earth at the leading edge of the Moon, a moving target a quarter of a million miles away, passing sixty miles in front of it three days after launch.

Buried in the dungeons of the auxiliary computing room was Hal Beck, an early entrant to the Space Task Group. Now he was the chief of the lunar mission design. His work of almost a decade was about to come to fruition. This was payback for the years of freezing at his desk, thermostats turned down to cool the computers in their office complex. Wrapped in sweaters with a heater at his feet in the midst of a broiling Houston summer, Hal represented the labor, the frustration, and the exuberance of almost eight years of work by the mission planners.

Chances are you have never heard of Hal Beck, who grew up, as many of us did, believing in Buck Rogers. He was one of the unsung heroes of Apollo, of whom there were many. It may not stretch the truth to say that without the likes of him we would not have made it to the Moon.

The next morning, shortly after dawn, I found myself in Mission Control, wearing a green vest hand-tailored by Marta (on occasion I wore a vest the color of the other leads for their flights). In the Trench, FIDO Jay Greene, RETRO Chuck Deiterich, and Gran Paules — the GUIDO — were racing the clock. The three had joined Flight Control after Gemini and had grown in their skills during the Apollo unmanned missions. Greene and Paules flew their first manned mission on Apollo 7. When the countdown resumed after the planned hold, Jay Greene finished configuring his displays for launch. After he gave the command, "Flight, FIDO is Go for launch," he muttered a silent prayer that it all worked. In Mission Control, for a few moments, time seemed suspended, everything happening in slow motion. Then in a collective fashion, the momentum built and Mission Control surged forward. Today we would go to the edge of the Moon.

Captain Refsmmat, the ideal flight controller and a member of the Trench through the Apollo and Skylab years. The graffiti on this Kilroy-type poster allowed the controllers to let off steam and express their opinions on virtually every topic. (Courtesy of Ed Pavelka)

It was at moments like this that we counted on "Captain Refsmmat," our imaginary mascot. In the Trench a "refsmmat" is shorthand for "reference to stable member matrix," a set of equations used among controllers, crews, and flight designers as the mathematical means to determine angles with reference to navigational stars. It is the one constant that ties together all of the other reference systems used during a mission, often as simple as a line drawn from the center of the Earth through the launch pad. With data from navigational stars and a refsmmat, the crew can determine the spacecraft's position and velocity in space with the spacecraft computer. The guidance officer at the control center is the keeper of the refsmmats during the mission, synchronizing the ground and spacecraft updates so that the computations will always agree.

The Captain was born during a discussion between John Llewellyn and a newcomer to the Flight Dynamics Branch. Standing by the coffee pot, the rookie asked Llewellyn the name of a controller who had just placed an IOU in the cup next to the pot. Llewellyn responded instantly, "Sheeet, man — that's Captain Refsmmat, the ideal flight controller! He's the best we've ever had in the Trench." The new guy nodded knowingly, glad to pick up the name of his new working partners, especially one considered the model for the Trench.

Ed Pavelka, a gifted FIDO of the Gemini era, heard of Llewellyn's joke and decided to sketch a picture of Captain Refsmmat for the branch. Within days, a two-foot cartoon was hanging in his office. Almost immediately, ideas from other Trench inhabitants poured in and Captain Refsmmat was outfitted in the tools of his trade. He wore a pot helmet with a hinged top opening to a radar antenna and truth-seeking glasses with a black line inscribed across the lens showing the correct deorbit attitude. He had a supply of refsmmats in a pouch on his belt and a variety of awards and decorations, consistent with his august status as the ideal controller.

The Captain was a patriot. He wore a crisp military jacket with captain's bars on the lapel and a pair of khaki shorts. With knobby knees, tennis shoes, and a broad military brassard, he was an apt replica of the ideal. For all of the things wrong in the world, Captain Refsmmat stood for what is right. Pavelka hung the cartoon on a gray metal locker in the hallway, and within days graffiti started to appear, expressing the various controllers' thoughts, opinions, gotchas. Over the weeks and the months, the graffiti provided an outlet for the working guys' feelings about their work, their bosses, and life in general.

Captain Refsmmat lived in Flight Control during the Apollo and Skylab years. He flew our missions, earning his medals for tough assignments and new worlds he had conquered. Today he was the fourth member sitting with the trajectory team in the Trench.

At 6:51 A.M. Central Standard Time, less than an hour after dawn, Apollo 8 lumbered skyward with Frank Borman, Jim Lovell, and William Anders on board. The earth orbital check-out and maneuver injecting the spacecraft onto a lunar trajectory was uneventful.

I was just an observer for this mission but I remember the feelings of pride and relief the instant the Apollo 8 crew left Earth for the Moon. After the launch, my feelings hit a new peak when the translunar injection time came and went. Then when I heard the crew report the maneuver's completion, it really hit me. I had to get up and walk outside because I was so happy I was crying. Being a bystander in Mission Control is tough — and doubly tough on a flight director and new division chief who had nothing to do except wait and hope and pray.

Observer status for me was a living hell. I never liked the viewing room, which was reserved for families, politicians, and the wheeler-dealer contractors. I have never felt comfortable with the high rollers, so if I was not on a console I roamed the back rooms, looking for a place to plug in my headset.

Apollo 8 was one of the best spacecraft ever produced by North American. The systems controllers continually searched their telemetry for the slightest fault, constantly reassuring themselves and their flight directors that, "No, Flight, we don't have any funnies." On the second manned flight of the CSM, only seven spacecraft discrepancies were noted. None was major.

In the early afternoon of December 23, after a brief countdown, a Mission Control wall clock clicked over to 00:00:000 — "all balls" in the controllers' idiom — and civilization crossed another boundary. Now only 30,000 miles from the Moon the Apollo 8 crew had left Earth's gravity field. At 2:29 P.M. Central Standard Time, mankind for the first time was captured by the Moon's gravity. The celebration was brief; the pressure mounting, the controllers were already computing the critical lunar orbit insertion maneuver to be executed in fourteen hours.

During their journey, the crew had not seen

the Moon, trusting the computations developed by the Trench. The two midcourse corrections set the conditions for the precise point and time to enter lunar orbit. The numbers were flawless, and as the early afternoon passed in Houston the tension mounted.

The two midcourse maneuvers had nailed the final trajectory and CapCom Gerry Carr's call, "You're Go for lunar orbit insertion," did not surprise the crew. The spectators in the viewing room hunched forward and the usual buzz of communications ceased. Borman had maneuvered Apollo to the burn attitude. In the control room, the computers had been rechecked and the pregnant waiting continued, with brief moments of banter. As the final minutes counted down, cigarette smoke hovered above the consoles, the room silent.

"Apollo 8 you're looking good . . . good all the way . . . ten seconds to loss of signal."

After a quick "atta boy" from Bill Anders, the final words came from Jim Lovell: "We'll see you on the other side." To the split second, a burst of static marked the expected signal loss. The first humans to see the "far side" of the Moon were now on their own. It would be thirty-two minutes until we saw the crew again and we would know the maneuver result.

After the time passed for the first of the lunar orbit injection (LOI) maneuvers the controllers scattered to the rest rooms. On their return the Trench changed the ten-by-twenty-foot projection display stretching across the front of the mission control room. For all previous missions the display was the ever familiar track of the spacecraft tracing its path across Earth's continents and oceans. During the translunar coast period the trajectory from the Earth to the Moon had been depicted as a skinny, stretched-out horizontal S. Now the display screen for the first time in a mission showed the pockmarked lunar surface.

Unable to bear the tension, Cliff Charlesworth stood and muttered to Lunney and Kraft, "I gotta get out of here." Walking down three flights of stairs, he emerged from Mission Control, lit a Lucky Strike, and began a brisk walk around the two duck ponds in the central plaza of the Manned Spacecraft Center. Frustrated at his inability to control his emotions, he finished his second cigarette, and then purposefully strode back to Mission Control.

The controllers sat in profound silence, watching the clocks, waiting to see if the burn had come off, reviewing the few options available if it did not. Pavelka no longer checked and rechecked the data. He knew it was right. He also knew it was too late now to make any changes. Every controller's mind focused on the one event we could only now see in our minds. Was the Apollo engine burning? Did we get a full burn? Did the crew wave off the LOI maneuver and were they now on a return path to the Earth? The minutes never seemed to end. It was like one of those dreams where you have to fight to wake up. Two clocks were counting down to spacecraft acquisition, the moment when we would reacquire communications and data from the CSM and astronauts. The clock now approaching zero was the one all eyes were watching. If the crew waved off, and the maneuver had not been performed, Mission Control would have an early signal acquisition and it would come when the clock reached zero.

The time came and went, so we knew Apollo 8 had performed the LOI maneuver. The next question was, did we get the planned full burn? Eyes now switched to the second clock. Again, time seemed to hang suspended, unmoving. Suddenly the other clock's numbers were all zeroes, and within a second of the time predicted, the ground controller announced, "Flight, we've had telemetry acquisition." The controllers murmured in relief, and a brief cheer broke out in the room. Apollo 8 was in the planned lunar orbit.

While the spectators in the viewing room continued their buzz, Lunney's controllers heaved a collective sigh of thanks to the trajectory gods, then hunkered down to review the telemetry and tracking data, giving it a meticulous reading. Borman, Lovell, and Anders were in lunar orbit — another event in the sequence of firsts, a new plateau achieved.

With the tension and anticipation relieved, the Mississippi Gambler, Cliff Charlesworth, lit another Lucky, reached for his coffee cup, and said, "Anyone want any coffee? I'm buying!" I have never seen a broader smile on his face. As the lead flight director, he had pulled the planning, teams, and mission together and he had done it well.

The rest of us could only wonder, or guess, at how it felt to be the first humans to see the far side of the Moon, coasting silently, now barely sixty miles above the surface. (The Russians, it should be noted, had photographed the "far side" using an unmanned probe.) Kraft, Gilruth, and Low on the back row of the control center could hardly contain themselves. The viewing room was overflowing and the people gathered there stood and cheered wildly before making their own dashes to the rest rooms. Missions are tough on kidneys and bladders.

During the two revolutions after the burn, the crew excitedly described the craters of the Moon, giving them temporary names to honor the leaders who got them there: Low, Gilruth, Kraft, Paine, Slayton. Craters were named for Grissom, White, and Chaffee, then for Ted Freeman, Elliott See, Charlie Bassett, and C. C. Williams, astronauts who were killed in aircraft accidents, as the Apollo 8 crew called the roll of the courageous test pilots who with their lives provided the foundation for this mission on Christmas Eve 1968.

Frank Borman, Jim Lovell, and Bill Anders were like explorers from ancient days, seeing a new land for the first time and reporting almost constantly during the portion of the orbit on the Moon's front side, where we could communicate with the crew. Borman, concerned as the crew day approached its twenty-fourth hour, grabbed a two-hour rest break, then demanded that his compatriots get some rest before preparing for their final orbit and the critical trans-Earth injection maneuver that would conclude the lunar phase of the mission and start the homeward-bound leg of their journey.

I was sitting at the console, reading the flight plan, when, on their ninth orbit of the Moon, Anders began reading from the book of Genesis. It was a surprise, beautiful and timely for this achievement and this day. I felt a chill as Anders said, softly,

"In the beginning God created the heaven and the earth.

"And the earth was without form, and void; and darkness was upon the face of the deep. . . .

"And God said, Let there be light: and there was light."

Now I was really grateful that I was not working the mission. I was enraptured, transported by the crew's voices, finding new meaning in the words from Genesis. For those moments, I felt the presence of creation and the Creator. Tears were on my cheeks.

One orbit later, on the tenth revolution of the Moon, early Christmas morning, the crew left lunar gravity for their return to the planet Earth, forever changing our world, opening the door to a new generation of explorers.

We grabbed for the lunar prize and we got it on our first shot.

There was a postscript to this perfect mission. For years we had kidded the recovery team to stay away from the landing point or else we would hit the aircraft carrier. As the guidance system performance improved, this actually became a possibility. The Trench did such a good job for Apollo 8 that Bill Tindall dispatched a letter to the head of the Recovery Division: "Jerry, I've done a lot of joking about the spacecraft hitting the carrier, but the more I think about it the less I feel it is a joke. The visual reports of the landing indicated the spacecraft flew right over the carrier and landed only 4,572 meters [2.8 miles] away. This really strikes me as too close. The consequences of hitting the carrier would be catastrophic. I seriously recommend that you relocate the recovery forces at least 8 to 16 kilometers [approximately 5–10 miles] from the target point."

14

1969 — THE YEAR OF APOLLO

There have not been many years in American history to rival 1969. Richard Nixon moved into the White House and across the globe from Northern Ireland to Southeast Asia it was a dangerous world. Senator Ted Kennedy drove off a bridge at Chappaquiddick, probably ending his chances for the presidency. Even more amazing, the Mets won the World Series. Yet it really was the Year of Apollo. The ecological movement kicked into high gear. Apollo 8's stunning images of the Earth in vibrant color, images never before seen by man until we pushed our way into space, brought home the reality of what we had accomplished in sending men to the Moon. It provided the environmental movement a powerful visual expression of the concept of "Spaceship Earth."

Now the images indeed seemed real to those of us who had helped send this craft to the Moon. For a brief moment in December 1968 we had united all humanity. In the coming months, in the greatest adventure of mankind, we would attempt to place two Americans on the surface of the Moon.

The fast-track effort for Apollo 8 put us behind for the Apollo 9 launch, now just two months away. Our holiday celebrations were brief. My White Team began training two days before the new year. We faced two missions before we could make the landing attempt. I had the Earth orbital flight test of the lunar module, and then on Apollo 10 Lunney would pull the pieces together in a full dress rehearsal of the lunar landing. These were the final months of the campaign to reach the Moon. Although we had few details, the death of the cosmonaut Vladimir Komarov in the flaming crash of the Soyuz 1 capsule in Central Asia in 1967 indicated that the Russians were having problems with their space systems. After an eighteen-month hiatus, the next Russian manned missions, Soyuz 2 and 3, in October 1968, accomplished a rendezvous, but the spacecraft were unable to dock. Docking, an essential technique for space operations, was finally accomplished in January 1969 on the Soyuz 4 and 5 mission. It looked like the Russians were almost three years behind us in operational manned capability.

My staff kept the division humming as my team prepared for Apollo 9, the last in Earth orbit. It was like Wally Schirra's flight in many ways, but this time it was a shakedown cruise of the lunar module, the last test before we went to the Moon. The lunar module had no heat shield; it could not return to Earth. The test required the CSM and LM to separate to test the LM engines and practice rendezvous, then the two spacecraft had to re-rendezvous and dock for the crew to transfer back from the LM to the CSM for return to Earth. We practiced solo rendezvous with Dave Scott in the CSM in case we had to rescue the LM crew. We had two manned spaceships to operate, a lengthy rendezvous, and a lot of engine testing. As the flight director I would be working with a team of twenty-one personnel, the largest MCC control room team in history. I was concerned that the span of control might be too large for rapid and correct decision making.

From now through the lunar landing, the missions were on two-month intervals with both MCC control rooms on the second and third floors operating simultaneously. We had a great Apollo crew, and the delay due to moving Apollo 8 into our slot let us get much better acquainted. Our crew consisted of Jim McDivitt, Dave Scott, and Rusty Schweickart. McDivitt and Scott were Gemini veterans and had spent time as CapComs in early Gemini. With the advent of dual spacecraft missions, we referred to the commander of the mission, McDivitt, as CDR; the lunar module pilot, Rusty Schweickart, as LMP; and the command module pilot, Dave Scott, as CMP. Since Scott, like McDivitt, had come up through the program as a CapCom in Mission Control, he was close to the controllers and sent them his pilot's notes drawn up during preparation for the mission. The controllers would check them for accuracy. Their review helped him create a damn good handbook for future CMPs.

Rusty Schweickart was a virtual unknown to us. He had spent little time in the Mission Control Center and, while involved and friendly, was deferential about taking a position or arguing a policy, an understandable attitude for a newcomer.

McDivitt established another first by designating astronaut Stu Roosa as the full-time representative of the crew for all topics. Given the complexity of this mission, McDivitt felt there was too great a chance for something to slip through the cracks. Roosa acted as a sounding board to keep policy issues constantly in the forefront. These steps seem elementary, but in the rapidly moving flight program, with the constant parade of crews and expanding control teams, we were hard pressed to do anything but the fundamentals for each mission. We were learning by doing, with little time to reflect, only to respond.

February 1969

My Apollo 9 teams were a mixed bag in terms of experience. One of the flight directors, Gerry Griffin, had been a superb Gemini GNC flight controller. Now, fresh from his first mission as Apollo 7 flight director, he was raring to go. Pete Frank, the rookie flight director, had been carrying most of the flight planning and mission rule work while Griffin and I were working with Lunney on Apollo 7. The flight control teams always stepped into the breach to help the flight directors, especially the rookies.

The primary objective of the mission was to flight-test the lunar module. The testing of the LM began shortly after getting to orbit. During launch, the LM is bolted into a long tapered adapter atop the forward end of the Saturn IVB rocket stage. The CSM sits atop the adapter and once on orbit, the command and service module separates from the adapter, turns around, and flies formation with the rocket. The four sections of the tapered adapter open like huge flower petals before being jettisoned, exposing the docking mechanism at the top of the ascent stage of the booster. It looks like something out of a James Bond movie. The crew maneuvers the CSM to carefully dock to the LM and then extracts it from the adapter.

The workload got so heavy in the final six weeks that Cliff Charlesworth, coming off Apollo 8, was drafted to fill in on the shift schedule and give us a hand. There were only so many trained and experienced people to go around. Once again, the clock was our enemy.

You learn never to relax during simulations. About the time that you think your team is really humming and ready to launch, SimSup pokes a hole in your bubble. We were about halfway through the training for Apollo 9 when Gerry Griffith, the SimSup, taught me a lesson I never forgot.

I had worked with my FIDO Dave Reed only once before, on Apollo 5. This mission was Reed's first manned launch. The Saturn launch period is extremely complex. The flight director and FIDO have four major abort options during launch and several of the abort options overlap. At any one second you may have two methods to terminate the mission, so the timing of executing aborts between the FIDO and flight director is critical. The FIDO uses five large-screen displays and his TV data to select the abort options. The flight director uses the same displays for launch phase timing and as a memory jogger on the abort mode boundaries. Griffith had observed that the handoff between Reed and me was not going smoothly and decided to give us a test.

During training, a FIDO has to learn to communicate with the flight director so Flight knows his intentions before critical points are approached.

The goal during launch is to get to orbit if possible. Orbit represents a stable point where we can gather our wits and figure out what to do next, or if necessary, how we will get home. An abort involves moving from the time-critical launch sequence to an even more time-critical abort sequence. This is an irreversible process, and doing it with problems on board the spacecraft is pretty tricky. The control team and crew spend a lot of time during training to get the decision process and timing right.

The launch simulation got off to a good start. When the S-IVB stage engine ignited, SimSup started a leak in the CSM propulsion system. My GNC recognized the leak a few seconds later and started monitoring the pressure decrease to compute the leak rate. So far the team was responding smoothly and we had done everything right. SimSup then shut down the S-IVB engine at the precise moment when we had an abort mode overlap. Reed had two choices, light the CSM engine and continue to orbit, or turn around and use the engine to deorbit the spacecraft in the Atlantic Ocean near the west coast of Africa. The GNC, however, had not computed how much fuel was available with the leaking tank to accomplish the maneuver.

The CSM, now near orbital velocity, was racing toward the African coast, covering five miles in each second we delayed. With no data from the GNC, Reed could not make up his mind which abort option to select.

I had been monitoring the FIDO-GNC communications and, at the same time, watching the giant plot boards. Reed was hesitating. With no call from FIDO I stepped into the breach, saying, "CapCom, Mode III abort . . . Mode III." The crew executed the abort but by the time I made my call it was too late to land in the Atlantic. All the crew could do was prepare for a land impact.

There is no feeling in the world to compare with the feeling you get when you know you blew it, and you have to explain in excruciating detail during simulation debriefing why you acted as you did. There are no excuses. The astronauts, controllers, training team, and MCC staff listened to the debriefings. When I finished mine, SimSup came up on the voice loop and rubbed the final salt in the wound. "Flight, the crew was killed. The landing point was in the Atlas Mountains in western Morocco. Those mountains are 14,000 feet high, the parachutes don't open until 10,000 feet."

I had blown it. I had killed the crew . . . the astronauts knew it . . . my controllers knew it. I knew it. I had acted like a rookie.

Marta does a lot more than make my mission vests. In the final few days before a launch, usually after supper, she will say, "Gene, I think it is time for your mission haircut." After leaving Langley I could not find a barber who would clip my hair short enough. Frustrated, I bought a hair clipper, and standing in front of the mirror I could cut the sides and top the way I liked it. Marta then stepped in and finished the top and shaved my neck. In December 1972 Carmen and Lucy at a family meeting told me, "Dad, you scare the boys away. They see you with the short haircut and they are afraid to come to the door. Couldn't you let it grow a bit longer." Marta nodded her agreement. Their plea was so earnest that I had to acquiesce. For the next seven months I sported somewhat longer hair, combed in a 1950s style. I finally rebelled when they asked me to get it styled. In August 1973 the kids asked me what I wanted for my birthday. My response was "I want to cut my hair!" The next day Marta and I collaborated on a crew cut, and I was happy again. The kids found out what Marta already knew: my hair length made no real difference. The boys were still afraid of meeting me at the door.

March 3, 1969, Apollo 9

Surveying my launch team as the countdown progressed, and looking at the enormous beast we were about to launch, I felt a disconcerting mixture of confidence and humility. I am sure that the pad team did also. The Saturn V on the television screen in front of me was the world's most powerful machine, towering 363 feet above the flat Florida shoreline. My team, whose average age was twenty-six, just a few years out of school, had within its hands the power to change the direction of history.

On the launch pad, ice from the liquid oxygen tanks' condensation glistened in the searchlights; mist swirled around the umbilical tower and platforms. At the top was the CSM, with the detachable escape tower for the command module at the very tip of it all. Buried in the tapered adapter section below the CSM and atop the launch vehicle was the lunar module, the spacecraft we would shortly test. Weighing over 6.5 million pounds, the Saturn rocket consumed twenty-three tons of kerosene and oxygen before it started to move. As it climbed along the launch tower, a ton of frost was shaken loose from the tanks, falling past the swing arms into the flame bucket. When the rocket exhaust hit the streams of water pouring into the flame bucket to absorb the intense heat, steam billowed along the flame trench that directs the exhaust heat away from the launch complex. By the time the Saturn booster shed its first stage, two minutes and forty-one seconds into flight, it had consumed almost five and a half million pounds of fuel.

When you turned loose the energy of a Saturn rocket, you simply had to have trust in your crew, your team, and in yourself. Through trust you reach a place where you can exploit opportunities, respond to failures, and make every second count. As gigantic as the machine was, and as puny as we humans were measured against its towering bulk, the human factors balanced the technology on the scale. It would be this balance that would be, indeed had to be, maintained successfully throughout manned spaceflight operations.

The control room contained twenty-one team members, but the decision process during a Saturn launch focused on ten: the three Booster engineers, FIDO, RETRO, GUIDO, two CSM systems engineers, the CapCom, and myself, the flight director. We had a bewildering set of options facing us during the twelve minutes of powered flight. My mission rules were perched on the right corner of the console, a multicolored, two-inch-thick document containing several thousand rules for the conduct of the mission. These rules had been whittled down to less than a hundred for launch. We knew from the pre-mission studies and simulations that a launch abort was the final and often risky option to terminate a mission. The nighmarish scenario we faced was making a wrong decision and placing the crew into orbit with no way to return to Earth. An equally nightmarish outcome was executing an abort that either was not necessary or that, if executed improperly, might also kill the crew. With only seconds to assess a situation and then pick a path, we had to determine clearly the course of action before we launched. Except for trajectory problems that allowed no alternatives, our judgment was that things had to be going to hell in a handbasket in the spacecraft or booster before we would abort the launch.

The count progressed. In the final fifteen minutes, you could feel this incredible pressure build; all controllers felt it. Once the Saturn was launched, we would be tied to our consoles for at least a half hour. I gave the controllers their final chance for a pit stop before the doors were locked. We made a final rush to the rest room, standing in line, then sprinting back to the consoles. When I returned, I put on my white vest while inwardly I was marching to the cadences of Sousa's "Stars and Stripes Forever."

During most of powered flight, our decision time frame was about twenty seconds, sometimes less. With our training, twenty seconds was a lifetime. In that time you can detect a problem, hold several crisp conversations, select displays, make a decision, and issue the command/voice instruction — all in less time than it takes to air a short television commercial.

Nearing launch, an internal clock kicked in as auto sequence started. I could feel the sweat on the palms of my hands. This was, after all, my first *manned* Apollo launch as flight director. At launch minus fifty seconds, the electrical power transfer from the launch pad to CSM fuel cells and batteries was complete. This brief period was the time that I hated; I always hated it. I had a long list of ground equipment I needed for launch, scattered around the world, much of it mandated by the mission rules. I prayed it all held together for the next twenty seconds. I established my personal cutoff for killing auto sequence at launch minus thirty seconds. My risk judgment told me that the MCC must suffer a crippling failure before I would call the launch team with a NoGo at this point, terminating the automatic launch sequence. I bowed my head briefly and made the sign of the cross as the engines roared and the crew called, "Liftoff; the clocks have started."

The Apollo 9 mission was sheer exhilaration for both the astronauts and Mission Control. While docked during the first four days we thoroughly checked out and tested the propulsion systems of both spacecraft. Then Rusty Schweickart and Dave Scott performed an EVA between the docked spacecraft before starting the rendezvous. The LM maneuvered to rendezvous with the CSM just as it would have to do in lunar orbit.

No Apollo crew was better prepared than McDivitt, Scott, and Schweickart. They had been in training since 1966, when they were initially assigned as the backup unit to Apollo 1 until replaced by Schirra's crew. If anything went awry during the lunar module testing, Jim and Rusty could have found themselves adrift in space in a machine with no capability of returning safely to Earth. In that event, Scott, flying the command module, would have been called upon to make the first space rescue in history. Scott knew his spacecraft better than any prior command module pilot. If he had had to initiate a rescue attempt, you could be damn sure he would have succeeded.

Combining two rendezvous techniques developed on Gemini 9, we flew the most complex rendezvous to date. The LM rendezvous radar, computer, and propulsion systems passed every flight test. LM emergency power-down checklists and techniques for using the lunar module engine while docked to the CSM were developed and tested. These rudimentary "lifeboat" techniques would provide get-home alternatives for certain CSM failures and would be vital during later missions. The mission debriefing proved that the lunar module, in a zero-G environment, was a remarkably sturdy space buggy. The only problem we had was that Schweickart was space sick for four days. Overall, it was a damned good mission with a great crew.

Several times during the mission I reached task saturation with the control of two spacecraft, while planning and executing the mission. I was convinced that the only way to ensure effective support for spacecraft operations when the LM and CSM were separated in lunar orbit was with two separate teams in the MCC, one following the CSM and the other the LM. Two complete communications sets were available at each console to support shift handover. When the CSM and LM were operating independently, I believed that one flight director could work with the GNC and EECOM, the other with CONTROL (lunar module guidance, attitude control propulsion, and navigation) and TELMU (LM electrical, environmental, and EVA systems). Lunney and Charlesworth were both skeptical that two flight directors could work side by side on the same mission from the same console.

A decision on the dual flight director arrangement was deferred until we could test the concept on Apollo 10, which would set the stage for the final push to the Moon.

Lunney led the team for the Apollo 10 mission, with Milt Windler and Pete Frank on the wings. I closely followed every event in the mission. The procedures, plans, and rules were virtually the same ones we would use on Apollo 11.

Tom Stafford and Gene Cernan took the Apollo 10 LM to within ten miles of the lunar surface, then staged a rendezvous with John Young in *Charlie Brown*, the mother ship. After thirty-one orbits of the Moon, Apollo 10 left for home. The flight plan, navigation and tracking techniques, and the exact procedures were used in an end-to-end basis to shake out any problems and further reduce the risks of the landing mission. Every MCC maneuver computation, every controller display, and even the team shifting scheme was tested.

By the time of the Apollo 10 splashdown we knew that the total mission system — the crew, controllers, and spacecraft — was ready to go.

The only remaining uncertainty was the lunar landing and the subsequent liftoff. Lunney's and Charlesworth's experience at the flight director's console during the mission persuaded them to agree to the dual flight director approach once the LM was on the Moon's surface.

15

SIMSUP WINS THE FINAL ROUND

There are luminous points in memory that are as fresh and vivid as if they had just happened. This is the way I remember the day I was commissioned in the Air Force, the day I got my wings, and the day I was married. I remember the first time I flew a Sabre and the hour of my bittersweet last flight. I remember meeting Marta, our joyous reunion upon my returning from Korea, and the births of each of our children.

In space, I remember the "Four-Inch Flight," my first mission as flight director, Ed White's EVA, and Apollo 1. Some of these moments tore my heart out while others were pure joy, an opportunity to share an instant with Marta or my team. Still, others were purely visceral, the thrill of doing something for the first time, being involved in a great event or leading in a great cause. Between the flights of Apollo 8 and 9, a brief meeting joined the list of those moments I will never forget.

As 1968 came to a close, I realized that if all the remaining missions went well, I would be a flight director for the first lunar landing. Every controller wanted to be a part of this historic mission and all had been jockeying for some position since early in the Apollo program. Although we were incredible team players, each of the flight directors wanted a challenging and historic mission. A position in Mission Control was the next best thing to being in the spaceship. More than just working the mission, however, I wanted to lead the team that would take the first Americans to the Moon.

The roll of the dice put Glynn Lunney, Cliff Charlesworth, and me on the schedule for the first lunar mission. This was a reunion of sorts, the first time the three of us had worked together since Gemini 12. Since the Apollo 1 fire, Hodge had rotated the lead assignments among Glynn, Cliff, and me. When I became Flight Control division chief, I saw no reason to shuffle the lead responsibilities, so Charlesworth retained the lead flight director role for Apollo 11.

One of Cliff's initial tasks was assigning the flight directors for the mission phases. There were eight major phases of a lunar mission; five of the eight would have been demonstrated on either Apollo 8 or 10. The three new mission phases were the lunar landing, surface EVA, and lunar ascent.

Our experience made it a toss-up for the phase assignments. Cliff had launched the most Saturn rockets; Glynn would have been to the Moon twice with Apollo 8 and 10, and I would have the most lunar module experience. We had all worked manned command modules. Cliff and

Glynn were considered by the controllers as trajectory biased, favoring work on the trajectory aspects of the mission, while I was considered as systems-biased, favoring work with the CSM and LM control teams. I hoped that when Charlesworth put the pieces together, he would give me the assignment for the landing phase of the mission.

The meeting where that actually happened was almost anticlimactic. Cliff walked into my office, stood at the window, stared out for a moment, and then turned with a smile on his face. He knew that I wanted to get down to business, but he just toyed with me for a few moments, passing the time of day, then abruptly said, "I think it's time to decide on the Apollo 11 phase assignments."

With no further preliminaries, he continued, "I think I should launch Apollo 11 and do the EVA. Milt [Windler] will take the entry. This leaves Glynn for the lunar ascent . . . and you with the landing. Is that okay with you?" I nodded, and the meeting was over. The entire session had taken less than sixty seconds.

I had drawn the flight director assignment to put the first man on the Moon.

I had to tell someone and the list would start with Marta. Seldom would I call Marta bearing glad tidings; it was usually, "I'm going to be late," or "I'm not going to be home to help one of our kids with a math test preparation." This time I couldn't wait to tell her. "Marta, guess what? Cliff gave me the lunar landing assignment!" When I called the staff in, they could tell from the way I was ricocheting around the office and from my ecstatic expression that I had gotten the glittering prize — the Moon landing. They were as happy as I was.

My landing team formed up quickly. Each team was constructed on a mission-by-mission basis. When the flight directors received their mission phase assignment, the branch chiefs carefully matched the personalities and strengths of controllers to those of the individual flight directors and their capabilities to handle the mission events.

Bob Carlton, nicknamed "The Silver Fox" for his prematurely gray hair, had the responsibilities for the LM navigation, control, and propulsion systems. His call sign was CONTROL. During the last seconds of landing, his slow, deliberate Alabama drawl would be the only voice on the intercom, calling out the seconds of fuel remaining.

Don Puddy, the tall, intense Oklahoman who never wasted a word, was the self-appointed leader of the lunar module team, responsible for communications, power, and life support systems, call sign TELMU.

Steve Bales, filling the GUIDO (guidance) position, was one of the first of the computer whiz kids. Steve in many ways was more like the systems controllers. He had not yet developed the arrogance so characteristic of the typical Trench inhabitant. He spoke rapidly, running his words together occasionally, then letting them stumble out. You could tell how he felt by his voice inflection. His large, round black-rimmed glasses set him apart from most of the controllers.

FIDO was Jay Greene, the pipe-smoking New Yorker, a rabble-rouser who did not like it when things got quiet. He liked to coach the flight director along a decision path. He was elite in the ranks of the FIDOs, cocky and crisp with his calls. Bostick, my branch chief with responsibility for the Trench, knew that I was the weakest flight director when it came to the trajectory, so he gave me an experienced FIDO who would teach me the new stuff I needed to know for the landing.

To Greene's left sat the RETRO, Chuck Deiterich. Like Greene, he had a classic disdain for any controller who did not immediately surrender to the wisdom coming from the Trench. Chuck would either bury you with more data than you needed or cut you at the knees . . . once telling me during a debriefing, "You don't want to hear about that, Flight . . . it's too technical!"

The combination of Chuck's Texas and Greene's New York accents during the rapid-fire exchanges on the voice loops made for interesting listening during time-critical operations. Their voices were unmistakable.

The final member of the Trench was Gran Paules, who would work with Bales. Gran was a tall, blond, taciturn controller who had the habit of turning to look at you when you called. His nasal inflection reminded you of someone constantly suffering from an allergy. Like Steve, he was typical of the next generation in the MCC.

Slayton selected Charlie Duke from the astronaut class of 1966 as my CapCom. Duke was well experienced in the operation, having worked on Glynn's team for the previous mission. For flight directors and CapComs, the principal tools used during the mission were the MCC intercom and crew voice loops. Our common job was to listen, integrate, communicate, and act.

During mission preparation, Duke provided the communications conduit. Aware of both crew and MCC concerns during meetings, he brokered and summarized the resulting actions with the crew. Watching him operate, I knew why the Apollo 10 and 11 astronauts wanted him in the CapCom's slot. Of all the astronauts, he would have made a hell of a good flight director. I had a great feeling about the easy confidence Duke showed during planning sessions. He contributed to making my mission preparation successful, helping to bring the controllers to the highest pitch of readiness in the three months before the lunar landing mission. The SimSup and his team came from the Flight Control Division's Mission Simulation Branch. SimSup was my other partner in team building. He worked with the flight directors and the branch chiefs in carefully monitoring controller performance during training and certifying them suitable for mission support. There was no way one flight director could do this job by himself.

My lunar module team, the four controllers in the Trench, and Charlie Duke were the core controllers for landing. Our job was to get the LM, an odd-looking contraption — like a praying mantis, Mike Collins said — close enough to the surface to let the crew take over and attempt landing. "Close enough" was subjective. Only the crew in the LM would know whether to land or abort in the last few hundred feet. It was our job to get them to their decision point.

There were three members of the command module team in the control room, looking over Mike Collins's shoulders in the spacecraft. Ed Fendell was making the transition to a consolidated (CSM and LM) communications position. Ed was intuitive in responding to problems and he developed great young controllers. He was also noisy, poked fun at every controller, and could be disruptive. You had to earn his respect and keep him on a short leash. I worked with Ed many times during Gemini and liked his spirit, commitment, and willingness to step up to responsibility. Above all, I liked him because he never left the flight director hanging and never catered to anyone. His independence irked many of the controllers, but they respected him. Glynn and Cliff considered him a pain in the ass. The rest of the team consisted of Buck Willoughby and John Aaron. During the landing phase their job was to monitor Mike Collins's solo operations in the command module. During the descent, if needed, we would use the Command and Service Module as a communications relay point and possibly an orbital rescue vehicle.

With the assignments completed, I called the first meeting of the White Team to finish working out the detailed landing strategy. Personal and team readiness would emerge from our study and the team working sessions on the trajectory, flight plan, and the mission rules. Then the simulation training would integrate the ground team with the astronauts and test our mission planning. The White Team had a total of eleven days of simulation to get ready for the landing. Only seven of these were with the crew. Four were with math models and a simulated astronaut.

Bill Tindall had started weekly meetings on the descent phase in April and had released a barrage of "Tindallgrams" and assorted notes. "Tindallgram" was the name given to Bill's comic and highly treasured memos of the techniques meetings he conducted from 1966 to 1970 to document key engineering and operational decisions. In May 1996 the memos were bound in a single volume and distributed to "Bill's many friends." Tindallgrams were converted into new procedures, flight plan entries, and the jargon used by the controllers in their Go NoGo.

One of the Tindallgrams really grabbed our attention and also gave us a few laughs. It began, "There is another thing about powered descent crew procedures that has really bugged me. Maybe I'm an 'Aunt Emma' — certainly some smart people may laugh at my concern, but I just feel that the crew should not be diddling with the computer keyboard during powered descent unless it is absolutely necessary. They will never hit the wrong button, of course, but if they do, the results can be rather lousy." The next day we started a review of every crew computer keystroke and its effect throughout the descent phase.

Another "Aunt Emma" note challenged the terms used by my flight controllers after landing, "Once we get to the Moon does Go mean 'stay' on the surface, and does NoGo mean abort from the surface? I think the Go NoGo decision should be changed to Stay NoStay or something like that. Just call me 'Aunt Emma.' "

We changed the procedures for the entire after-landing process into a series of Stay NoStay decisions. Tindallgrams, spiced with humor, idiosyncratic grammar, and personal "revelations," got the job done. After Apollo 11, at a post-mission beer party, Flight Control made Tindall an honorary flight director, with the team color Gray. His color is retired, like that of many flight directors, and now hangs in the third floor of the Mission Control Center.

I asked him to sit next to me at the console for the lunar landing. Tindall, ever modest, declined, but I persisted in my request and he finally agreed. He was one of the great pioneers of manned spaceflight.

During the mission rules sessions, Buzz Aldrin was the crewman usually involved, demonstrating his knowledge of a variety of subjects, and generally dominating the crew side of the conversations. Neil Armstrong seemed more the observer than the participant, but when you looked at his eyes, you knew that he was the commander and had all the pieces assembled in his mind. I don't think he ever raised his voice.

He just saved his energy for when it was needed. He would listen to our discussions and, if there were any controversy, he and Aldrin would try out our ideas in the simulators and then give feedback through Charlie Duke to the individual controllers. Buzz and Neil seldom took a strong position during the meetings, but they were good listeners. They knew enough about us to trust us, to give us the benefit of the doubt. Mike Collins used a different tactic. He worked directly with the Trench and systems guys. By the time we got to the rules sessions, all the problems were ironed out.

We published our first complete set of rules for the Apollo 11 mission on May 16, two months prior to launch. With no landing simulation experience, this first set of rules represented the sum total of our knowledge from our meetings. With the simulations starting in early June, the learning curve would be steep, resulting in planned rules updates weekly until launch.

Simulation training is broken into two parts, nominal and contingency. The nominal training occurs early in the simulation period. It lasts only two to three days and is used to establish crew–controller action timing, locate the Go NoGo decision points, and exercise the procedures for the planned mission. The contingency training tests the crew–controller decision process in a mission environment while solving complex trajectory and systems problems. Training scripts are developed by SimSup's team, and problems are programmed into the simulators without the crews' or controller's knowledge. The training environment becomes as close to the real thing as possible, with the training team testing the flight teams' strategy, knowledge, and coordination while probing into the psyche of the crew and controllers. Nothing is sacred; no quarter is given and none asked. Training for a lunar mission was a daunting task. Training to cover every conceivable aspect of the first lunar landing bordered on the impossible.

To get a handshake on the unwritten rules for the landing, I had a final strategy session before simulation start-up with Neil, Buzz, Mike, and Charlie Duke. It was in this session that I outlined the landing strategy. We had only two consecutive orbits to try to land on the Moon. If we had problems on the first orbit we would delay to the second. If we still had problems, we would start the lunar descent to buy five additional minutes to solve the problem. If we couldn't come up with answers, we would abort the landing and start a rendezvous to recover the LM, then jettison it and head back home. If problems surfaced beyond five minutes, we would try to land and then lift off from the surface after a brief stay. We would try for the landing even if we could only touch down and then lift off two hours later when the CSM passed overhead in lunar orbit with the proper conditions for rendezvous. (The LM had limited electrical power. The lunar liftoff time had to be precisely established to allow rendezvous and docking before the LM batteries failed.)

I knew Armstrong never said much, but I expected him to be vocal on the mission rule strategy. He wasn't. It took time to get used to his silence. As we went through the rules, Neil would generally smile or nod. I believed that he had set his own rules for the landing. I just wanted to know what they were. My gut feeling said he would press on, accepting any risk as long as there was even a remote chance to land. I believed we were well in sync, since I had a similar set of rules. I would let the crew continue as long as there was a chance.

The SimSup presided over some of the most complex technology of the space program. The only *real* things in the simulation training were the crew cockpit and the consoles in Mission Control. Everything else was make-believe, created by a bunch of computers, lots of formulas, and a skilled bunch of technicians. The business was more art than science because we were training for something that had never been done before.

Every time an astronaut threw a switch or a controller sent a command, a series of equations was triggered in the simulation computer. This pulse started a chain reaction: the data cascaded through sixteen other computers, sending impulses to the cockpit or the controllers' displays. SimSup needed a thick skin. When the simulation hardware and computers misbehaved, he had an unhappy crew and control team while we waited hours to restart the training. If they balked too often, Slayton, Kraft, and the flight directors would get involved. The squeaky wheel invariably got the grease, priorities were rearranged, and someone downstream, another flight director and group of astronauts, would be stuck with less and would start screaming about being cheated out of their training. Simulator training was routinely conducted sixteen hours each day. Both primary and backup crews were being trained for the subsequent two or three missions on two sets of training hardware. For the coming mission in the sequence SimSup had a prime and backup crew and four teams of controllers to get ready for a mission in ninety days, and he always had the priority.

The simulation team, like the flight director, worked to the launch day deadline. When the controllers clocked twelve-hour days, the training team worked fourteen. When he was not training, SimSup was studying the controllers, crews, and mission strategy, looking for the holes and developing new training runs to exploit the perceived holes.

SimSup for the descent phase was Dick Koos, who struck me as a quiet young academic. In fact, he was a discharged sergeant from the Army Missile Command at Fort Bliss, Texas. Dick was an early hire into the Space Task Group, assigned to train the crews and control teams. His background was in computer guidance for the early ground-to-air missiles. Simulation was an entirely new field, but with his computer experience he took to his work with a passion.

Koos went up through the ranks rapidly and developed into an excellent SimSup. Dick was a thin guy, wore wire-rimmed glasses, expressed himself in incomplete sentences, and seemed unsure of what he was trying to say. His external demeanor set you up for his training sessions, which were like a rapier, cutting so cleanly that you did not know you were bleeding until long after the thrust. Koos was a worthy adversary and an excellent choice for training my White Team for the Apollo 11 landing.

SimSup had months to design the training sessions; we had but seconds to minutes to solve problems he posed. The early sessions were rugged. SimSup's team attacked every aspect of our knowledge, even the relationships between space and ground teams. They pounded away at the strategy and timing. When we were through, we were in a place beyond exhaustion.

Landing training kicked off at launch minus eight weeks. My team, fresh from Apollo 9, had a "hot hand," and from the beginning it seemed we could do no wrong. The first two training days were used to nail down the timeline for the landing preparation, establish the Go NoGo points, give the controllers a walk-through of the landing sequence, and become familiar with the three-second communications delay when working at lunar distances. Initially the three-second delay didn't seem like much, but if things start to go wrong in the final seconds before landing you could quickly find yourself in the corner of a box with no more options. The mission rules and procedures were refined during the initial two days of training and I felt well prepared to start the landing abort simulations.

Mental preparation was key to getting through a simulation. Each individual on the team had to find his very own way to be up for the challenge. Marta always sensed when I had to start working on it; she would say, "Isn't it time for you to get ready?" She would then round up the kids and give me the time and space to start my internal preparation for each day. She made sure that I had some internal peace and was centered as I left to face whatever the day would throw at me. We had so much to do — and so little time.

Dick Koos was as concerned about the adequacy of the training schedule as I was. Putting all of the pieces together in eleven days of training scattered over two months was tough. After the first two sessions Koos's judgment was that we were too damn cocky, and a bit of humility learned early in the training might make us more receptive. Looking at my team through the glass wall in the control room, Dick gave orders to increase the pressure. Smiling confidently, he thought, "Kranz's team will remember June 10 as the day that started them down the path to the Moon." Koos's team leaned forward at their consoles, savoring the coming battle. Today only the fittest would survive.

The first session was the warm-up. Seconds after we started the descent it seemed every controller had problems. The voice loops were jammed by controllers voicing instructions through Charlie Duke to the crew. Seconds after the crew responded another problem surfaced, then another, until Bob Carlton advised he had problems in the ascent stage. If we continued we'd leave the crew on the surface without a way home. Once the abort call was made and the engine throttled up, Koos called on the loop, "Good one, Flight. You nailed it. Let's start the turnaround for the next run." Through the glass wall I could see Dick standing behind his console. If the first run was an indication, today was going to be mano a mano. By the third run of the day the time criticality and complexity of each training run was peaking and my team was barely holding its own. Koos was having his own problems trying to keep the simulation computers from crashing.

The fourth run ended in a crash. In the Trench Jay Greene got behind on his calls, allowing the LM landing speed to build up. Our final instruction to abort was too late and Greene's large plot board in the front of mission control was mute testimony to the futility of our action. With the three-second delay in communications to the Moon, the crew was splattered across the Sea of Tranquillity. This was our first crash, the result of a few seconds' delay in our communication and decision process. After a tough and very frank personal debriefing, Jay and I dared Koos's team to get us again.

On the next session Koos delivered the coup de grâce with a virtual repeat of our previous crash. This time, with the crew approaching the lunar surface, the LM primary computer failed while I was working an LM electrical problem with the systems controller. The distraction caused by troubleshooting an electrical fault resulted in a late switch to the backup computer system for the abort. It seemed that no matter what we did, we just were not fast enough. We were learning the hard way about the *deadman's* box, the seconds-critical relationship of velocity, time, and altitude where the spacecraft will always impact the surface before the MCC can react and call an abort. In flying terms we were behind the power curve. The debriefing was long and intense, focusing on the need for some new rules. Approaching the Moon at a high rate of speed the LM can go a hell of a ways in three seconds. Greene took the action to change his mission rules and plot board limit lines to add a bias to accommodate the communications delay.

The next two runs were a washout. I felt like a novice flight director, the sweat soaking my shirt at the armpits. There was something in the air, something I could not put my finger on. I felt unprepared, edgy. My moves and calls became hesitant and unsure, and I believe my voice betrayed my unease and passed it to my team.

Koos never backed off; his pressure was unrelenting. We were just hanging on, and our performance was in a downward spiral. Every team member, frustrated, tried desperately to get the team on track. By the final training run I felt like the coach of a sandlot ball club behind 21–0 in the third inning. All this had taken place in one day. I had just had my worst day of simulation ever as a flight director. But when the LM headed for the lunar surface, I would be working in precious seconds. We had to work out the bugs now.

During training runs it was customary for the

big bosses, Kraft, Slayton, and even George Low, to listen in on the flight directors' loop from their offices. We would cut the loops during the debriefings, so that we had some privacy for soul searching or a plain old-fashioned ass chewing. After the final busted training run, the telephone behind the console rang. Frustrated, I picked it up with my customary "Kranz here." I heard the familiar voice of Kraft. "I listened to your runs today," he said. "Sounds like you had a tough time. What's going on?"

I think he really wanted to take a reading on my frame of mind by listening to the sound of my voice. He knew the business, and he knew the job, so my response was simple. "Chris, you've had these types of days. It is just a matter of time and training, we'll work it out." After Chris hung up, I switched off the ringer on the phone so I would no longer hear if he called.

SimSup was winning the battle and there was little we could do except hunker down, study some more, get more training under our belts, and come back and do it again and again and again. This was the time where "*Tough* and *Competent* . . . *Discipline* and *Morale*" took on a real meaning for me. "Morale" was not a new word in our vocabulary. The belief in our mission, our team, and ourselves was the key to our eventual success in Gemini. Morale sustained us during the difficult EVAs and when the Agenas failed to reach orbit. I had to practice what I preached.

Sam Phillips set up a preliminary telephone conference the week before the Flight Readiness

Review with George Low, flight surgeon Chuck Berry, and myself at the Houston end; Kennedy Space Center director Rocco Petrone and Deke Slayton were at the Cape. I was surprised at the turn of the meeting when Phillips asked if we each felt comfortable with the schedule. He indicated a willingness to push launch into August if we needed more training time.

We each carefully measured out the time we had remaining to train and figured the few extra days would not buy us that much. Then it came down to Chuck Berry. Chuck was concerned about the crew workload but after stammering a bit about the crew schedule, he also gave a Go. My team was coming up to its peak very smoothly, and I did not want to back off. This was a time when the pressure was good. I think that Phillips also talked to Neil that day, and he got a Go from the crew.

The Flight Readiness Review was conducted on June 17, and there were no major open items. The review went well until Kraft made a few comments about the landing data rules. A free-for-all started, and I was called on to write some specific rules on the communications and data requirements for landing. This issue continued to be debated until the week before flight, and it appeared that some of the folks at headquarters were getting damn nervous about the consequences of a crash, if one occurred. Chris, Cliff, and I agreed on the real rule: that "we must have enough data to reconstruct what went wrong." This rule left me the maneuvering room to take it right down to the surface before I had to make a land or abort call. Once we were close, I intended to let the crew go if everything appeared okay to them. I considered a low-altitude fire-in-the-hole abort more risky than landing without data. I always looked at a fire-in-the-hole abort the same way I looked at a parachute when I was flying jets. You use a parachute only when you have run out of options. The day before the launch, I processed a write-in mission rule change that legitimized this landing philosophy: "The flight director will determine if sufficient data exists to continue the landing." No computer could make this call — it had to be a human decision.

July 1969

During the last two weeks of training, the individual and team confidence were restored, thanks to the superb efforts of the simulation supervisor and his team. We were approaching readiness — so we took the day off on July 4, 1969.

The MCC final training day always had been a confidence builder, with most of the training runs focused on achieving the mission objectives. However, this wasn't the case when we returned on July 5, and by midday I was doing a slow burn. Since Armstrong and Aldrin had deployed to the Cape, Koos was running us through the paces with the Apollo 12 crew. We were in top form, having aced six tough landing aborts in a row. As we continued to work through the final training exercises, the Apollo

12 backup crew moved into the simulator. SimSup often did this in the last day of training, giving us a less experienced crew in the simulator that forced us to do more prompting and work harder. This way we would not take anything for granted. We had started the day with Pete Conrad and Alan Bean. By midday, however, their backups, Dave Scott and Jim Irwin, joined the simulation.

The final simulation before a mission was much like a graduation ceremony, except that instead of going out into the world to get a job, we had the task of landing an American on the Moon. Sitting at the console, late in the afternoon on July 5, my mind had closed the books on training and was racing ahead to the thousand items to be closed out in the final two weeks before launch. Mentally I had made a fatal misjudgment.

Dick Koos had been monitoring my team and was not about to give us our diploma. He made up his mind that we would have to earn it. Dick quickly scanned the simulation scripts, then called out to his team, "Hey guys, open your books to Case No. 26 and have them load it in the simulators." The technicians coordinating the simulator setup responded, "Case 26 is loaded." Dick smiled and turned to his team. "Okay everyone, on your toes. We have never run this case, so it is going to take a helluva lot of precise timing on our part. This one must go by the numbers, so stand by for my call-outs. If we screw it up I hope you got a bunch of change 'cause we'll end up buying the beer!"

The simulation picked up with the crew performing the final systems checks before starting the descent. I polled my controllers for the start engine Go NoGo, and Charlie Duke called to the LM crew, "Eagle, you are Go for powered descent." Five minutes later, the descent engine started and we were on our way to the surface. I thought, "This is going to be a good one to wrap it up."

Three minutes into the landing sequence Koos nodded to his team. "Okay gang, let's sock it to them and see what they know about computer program alarms." The LM computer provides a series of five-digit alarm codes. The computer alarms signal crew or ground procedural errors, computer hardware or software problems, or out-of-limits conditions. An ominous note states "30000 series alarms indicate a computer abort code that results in a software restart. 20000 series alarms are more serious and will result in the computer going to idle."

In the Trench, Steve Bales, the GUIDO, was busier than hell. He had done well so far today in training and was damn glad it was all about to end. Steve was responsible for the LM computer. He had to make sure it got the right data from Earth and then had to be certain that guidance, navigation, and control functions during the landing were being executed properly.

Within seconds of Koos's malfunction entry, Steve was peering intently at his television display. He was seeing a 1201 alarm code indicating a computer restart. This was the first time he had seen this code except during computer ground testing. An equally perplexed LM pilot in the simulator called up data on the LM computer display. The code was meaningless and he decided to wait for a Mission Control call to enlighten him.

At Bales's fingertips was a small one-quarter-inch-thick blue handbook containing a glossary of the LM software. Quickly paging through the index he read "1201 — Executive overflow — no vacant areas." This meant that the computer was overloaded. The LM computer was unable to complete all its jobs in the course of a major computer cycle. Bales had no mission rules on program alarms. Everything still seemed to be working; the alarm did not make sense. As he watched, another series of alarms was displayed. Punching up his backroom loop, he called Jack Garman, his software expert. "Jack, what the hell is going on with those program alarms? Do you see anything wrong?" Steve was counting the seconds, waiting for Garman's response, happy that the crew had not yet called for an answer.

Garman's response did not help. "It's a BAILOUT alarm. The computer is busier than hell for some reason, it has run out of time to get all the work done." Bales did not need to consult the rules; he had written every computer rule. But there were no rules on computer program alarms. Where in the hell had the alarm come from? He felt naked, vulnerable, rapidly moving into uncharted territory. The computer on the LM was designed to operate within certain well-defined limits — it could only do so much, and bad things could happen if it were pushed to do things it didn't have the time or capacity to do.

Staring at the displays and plot boards, Steve desperately sought a way out of the dilemma. The computer was telling him something was not getting done and he wondered what in the hell it was. After another burst of alarms Steve called, "Jack, I'm getting behind the power curve, whatever is happening ain't any good. I can't find a damn thing wrong but the computer keeps going through software restarts and sending alarms. I think it's time to abort!"

Seconds later, oblivious to the problem, I was startled by Bales's call, "Flight, Guidance . . . something is wrong in the computer. I've got a bunch of computer alarms. Abort the landing . . . ABORT!"

Charlie Duke picked up the call. "We gonna call an abort, Flight?" My response was curt: "Abort, CapCom, abort!"

If there was one word guaranteed to get your attention in Mission Control it is the word "abort." This word is never used casually and literally rings across the voice loops as the word is passed to the crew, computer controllers, and support personnel. An abort is an intensely time-critical effort where every action must be perfect and perfectly synchronized. In an abort, your chances of getting out alive are good *if* the abort is done at the right time. If you are off the timeline, your chances are not good 200,000 miles away from home. An abort is the last option, one that must be perfectly executed with perfect timing if you're going to pull it off.

The crew confirmed the abort call as they throttled up the descent engine, then staged. The ascent engine ignited and moments later they set up a rendezvous with the command module. I felt that we had made the right, necessary call, but I was really unhappy with Koos. Dammit, we should have finished our training with a landing on the surface.

The flight controller debriefing was extensive. After listening to the confession of the team members, Koos gave his evaluation of our performance. Slowly, methodically, Dick took us through the problem, then plunged in the dagger: "THIS WAS NOT AN ABORT. YOU SHOULD HAVE CONTINUED THE LANDING."

Koos had grabbed me by the throat; I wondered where the hell he was going. Half dazed, I was anchored to my chair as he continued: "The 1201 computer alarm said the computer was operating to an internal priority scheme. If the guidance was working, the control jets firing, and the crew displays updating, all the mission-critical tasks were getting done." Koos's voice then became almost fatherly as he continued, "Hell, Steve, I was listening to you talk to your back room and I thought you had it nailed. I thought you were going to keep going, but then for some reason you went off on a tangent and decided to abort. You sure shocked the hell out of me!"

Then Koos made the final cut with his knife: *"You violated the most fundamental mission rule of*

Mission Control. You must have two cues before aborting. You called for an abort with only one!"

Bales, the proud, capable young computer whiz kid, was devastated by the simulation. The controller's world, however, is black and white, Go or NoGo, right or wrong. A controller can never make an excuse. His only answers when he fails are either "I was wrong" or "I don't know, but I will find out." Bales was frustrated and mad, damn mad, and his response was short. "Flight, I'm gonna pull a team together after we finish the debriefing. I'll tell you what the hell went on when we figure it out."

Every controller has experienced the bitter taste of failure. A single busted training run is abysmal; a busted run on the final day of training is unacceptable. Slowly, we took off our headsets and packed up our gear. We had run the last race and SimSup had won the battle. We would just have to get on with our job.

Later that evening, I got a call from Steve. "Koos was right, and I'm damn glad he gave us the run. The computer whizzes at the MIT labs, and our own assessment, said we could have continued. I'm going to stay with the team tonight and get out some rules. I've talked to Koos, and he is going to set up some training runs in the morning, if that's okay with you."

Koos scheduled four hours of training on program alarms the next day. The runs were scheduled with the Apollo 12 backup crew as well. SimSup triggered various alarm types during several intense training sessions while Steve Bales and Jack Garman collected computer performance data and response times during alarm conditions.

On July 11, nine days prior to launch, Bales modified his already lengthy listing of reasons to abort the lunar landing, adding a new entry to the trajectory and guidance section of the rules book.

> Rule 5-90
> Item 11, "powered descent will be terminated for the following primary guidance system program alarms — 105, 214, 402 (continuing), 430, 607, 1103, 1107, 1204, 1206, 1302, 1501, and 1502."

Steve did not put program alarms 1201 and 1202 in the mission rules listing requiring an abort.

The intense training period prior to flight had found our Achilles' heel, something that could have distracted the MCC team and crew at the wrong time. Something that could have been a mission-buster.

SimSup had won the last round.

16

"WE COPY YOU DOWN, EAGLE"

On the day before launch, I feel like I am going into the seventh game of the World Series or playing for the Stanley Cup. The energy starts flowing, and my mind is filled with thousands of bits of information that I will need soon. I am impatient, eager to get on with the mission. Even at home I pace in endless figure 8s like a large cat in a small cage, as I frequently do behind my console.

Marta has been through this before and knows there will be no relief until launch. She keeps the conversation light, but she knows I am starting to feel the pressure. This wasn't unique to the lunar landing; it happened every mission.

July 16, 1969, Apollo 11

I am up at 4:30 on the morning of the launch, wide-eyed alert, and thinking about the countdown. There have been no phone calls, so it must be going well. I can't wait to get to Mission Control and find out for sure. I fire up my psyche and crash around the house like the proverbial bull in a china shop. Marta tries to keep me quiet, since the kids are sleeping. As usual, she makes me an enormous lunch, generally two of everything. We say goodbye in hushed tones. I'm sure she's glad when I leave.

Prior to launch, the pressure I feel asserts itself through nervous kidneys, until commitment of the final Go. Then I become icy calm. Other than that, I never have any problems. I sleep well. My only other on-console symptoms are sweaty palms, a tendency to engrave words in the log, and the endless clicking of the ballpoint pen. The other flight directors kid me when the sweat-soaked paper curls as I write.

As I drive to the MCC, I wonder what Neil Armstrong, Buzz Aldrin, and Mike Collins are feeling as they prepare for this day. How do they feel as they enter the transfer van, go up the elevator, and across the platform to the command module? I believe we share the same feelings when it is time to get the show on the road. There is anticipation of the countdown reaching zero, the point at which there is no turning back. It is the final commitment.

The Black Team led by Glynn Lunney began support of the Apollo 11 countdown twelve hours before the predicted launch to support the Cape checkout of the CSM and booster systems. (The LM is not checked out during the launch day countdown and will not be powered up until shortly before the lunar landing.) In this way the teams can start working into the mission shifting cycle.

I arrive shortly after Charlesworth and Lunney have completed their handover. When Lunney goes off to get some coffee, I search for a chair. We tried labeling the chairs, but on launch day they have a habit of moving around the room and losing their labels. The back row is filled with the brass — Kraft, Bob Gilruth, and George Hage, the mission director who represented the NASA headquarters mission policy interests. As the count progresses, Charlesworth lives up to his Mississippi Gambler image. He is his usual cool self, saying little and wearing a smile across his broad face. He is ready to play any hand that is dealt him today during the Saturn launch.

There is no external indication that today is any different from any of the other days in his life, although Cliff seems to be keeping closer tabs than usual on the Trench. He likes to play mental gymnastics with his people, asking questions to which he already knows the answer, showing his guys that he has not lost his touch. Today he is pressing them harder. I think this is how he relaxes. With the uncertainties and the fast decisions we face, I think all launch flight directors search for something to feel comfortable with and hold on to. I sit to his left and enjoy watching him do his thing.

Kraft, seated on the row above us, is also having his problems. He left his heart at the flight director's console after Gemini 76. Since that time he was faced with the formidable task of leading his four divisions into Apollo. As the count progresses toward liftoff, he becomes nervous and fidgety. He asks Charlesworth questions about the countdown. Cliff turns, frustrated by the interruptions, and in a mock serious voice, says, "Chris, if you don't settle down, I'm going to have to ask you to leave the room. You're making me nervous." I smile; this is one of the few times we can tell our boss to "cool it!" Kraft hesitates, gives a thumbs-up, reluctantly settles in his chair, and then mutters at the console.

The countdown progresses smoothly. It is hard to believe that this is the day we are going to launch the mission that will land on the Moon. Charlesworth gives the Cape the Go for the start of the terminal count and advises the controllers of his intention to lock the doors at launch minus nine minutes. The controllers scramble in the usual last-minute rush to the rest rooms. After the completion of the final communications checks, everyone hunkers down and I mumble a silent prayer for the crews and controllers as we start the voyage.

The launch is flawless, as if this is just another simulation on a very good day. The only indication that this is the real lunar mission is the muffled commentary of the public affairs officer, Jack Riley. Riley is a neat guy, trains as a member of the team, and covers his flight directors' flanks just like a good wingman. Sitting next to Charlesworth, I hear Riley's voice over the air path. He is speaking so loudly into his microphone that his words penetrate the background buzz of the room. I pick up his words . . . "lunar landing mission." Then it sinks in. Today is different. We are launching the mission that will try to land Americans on the Moon. On this flight, America will go the final 50,000 feet.

Charlesworth continues his chatter with the controllers, giving the crew their Go's periodically throughout powered flight. All eyes in the control room are on the plot board, as the markers plotting the radar trajectory streak along the flight path and into the cutoff box. Collins, the command module pilot, calls out "Cutoff!" at 00:11:42 MET (mission elapsed time), and the controllers scramble to call up their displays for the orbital Go NoGo decision. After a rapid conversation with his controllers in the trench, FIDO Dave Reed shouts, "Go, Flight. We are Go!" We are committed. We are in Earth orbit and there is no turning back.

I pick up the second shift after Charlesworth has guided the mission from launch through translunar injection and has extracted the lunar module from the Saturn IVB. There is little for me to do after shift handover except track the spacecraft and get the crew to sleep. This is my first experience with translunar coast, and for the first time I enjoy continuous communications while the spacecraft is en route to the Moon. Since there are no problems, my team spends the entire shift studying and noting any funnies. Each spacecraft is unique and has its own personality. Learning these characteristics is essential if the controller is to make the right calls and not get fooled under pressure.

Buck Willoughby, the CSM GNC, John Aaron, Ed Fendell, and I go over each measurement, discussing everything that is in any way different from expected. As we talk, I make notes on my spacecraft schematics and in the mission rules. SimSup has taught the controllers many lessons about data integrity. At one time or another, every controller has been faked out by his data and has made the wrong call.

The mission progresses without a glitch, and shift rotations go smoothly. By the time of my third shift rotation, my White Team is well into the groove and, for the first time, my lunar module people have something to do other than sit and fret. Except for a brief communications check on the fourth day, there is no power margin to allow us to look at lunar module data until the final check-out for landing.

This makes it tough on my LM team and support staff rooms. The first time they will see data is when they are giving their Go NoGo's at LM power-up, six hours before the lunar landing attempt. Their learning curve has to be near vertical, and I expect surprises as we go along. The third shift is a welcome break for my controllers as the crew pressurizes the LM and, during the middle of the shift, climbs into it to make the first in-flight visit for a visual inspection.

For the next hour and a half, the crew takes the world on a TV tour of the spacecraft, describing displays and providing a stark view of the cockpit. It's an obvious tight fit for two crewmen. Although the LM controllers do not see data, at least they know that their spacecraft has arrived in space okay. They are finally getting a piece of the action. This is our last shift prior to landing.

After finishing with the post-shift press conference, I go over to the Singing Wheel to have a beer with the team before going home. Mission events never fall into neat, equally spaced increments of eight hours. My team must take thirty-two hours off to synchronize with the lunar trajectory for landing. During this thirty-two-hour period Charlesworth will get the spacecraft into lunar orbit, then Milt Windler, the Maroon Team flight director, will have the crew trim the orbit and then perform another interior inspection of the LM. Four flight control teams are being used for the lunar phase of the mission to provide flexibility and, once the LM is on the lunar surface, to support the CSM solo orbital operations. Lunney will come in for the shift preceding mine, presiding over the crew sleep and, with the assistance of the Trench, nailing down the final trajectory for landing. This "whifferdill," as we call it, sets up the shift sequence for my shift for landing. (A "whifferdill" is the controllers' term for an adjustment to a shift schedule in order to accommodate events that are going to take place in the lunar phase.)

The pre-mission flight plan has the crew in the LM going to sleep after landing, but no one believes it will happen. During the whifferdill Charlesworth moves into a shifting position so we can give a Go and be ready for an EVA shortly after the Stay NoStay decisions. Whifferdills happen every mission and are pretty messy.

Sometimes you come on shift with only an eight-hour gap with the previous one, other times the adjustment is as much as thirty-two hours. You just have to tell your body to ignore how it feels and get on with the job.

The few patrons in the Singing Wheel are watching the TV news as my team orders a couple of pitchers of beer. I glance up as I hear my voice coming from the TV in the bar. The commentator quotes me saying, "The lunar mission is on schedule; there are no problems impacting the planned landing." When the pitchers of beer are drained, I bid a muted farewell to my teammates and then drive home. I read the newspapers, watch television, and try to force myself onto my new shift schedule. It does not work, and I fall asleep on the sofa. When I wake up, and face the extra sixteen-hour gap, I finally come to terms with the realization that the next shift is the "real one."

I go to Saturday evening mass. Blessed by my mother with strong faith, during almost every mission, I find a way to get to church and pray for wise judgment and courage, and pray also for my team and the crew. Our pastor, Father Eugene Cargill, knows the risks and the difficulties of our work and the need for extra guidance. He knows that tomorrow is a special day, and he says a few words about it in his sermon. After mass, he talks with me briefly, finishing with a thumbs-up. Then I go home, have a great supper and a couple of beers, and Marta keeps the kids quiet when I go to bed early. I sleep well.

July 20, 1969

I wake up feeling refreshed and have a quick breakfast. The eastern horizon is just starting to show a bit of light as I hit the road. I arrive at the control center without any memory of passing through League City and Webster, small towns along the way. In an instant, it seems, I am pulling my '67 Cougar into my parking space on the north side of the building, just as I have done hundreds of times before.

Today a guard approaches me and instantly recognizes me. He says, "We gonna land today, Mr. Kranz?" His teeth flash and I see the gold cap on his tooth. It is Moody. I don't know his first name. He is ageless, always standing proud in a crisply pressed uniform at the MCC entrance. The name on his badge just reads "Moody." His cheerfulness makes him as effervescent as usual, a favorite of the controllers. Moody's greeting snaps me back to reality. I smile, give a thumbs-up, and respond, "Today's the day. We are Go." Additional guards are present on mission days to patrol the building and limit access to the control room. They learn to mirror our feelings, and we feel a closeness, a kinship with the MCC guards.

As I walk to the MCC, I note the "egg crate" facade over the entrance. It always sticks out as an anomaly in the four-story, featureless, windowless, boxy, pea gravel and concrete structure. To the left is our office area, its windows well lit and filled with engineers moving deliberately between offices. Approaching the MCC lobby, voices echo like in a canyon as a small group moves past me to the cafeteria for breakfast. We've come a long way since the roach coach back at the Cape.

The guard at the entrance nods as I pass through the lobby. He checks my badge and waves me through. The elevators are hydraulic, like a car lift, and have a habit of getting stuck between floors. Today is not the day to get stuck in an elevator so I take the stairs to the third floor, passing controllers wishing me good luck. Other than that simple statement, everyone avoids unnecessary conversation and does not intrude on my privacy. Usually there is a lot of chatter and kidding among controllers.

My footsteps echo as I walk down the high, narrow gray hallway to the control room. I have the same feeling every time I walk into the MCC. It is a place where history is being made, day by day. It is the home base, the control center for our explorers. As I continue down the hall, I get my usual vague feeling that somehow my entire life has been shaped by a power greater than me to bring me to this place, at this time.

Our target today is the Moon, traveling 2,287 miles per hour in its orbit. Mountainous, pelted by micro-meteorites, and with craters 180 miles across, it is about one fourth the size of Earth. With only one sixth of Earth's gravity, it has no air, no moisture. The temperatures range from plus to minus 250 degrees Fahrenheit during the two-week-long lunar days and nights. This heavenly body has never seen an earthling, never felt a footstep. But, as the scientific evidence from

Apollo will help confirm, Luna is our geop
ical sibling, separated from us in the violen
mation of Spaceship Earth.

The mission operations control room door is heavy, and entering the room, I again realize how small it really is considering the magnitude of operations that take place in it. My eyes have difficulty adjusting to the heavy gray-blue gloom cast by the world map and the dimmed lights over the Trench. I listen to the ambient voice level of the room. It is always the first indication of what is going on. Today it is quiet. Lunney's team is busy closing out its shift, and a lot of messages are being read by the CapCom.

I glance at the TV of the flight plan to the right of the room. The astronauts are awake and well into post-sleep activities. Many of my White Team controllers are on the console and already starting handover. Jerry Bostick, chief of the Flight Dynamics Branch, is standing behind the Trench, listening to his controllers. He is tall, thin, wears a coat, and has allowed his black hair to grow long; he used to have a crew cut like mine. He is in the process of taking a pulse check on his people. Bostick is like some permanent fixture in the MCC; I wonder if he ever sleeps because he is always there, standing behind his controllers, head cocked, coaching his people.

The coat rack is overly full. It swings like a pendulum and it threatens to tip over as I hang up my sport coat. The trip to the flight director console is like walking through a minefield, dodging books, lunches, and the spaghetti of headset cords. The room smells of cigarettes, with an overlay of pizza, stale sandwiches, full wastebaskets, and coffee that has burned onto the hot plates. The smell has never changed since we opened the control center four years ago.

A bouquet of roses glows red against the gray wall of the Mission Control room. The bouquet always arrives as we near launch day for the Apollo missions. The accompanying card simply states "from an admirer." Initially they came from Dallas, subsequently from various Canadian cities, and then the eastern United States. The sender became known among the controllers simply as the "flower lady." For us they were a tangible link with someone who represented the hopes and good wishes of the millions who cheered us on as we pushed deeper into space. We would not know the name of this anonymous supporter until the end of the Apollo mission, when we received, for the first time, a card signed with only the sender's first name, Cindy. It became almost a talisman; the launch flight director always wanted to know that the flowers had arrived — and they always had every time. We placed the flowers in a vase on a small table to the right and beneath the operation room's ten-by-ten-foot TV screen. This was in the area where we normally congregated to celebrate a successful mission. We knew that the TV cameras would pick up the roses sitting there in the background, thus showing our appreciation to the unknown well-wisher.

I talk briefly to each of the controllers, touch them on the back, and say, "How's it going?" I snag a brownie to go with my coffee as I pass by the procedures console. I hang my silver and white brocade vest behind the flight director console, deposit my lunch in a drawer, place my Cokes in the refrigerator, and continue across to the exit, to the spacecraft analysis room. The SPAN room log tells me what the engineers are working on, and today it indicates that everything is normal.

Mission Control during critical events is like a magnet, drawing controllers and astronauts close to the action. Every person not working a shift tries to find a place to plug in his headset. Each console has four communications outlets and as the landing time approaches, every outlet is filled. Since the astronaut observers do not have support staff rooms to hang out in, they burrow into obscure corners to find a place to plug in. Today, a bunch of them have found their home in the SPAN area. I do not blame them. Today, we will watch Armstrong and Aldrin open a new chapter in the history of exploration.

Once we start the descent, we will have very little time to avoid disaster if things go wrong.

We have a ringside seat; the only better one is in the spacecraft. My final stop is the simulation control area; I want to say thanks to SimSup Dick Koos and his team. To my surprise, he is not there. Having finished my circuit of the control and support areas I return to the flight director console. Lunney would never be awarded the Good Housekeeping award for the condition in which he leaves the console, so I set about to clean up the debris and make room for my flight books. Lunney and Windler are not cigarette smokers, so I dig around to find an ashtray. Lighting a cigarette, I pull my headset from the pouch and plug in on the left side of the console, punching up the intercom loops to listen to my team as they conclude handover.

Everything is going smoothly, so I start reading the logs for all the shifts since I was last on console. Kraft arrives, and as he passes behind me, he pats me on the shoulder and says, "Good luck, young man!" He does not have to say more. He occupied this chair from Mercury to the Apollo 1 fire. I wonder how his stomach is today and whether his customary supply of milk is safely tucked away in the refrigerator. The last person to wish me good luck as he leaves the control room is John Hodge, who, like Kraft, has closed out his era in Mission Control and moved into the ranks of management. A new generation of controllers, many mentored by him, are now on the consoles.

Lunney finishes updating the log and indicates the crew has been ahead of schedule all morning. He, Bostick, and the FIDO have been trying to resolve a 500-pound weight discrepancy during much of the shift and have fine-tuned the maneuver times to an exquisite level. Lunney ends his log with the comment that much of the work is trivial . . . "all peanuts."

Preoccupied, I put on my vest and move my gear to the right side of the console. I notice that the space artist Bob McCall is seated on the console step to my right. (NASA had run a competition to select artists to document the program. McCall was one of those chosen.) Looking over his shoulder briefly, I marvel at his work as he rapidly makes a series of pencil sketches of people in the room. I select my TV displays, bring up my intercom loops on the panel, and make the first entry in the flight log: "95: 41:00 MET White Team — descent, crew in LM, pressurizing preps — all looks good." I adjust the intercom foot switch and call the controllers to give me an "amber" and check in. A small status light panel is at the top of my console. Each controller can signal me with the colored light to give his status. A green light signals "I'm Go!" Amber has several connotations, among them "console handover is in progress" or "I'm currently away from the console" or "I've got a problem, call me when you have time." A red controller status gets a flight director's immediate attention. It indicates that the controller has a serious problem or is preparing to call an abort.

The panel status lights at the top of my console instantly change from green to amber. Even the surgeon is listening in to the communications loop today. I advise them to go green if their handover is complete. One by one the status board returns to green. Andrew Patnesky, the NASA photographer at MCC, walks by and bends over. I rub his bald head for good luck, saying, "We're Go, Pat." He croaks, "Good luck." Then he moves to sit behind the CapComs. Pat and I established this ritual way back in Gemini, and it seems to work. Today is not the day to omit it.

One spacecraft revolution in lunar orbit takes about two hours. During the front-side pass, we receive data for about one hour and fifteen minutes, followed by forty-five minutes when the spacecraft is out of sight behind the Moon. At spacecraft acquisition of signal on revolution eleven, the crew is still ahead of the timeline by about thirty minutes, and controllers scramble to check displays to make sure they didn't miss any checklist items while the spacecraft was behind the Moon.

Spencer Gardner, my flight planner, brings everyone to the correct page, identifying checklist items the crew is currently performing. I like his crisp, businesslike call. Spence is on top of his job today. The crew works with the ground on voice checks, navigation updates, computer memory dumps, and docked alignments. Revolution eleven passes quickly, and at loss of signal the crew is working smoothly, still about thirty minutes ahead of the timeline.

The next revolution is equally smooth as the crew continues with LM landing gear deployment, autopilot checkout, and communications testing. They power up the steerable antenna, and for the first time we see the complete set of telemetry from the LM. The controllers quickly assess the data and happily give their Go. The Trench is scrambling to keep up with the crew, provide navigation updates, synchronize clocks on board the spacecraft, and, finally, give maneuver data to the crew.

The exquisite ballet of flight crew and ground

controllers continues. Each participant is in perfect harmony with the other, moving to a cadence dictated by the laws of physics and the clock. I reference my workbook, note all items completed, and at 99:24 MET poll the controllers. "Okay, all flight controllers, go amber and stand by for Go NoGo for undocking." The poll ends quickly with all controllers echoing, "Go!" Charlie Duke passes the Go to the crew. From now on, there is no getting ahead of the timeline. The CSM and LM are now flying in tandem around the Moon.

Over the air path in the control room I hear the voice of public affairs officer Doug Ward. He comments on the flight plan for the coming revolution. Ward is the youngest of the public affairs officers, does his homework thoroughly, and has what it takes to be a great flight controller if he wants. During press conferences, he is always ahead of the game and knows when to run interference for the flight director. He will pick up the ball and run with it if he thinks his flight director is about to get hung out to dry by the media.

Unknown to the MCC, Armstrong and Aldrin have not completely vented the pressure in the tunnel between the CSM and LM. When they undock, the pressure in the tunnel, like the cork in a champagne bottle or in a popgun, gives a slight thrust to the LM spacecraft. It is as if the crew had performed a very small maneuver with the rocket engines.

When we acquire telemetry and voice on revolution thirteen, the final revolution before CSM separation, the lunar module performs a pirouette while Collins, in the command module, makes a visual inspection. The two spacecraft continue flying formation, and the ballet enters its second act. Charlie Duke rattles off the long string of maneuver data for the landing and the abort and rescue options. The voice readbacks of the data by Aldrin and Collins are confirmed by the MCC team. After separation, the flight control team splits into two elements, each working with its own communications links and data stream to the two spacecraft.

This is the busiest time of my shift. I now have to keep logs on two spacecraft, each with its own plan, procedures, and timeline. The common link between the spacecraft is provided by Spencer Gardner's flight plan. I keep the separate groups in harmony as the intensity in the room increases, then poll the controllers for the separation Go NoGo. We have met all of the criteria, are on the timeline, and both spacecraft look good. We take a deep breath and give a "Go for the separation maneuver." Duke passes to the lunar module the Go for the maneuver that will bring it to a point 50,000 feet above the Moon.

I become aware that each of the controllers has reduced the crowding around his console. Duke and Slayton have cleaned house at the CapCom console. Astronauts Pete Conrad, Fred Haise, Jim Lovell, and Bill Anders have relocated to other parts of the room. Tindall maneuvers into the chair next to me, and I motion him to move closer to the console, where he can see the TV displays. This is his day and, I think, one of the happiest of his life. Of all our lives.

Collins acknowledges the Go for separation and, thrusting upward, performs a small radial separation maneuver. As the CSM moves away from the lunar module, Armstrong and Aldrin power up the radar and check it out by tracking Collins's departing CSM. I wonder when I will start to see signs of pressure in my team. So far, the reports are crisp, their voices almost the same as they were in training — the controllers are in a groove. I marvel at how well they are holding up, for no matter how hard they try to appear relaxed and cool, I know the pressure has to be building in them.

As the clocks continue their relentless progress, I can finally feel the tension mixed with excitement in the room. The air starts to crackle as we anticipate coming events. I notice that the paper in my logbook is damp from my palms, and the paper is curling in a tight roll as I engrave each word with my ballpoint pen. Although I am not really aware of it, I'm close to maximum stress at this point, even though mentally I am as cool as a cucumber.

When both spacecraft go "over the hill" and we lose telemetry and voice, I advise the controllers to take five. The rush for the rest rooms, led by the people in the Trench, is the first indication of the pressure the controllers are feeling. I thought that nervous kidneys were exclusively my problem. I follow the stampede and listen to their voices. There is no loud talk and no joking. Their faces reveal a level of concentration and preoccupation that I have never seen before. I do not want to look at my face in the mirror for fear that I might let my own feelings show.

As we reenter the control room, I am inspired by the controllers' mettle, for it takes courage to step up to our work. They are mostly in their mid-twenties. By comparison, I feel old at thirty-five. As I look around, it becomes real for me; in the next forty minutes, this team will try to take two Americans to the surface of the Moon. It will all be on the line. We will land, crash, or abort. In forty minutes, we will know which.

The emotion I feel in these final few minutes takes over. I have to talk to my team. I call on the loop, "Okay, all flight controllers, go to assistant flight director conference." The AFD conference loop is a private communications channel used principally for debriefings or for soul-searching discussions with an errant controller, and it is used exclusively by Flight or the AFD. No one outside the control room can listen in. After controllers complete their check-in, I begin to speak from my heart:

> Okay, all flight controllers, listen up.
>
> Today is our day, and the hopes and the dreams of the entire world are with us. This is our time and our place, and we will remember this day and what we will do here always.
>
> In the next hour we will do something that has never been done before. We will land an American on the Moon. The risks are high . . . that is the nature of our work.
>
> We worked long hours and had some tough

times but we have mastered our work. Now we are going to make this work pay off.

You are a hell of a good team. One that I feel privileged to lead.

Whatever happens, I will stand behind every call that you will make.

Good luck and God bless us today!

I pause briefly, then resume, "Okay, all flight controllers, return to the flight director's loop." I think my phrasing was a bit more emotional than this, but since there is no recording of this private moment we shared, I have put down my best recollection of what I said. I did something I thought was important, something the team had earned, in the good times as well as the bad, an expression of my esteem, my confidence. We were a band of brothers.

I am sure the people in the viewing room and the press corps wondered what in the hell I said. Those who are still around will know now.

I note the time in the log, and call, "Ground Control, lock the control room doors." I pause briefly, then say, "Take Mission Control to battle short." From now on, no controller can leave or enter the room. The main circuit breakers in the MCC are blocked and closed. We count the few minutes to acquisition and I light up another Kent, inhale deeply, and say a prayer.

Behind the Moon and out of contact with Mission Control, Armstrong performed a maneuver slowing the LM orbital velocity, allowing the lunar gravity to pull the spacecraft toward the surface. The LM is now silently coasting to an altitude of 50,000 feet. The final landing phase will take about twelve minutes.

Bill Tindall stirs in the seat next to me. There is an air of expectancy in the room. The clock hits zero and the ground controller says, "Flight, we've had acquisition." I do not know what the controllers are thinking at this moment, but it hits me; this is it — *landing day.*

We are too busy now to think about this being the first landing. We do not have to look for problems because they come right at us, like flies drawn to a picnic lunch. Voice communications are broken, and LM telemetry is unable to lock up. The noise on the air-to-ground communications loop is deafening. Every controller punches the loop off so he can hear communications among the flight control team. Don Puddy instantly swings into action with his back room and his CapCom to select an alternate antenna.

FIDO Jay Greene asks for a report on the descent orbit injection maneuver that the crew has performed behind the Moon. The maneuver sets up the conditions for landing, so the report is critical to Jay's evaluation of the tracking data. The lunar module is now coasting toward the point for descent engine ignition. While the LM is descending, Mission Control is checking the spacecraft systems telemetry, and Armstrong and Aldrin are performing landmark tracking to make sure they are in the landing corridor. We get communications with the LM briefly; just long enough to get the crew's maneuver report.

The communications problem has bit us, and

I am hard pressed to keep my frustration from surfacing in my voice. We have only two chances to get to the Moon and I sure as hell don't want to blow off one of them. Every member of the White Team is ready for the race. Now we're dead in the water. I have only five more minutes and then it's Go or NoGo. I say a brief prayer, "Please God, give us comm."

The mission rests now on Puddy's back. Charlie Duke works with Puddy to maintain voice communications with Armstrong and Aldrin in the LM so we can continue with the final preparation to start descent. Duke has to work around the comm outages and remember the controllers' instructions. He watches comm signal strength indications and suggests an LM attitude change to try to improve the voice comm. Neil Armstrong and Buzz Aldrin continue to roll through their checklists, while we try to sort out the problems.

We get a burst of telemetry data at the time for the Go NoGo for powered descent, and I poll my controllers. The controllers make a rapid Go assessment, and then we lose data again. Since Duke cannot communicate with the LM, he relays the Go through Collins in CSM *Columbia*, who passes the Go to the LM *Eagle*. The intensity of the effort is coming in waves, centering on Go NoGo points. Puddy has seen enough data to recommend a switch to the aft antenna, and Duke again relays the message to Collins. A minor attitude change of the LM helps clear up some of the communications, but we still have signal breakups to contend with.

This is not going the way it should, and I remember the mission rule on data needed for landing. It is up to me to decide if we have enough to continue. I am thankful that I swung the rule change to allow making the final call much later in the descent. Communications and data improve momentarily, and we listen to the final checklist items as the crew prepares to start the descent engine. All checklist items are complete both on the ground and in the air, then the mission smoothes out in the sixty seconds prior to starting down.

The intercom crackles, the voices diced and chopped, staccato. I listen for any hint of concern or confusion. The voices have the steady, cool response of a well-tuned aircraft engine on a sunny day aloft. We are now engaged; the battle has been joined. The communications problem is the opening salvo. I am sure there will be more as I listen for the word "Flight!" to trigger me into the action chain. On a couple of occasions I have to order everyone to keep the level of chatter down in the control room because I have to be able to hear all the controllers, sometimes two or more of them speaking at once on the comm loop.

As we approach the start of the burn, the noise on the air-ground voice comm starts to sound like bacon sizzling in the skillet, indicating another imminent loss of communications. As the wall clock hits zero, the crew calls out, "Engine start . . . 10 percent thrust." The lunar module uses a low thrust level to settle the propellants in the rear of the tank before going to full throttle. As the crew continues to throttle up, data is again lost. The team reacts swiftly to recover communications. Puddy requests the LM aft antenna. Duke relays the request, and Collins calls Aldrin to switch antennas. While communications are being restored, Bales indicates he now has a problem.

The landing target is in the center of a ten-mile-long and three-mile-wide oval area (the "footprint"). To hit the landing point the LM descent engine must be started at a precise velocity 260 miles before the target. The pressure in the tunnel at separation changed the planned velocity at the ignition point.

"Flight, [this is] Guidance. We're out in the radial component of velocity. I'm halfway to my abort limit. I'll watch it, and if it doesn't grow, I think we'll make it." Bales's concern is that the navigation system may be in error and that it will affect the trajectory during landing or if we abort late in the descent. I am also becoming concerned over the trajectory because Steve's words confirm FIDO's call that the altitude is a bit low.

Like a bolt of lightning, the data is suddenly restored. The controllers make a quick assessment, and all systems are Go. Radar data continues to be "ratty" and is frequently lost. We have just enough to provide the needed data comparisons between the ground indications and those on the *Eagle*. Bales continues his assessment. The downrange error is not increasing, so he determines that the navigation is good. With data steady for the moment, we verify proper thrust levels. Aldrin calls from the *Eagle*:

"I'm seeing some fluctuation in the AC voltages." We quickly confirm that the electrical system is looking good on the ground, and Aldrin concludes that the hitch is in the meter on board.

Everything is Go, and for the moment, the energy level of the room has settled down. We are running by the clock and are a quarter of the way to the surface. Bales again reports the downrange error is not increasing, and again states that he thinks we will make it through his mission rule gate. I have the fleeting impression that if it is close, he will bias his call on the Go side. After a visual position check against lunar landmarks, the crew indicates they also think they are a little long. I add up the pieces. Three data sources now say we are not going to hit the planned landing point. We are going to be long. I dig into my memory for the description of the landing site near the toe of the footprint. I know it is rough, full of craters and boulders. I hope Armstrong can find a landing site.

I relax prematurely, and once again the data gets ratty. I tell the controllers to make their Go NoGo decision based on the last data they think is valid. I trust enough in their judgment and the spacecraft to keep descending without data for a while. I go rapidly around the horn, and all controllers are Go, especially Bales. His "Go!" resounds clearly through the room like a cymbal. He does not need a voice loop today. I chuckle as I continue polling the controllers.

Duke advises the *Eagle* that they are Go to continue powered descent. Communications are noisy but usable, making it tough for me to pull the voices out of the background noise. It is the kind of workload that SimSup routinely put us through, and his training is paying off with my team today. Tindall must have been holding his breath, for he exhales noisily, and he knocks a book to the floor as he stirs in his chair.

Data returns just as the LM radar locks on to the lunar surface. Bales reports, "Radar, Flight," and his voice briefly quivers, betraying his true feelings. Steve has passed another of his Go NoGo milestones. Then, Aldrin reports, "Program alarm. It's a 1202."

Controllers are still studying radar quality prior to incorporating the data into the computers and do not immediately pick up the alarm call. Seconds later, activity at the guidance console comes to an abrupt halt as the implications of the alarm sink in. Bales calls, "Stand by, Flight," when his backroom support, Jack Garman, brings the alarm to his attention. Duke repeats, "It's a 1202 alarm," in a questioning voice. "It's a 1202 alarm" echoes across the loops for several seconds.

Aldrin requests a reading. It is like coming to a fork in the road where you're uncertain which direction to take. Many of the controllers are oblivious to the alarm and are continuing the decision processes related to accepting the descent radar. Bales, Duke, and I start work to resolve the program alarm. I don't think anyone outside the flight control team understands the real significance of the alarm, in the midst of the rapid-fire exchange of communications. Duke muses aloud on the flight director loop, "It's the same one we had in training." He audibly expresses our collective feeling, almost wonderment. These were the same exact alarms that brought us to the wrong conclusion, an abort command, in the final training run when SimSup won the last round. This time we won't be stampeded.

The significance of this is not lost on any of us. The alarm tells us that the computer is behind in its work. If the alarms continue, the guidance, navigation, and crew display updates will become unreliable. If the alarms are sustained, the computer could grind to a halt, possibly aborting the mission. Each alarm must be accounted for. They have the capacity to create doubt and distraction, two of a pilot's deadliest enemies.

Prompted by Gibson in the back room, Bales says, "We're Go on that alarm. If it doesn't recur, we are Go." Then, in the blink of an eye, he swerves back to the nominal mission and says, "He's [Aldrin] taking in the radar data." We pass the throttle-down times, continue the routine assessments, and a backroom controller inadvertently comes on the loop, saying, "This is just like a simulation." I smile and agree. There is nothing like working out a problem to relieve the tension on a team. Controllers always work best under pressure, and they are doing well during today's final exam.

The radar data smoothly corrects the altitude difference in the computer, and as we watch, we see another program alarm. Aldrin calls, "Same alarm, and it appears to come when we have a 1668 up." (1668 is the LM computer display of time, landing site range, and altitude.)

Bales quickly responds, "We are Go. Tell him [Aldrin] we will monitor his altitude data. I think that is why he is getting the alarm." This information is quickly passed to the crew. Above all, we have to prevent a rapid string of these alarms, which will cause the computer to go into an idle mode and abort the landing. With pressure mounting on the team, I get on the loop, "Okay, all flight controllers, hang tight. We should be throttling down soon."

Today we are gobbling up the alarms as they occur. I mentally thank SimSup for the final training run on program alarms. Throttle-down comes uneventfully. LM systems and trajectory are good, and Duke advises the crew they are looking great. As the crew continue to pitch over to the vertical for the start of the landing phase, they select the steerable antenna to assure continued communications.

We seem to gain strength as the problems mount. Again, I repeat, "Okay, everybody, hang tight . . . seven and one half minutes." I run out of breath with that statement. Bales comes on the loop. "The landing radar has fixed everything, the LM velocity is beautiful." Carlton, the LM CONTROL, has been watching the fuel gauging system, and he selects the fuel quantity measuring probe that will be used for giving the crew and control team voice call-outs on "seconds of fuel remaining." The voice loop calls are now back to the expected traffic levels, and eight minutes after checking with Bales that the landing radar has updated the computers, I start to close out my final mission rules for landing. We have met all of the mission rule requirements. It is time to make the final landing Go NoGo decision.

After the LM pitches toward the landing attitude the computer automatically completes the braking phase and switches to the final approach program. I know we are long, and the crew is now able to verify that the automatic system approach will take them into a large crater. My console telemetry display indicates the LM is about 7,000 feet above the surface, with a descent rate of 125 feet per second. Armstrong now selects a new landing point in the computer to overfly the crater, and Carlton reports that Armstrong has checked out the manual attitude control.

I start around the loop for the landing Go NoGo. I have met all of my rule criteria, and I am sure that controllers have, too. We are about to hand over the control for the final phase to Armstrong and Aldrin. Soon, we will be spectators just like the rest of the world. The controllers respond crisply and again Bales gives a "Go!" that rings through the room. I continue with my final polling. All controllers are "Go!"

With deliberate emphasis, I say, *"CapCom, we are Go for landing."* The voice exchanges become furious as Duke gives the Go to the crew, now busy trying to find a landing site. There is a brief pause, then Aldrin responds, "Roger, understand. Go for landing," and then continues,

"3,000 feet . . . program alarm . . . 1201." Duke acknowledges, "1201," and it echoes through the intercom loops with Bales advising, "Go . . . same type . . . we're Go." As it gets tougher, the team gets tighter. I am about to bust a gut with pride for my people.

The intensity increases and all calls become even swifter. They are emotional, but crisp and shorn of excess verbiage. Controllers now report what they are seeing, and Duke starts to choose data from the controllers to send up to the crew. As the LM passes through 2,000 feet, Duke picks up another alarm, this time a 1202, and he advises the crew we copy. Throughout the descent so far we have not seen any discernible effect of the alarms on LM computer performance.

Aldrin does not bother to respond. The control team has gone to a negative reporting mode as seconds become even more precious. Normal reporting stops and controllers report only NoGo conditions, with the exception of Carlton's fuel-remaining calls. The room is silent, expectant, listening intently to the crew calls: "700 feet, down [descending] at twenty-one . . . 540 feet, down at fifteen." During the descent, Buzz Aldrin, the LM pilot, selected landing data from the computer display and called out the critical data to Neil Armstrong, who was piloting the LM. The reports normally consisted of altitude, rate of descent, and forward velocity, although in many cases only the single most critical element was reported.

Carlton calls out in hushed tones, "Attitude

hold." I acknowledge, "ATT hold," then silence. The crew is searching for a landing site. Duke, in even more hushed tones, states, "I think we better be quiet from here on, Flight!" I respond, "Rog, the only call-outs from now on will be fuel." My voice loops become silent, the atmosphere electric as we hang on to each of the crew's words and wait for Carlton's call. We are within 500 feet of the surface and continuing the descent. We watch displays that the crew cannot see and listen for sounds yet to be uttered. If anyone so much as clears his throat, twenty other voices shush him.

Reflexively, I reach out and grip the handle on the TV monitor with my left hand and think, Damn close! I continue to keep up with my notations in the log. Again, I feel Tindall stir in his chair as he leans forward to look at the displays. It must be hell to be a spectator today. I have to break through the tension. I run a quick status check, "FIDO, are you happy!" "Go, Flight!" "Guidance, how about you? Are you happy?" "Go, Flight!"

The tempo picks up, the crew call-outs of altitude and descent rate increase in frequency. You can almost feel the crew in *Eagle* reaching for the surface. I look at my displays. The descent rate is almost zero. They are hovering now, and I try some body English in my chair to help them find a place to land. I look at the clock and my log. It is more than eleven minutes since we started descent.

In every training run, we would have put it down by now. It is going to be close, damn close, closer than we ever trained for. The voice loops are silent. Then someone unconsciously keys his mike, and for a few seconds you can hear him breathing, then he unkeys. It is quiet, no discussion at all, and in these last few seconds, I feel that every controller, our instructors, program management, and those in the viewing room are mentally on board the LM, feeling for the surface along with *Eagle*'s crew and aware fuel is almost gone.

The crew reports, "250 feet, down at two and a half, nineteen forward." Still near hovering, I think, but moving forward pretty rapidly. They are over a boulder field trying to find a landing spot. I write in my log, "Here we go," and advise Carlton, "Okay, Bob, standing by for your call-outs shortly." The crew continues, "200 feet," then, "160 feet, five down, nine forward."

"Low level," exclaims Carlton on the flight director loop. Propellant in the tanks is now below the point where we can measure it. It is like driving your car on empty. Controllers turn their mental clock on. We have 120 seconds or less to land or abort.

Carlton's backroom controller, Bob Nance, using a paper chart recorder, is mentally integrating throttle usage by the crew and giving Carlton his best guesstimate of the hover time remaining before the fuel runs out. During training, he got pretty good at it and could hit the empty point within plus or minus ten seconds, but I never dreamed we would still be flying this close to empty and depending on Nance's eyeballs. I wait for Carlton's next call.

Armstrong is flying and Aldrin is reporting, "100, three and a half down, nine forward." As the crew passes seventy-five feet, another call comes from Carlton, "Sixty seconds." I marvel at his calm voice and wonder if he feels the turmoil I am starting to feel. I mentally integrate the time. At seventy-five feet of altitude with a descent rate of two and a half feet per second, we will have about thirty seconds of fuel remaining at touchdown, assuming Nance's integration is good. It could be a lot closer!

Duke repeats Carlton's call on the uplink: "Sixty seconds." There is no response from the crew. They are too busy. I get the feeling they are going to go for broke. I have had this feeling since they took over manual control. They are the right ones for the job. I cross myself and say, "Please, God."

Carlton's voice again penetrates, "Stand by for thirty seconds, thirty seconds." Duke echoes the time on the uplink. The whole mission is now down to the last thirty seconds, assuming we guessed right on fuel.

It is quiet, damn quiet, the silence so great you could hear a feather hit the floor. All the air seems to have been suddenly sucked out of the control room as each controller gasps and then swallows a gulp of air, then holds it for Carlton's next call.

The crew report almost brings us to our feet: "Forty feet, picking up some dust, thirty feet, seeing a shadow." They are going to make it! It is like watching Christopher Columbus wade ashore in the New World. Carlton calls, "Fifteen sec . . ." then stops.

There is a lengthy pause in all crew communications, then, "Contact light . . . engine stop . . . ACA out of detent." It takes me a second to realize the crew is going through the engine shutdown checklist, just as they did in training. It really sinks in when Carlton, in a droll, almost bored voice says, "Flight, we've had shutdown." Duke responds, "We copy you down, *Eagle*."

Spectators in the viewing room and our instructors are drumming their feet on the floor, standing and cheering. We remain rooted in our chairs, but the sound seeps into the room. I experience a chill unlike any in my life. I am totally unprepared for the flood of emotion. This is the one thing that we never trained for — the instant of the actual landing. I am choked up, speechless, and I have to get going with the Stay NoStay. There is not one second to spare, and I just cannot speak.

While the world waits, Neil Armstrong sends goose bumps around the globe with the words: "Houston, Tranquillity Base here. The *Eagle* has landed."

Frustrated at my lack of emotional control, I slam my forearm against the console. My pen flips into the air, startling Tindall and Charlesworth. In a voice that cracks, I say, "Everybody stand by for Stay NoStay. Stand by for T-1 [Time-1]."

Charlie Duke, equally unsteady and in an emotion-filled voice, closes out this phase with a statement that expresses our feelings, "Roger, Tranquillity. We copy you on the ground. You got a bunch of guys about to turn blue. We're breathing again. Thanks a lot."

In case of an emergency after the lunar landing, three LM liftoff times were selected that would permit a CSM rendezvous within the electrical power lifetime of the LM. The T1 time was two minutes after landing, T2 was eight minutes after landing, and T3 coincided with the CSM orbital pass two hours after landing.

While the world is celebrating, each of the controllers overcomes his own emotional overload and proceeds to swiftly assess spacecraft systems. They start the process to check for an acceptable surface attitude, then verify the primary computer configuration and LM systems status for a possible immediate liftoff. Within thirty seconds of landing, I start polling controllers to commit to a surface stay of at least eight minutes. If I receive a NoStay we must lift off in the next sixty seconds. All controllers crisply state they are "Stay for T1," which Duke promptly relays to the crew.

Without a break, the White Team rapidly recycles and, minutes later, gives the crew the "Stay for T2." We then hunker down for the final Stay NoStay decision.

Sixteen minutes after the T2 Stay, Carlton jolts me with the call, "Flight, the descent engine helium tank pressure is rising rapidly. The back room expects the burst disk to rupture. We want the crew to vent the system."

My team doesn't have the opportunity to savor even a few seconds of the euphoria after the landing, as we watch the descent engine helium pressure rise, stabilize, then plummet. Carlton hovers over his telemetry display, anxiety coloring his drawl, then with a deep sigh he says, "Flight, we're now okay. The pressure has dropped and the system is stable." From that point on, the Stay decision is a piece of cake, Duke giving the crew the final, "*Eagle*, you have a Go for extended surface operations."

Windler and Charlesworth come to the console during the T3 Stay NoStay processes, and prepare to step in if the crew requests an early EVA. There is controlled confusion as elements of two teams circle the consoles, unsure of which team is in charge until I hand over. Tindall walks over, his eyes moist, offers a handshake and says, "Damn, Geno, good show!" The lunar landing techniques were his. I made no proprietary claim on whose show it was. It was a victory for the tens of thousands who worked on and believed in Apollo.

Before leaving for the press conference, I walk to the simulation control room to thank Koos and the training team. The instructors are unbelieving that the last problem given us in training is the one big problem during the landing. (I also learned that Koos, in his haste to get to the MCC for landing, rolled his new red Triumph — a TR3. I thanked God he came out all right.)

While walking across to the press conference with Ward, I finally have time to absorb the full reality of it, in a moment of silence when there is no busy chatter on comm loops and my mind can move into a reflective, rather than reactive, mode. We have just landed on the Moon. In a way, I feel cheated that I didn't have the chance to savor those seconds as deeply as those who watched. I thank God for being an American, and I think of my team and the way they performed during the landing. More than ever, I appreciate the great training, the unrelenting pressure put on us in getting ready for Apollo 11.

All I want to do is get the press conference over, so I can get back to Mission Control. Today we were perfect, devouring each problem and grasping for each opportunity. When I get back to the control room, Milt Windler is in the process of orchestrating the planning to get the EVA preparation started in about three hours. Cliff Charlesworth is already bringing his team on line. The world is about to witness an explorer setting foot on a New World. I sit with Charlesworth, awaiting Armstrong's descent from the lunar module to the surface. I am sure of it now — this is the best day of my life.

On July 20, 1969, at 9:56:20 P.M. Houston time, Neil Armstrong steps from the ladder to the surface and, as his boots touch the lunar dust, he declares, "That's one small step for man, one giant leap for mankind."

It was worth every sacrifice for this moment. I remember President Kennedy's words, "We choose to go to the Moon. . . . We choose to go to the Moon in this decade, and do other things . . . not because they are easy, but because they are hard!"

I was more teary-eyed in the months after Apollo 11 than at any time in my life. Every time I heard the National Anthem, or looked at the

Moon, or thought of my team, I got misty. On August 13, just a few days short of my thirty-sixth birthday, Marta and I were invited along with program managers, designers, controllers, and astronauts to a presidential dinner at the Century Plaza Hotel in Los Angeles. The Army Drum and Bugle Corps, in their brilliant red jackets, shiny brass, and blue pants kicked off the ceremony with the ruffles and flourishes. If ever I was on a high this was it. After a marvelous evening, a series of awards was presented by President Nixon. Steve Bales, my twenty-six-year-old, bespectacled guidance officer, accepted the Medal of Freedom on behalf of the entire team for flight operations.

I don't think anyone outside the program ever expected us to succeed on our first attempt at landing on the Moon. Now that we had set the standard we were expected to do it again . . . and soon.

With six children — some of the girls playing in the band, some cheerleading, and Mark playing — I spent a lot of time at high school football games every fall. As the harvest moon rose over the stands in the east, I never failed to stare at it with the binoculars, picking out the Apollo 11 landing site, hoping that someday we would continue what we started. I pray that my children will someday feel the triumph, the joy, and the shiver I felt the day we painted the Moon with our Star-Spangled Banner. Their day will come when we put men on Mars or accomplish some other feat where the human factor makes it possible to achieve something that technology, no matter how brilliant and advanced, cannot. We have "slipped the surly bonds of Earth" and our destiny will ultimately lead us to the stars that glow in our deep black night sky, like diamonds scattered on a field of velvet.

17

"WHAT THE HELL WAS THAT?"

Fall 1969

On the Moon there was a flag from the Earth, and on Earth there were pieces of rock and soil from the Moon. It seemed a fair exchange.

The days after a mission, and sometimes the weeks, are tough. The euphoria of the missions, coupled with the emotional intensity of the parties and the debriefing with the crew, is followed by a strange kind of emotional decompression, as if you are a diver who has come up from the pressure of the deep sea and has to gradually adjust to the sudden absence of that pressure. We did not have much time to decompress. We also felt like we had pushed our luck; solving the Apollo 11 problems and then landing with only seconds of fuel left was a lot tighter than any of us expected. After the string of successes that started with Apollo 7, I had a nagging premonition that we were about to break this lucky streak. When you get this feeling you keep it to yourself and move forward. That kind of feeling has no place in our business.

The lunar program now focused on pinpoint landings, extending the duration and complexity of the surface activities, and mapping the Moon. There was enough action for everyone. The systems controllers were the crew chiefs for the spacecraft. They were by nature and training tinkerers, mechanics. Living on government pay and raising a family was not easy, so a lot of us saved money by doing our own auto repair work, swapping tools and skills as necessary. You could identify the houses of the NASA guys who personally kept their old, but well-maintained, cars and motorcycles humming along by the oil stains on their driveway. Smooth-running engines and the harmonic rhythm of the valve train were music to their ears.

Whether it was a car or a spacecraft, the systems guys were the experts in diagnosis and providing quick fixes, using the materials and tools at hand. They had a gut knowledge of why things worked and why they broke down. They grew up with the legacy of Aldrich, Brooks, Hannigan, Fendell, and Aaron, the taskmasters who learned their trade in Mercury and Gemini. They were the kind of people you liked to have around when things unraveled. They worked like detectives, suspicious of anything that did not seem to fit, doggedly tracking down every glitch, relishing the opportunity to explain what was, to the rest of us, inexplicable. They would soon prove themselves the ultimate backup when their own systems let them down.

November 14, 1969, Apollo 12

After the successful lunar landing, it did not take the Apollo program office long to establish more, and more demanding, objectives for the next lunar flight. Gerry Griffin was teamed with Pete Frank, Glynn Lunney, and Cliff Charlesworth for the second Moon landing. This would be Charlesworth's last mission; he was moving on to a role in management.

The lunar module was targeted to set down next to Surveyor 3, an unmanned spacecraft that had landed three years earlier. The craft was sitting in a 700-foot-wide crater in the Ocean of Storms. Apollo 12 had an all-Navy crew, Pete Conrad, Dick Gordon, and Alan Bean. Conrad and Bean would perform a landing that required a manual guidance update. Landing near the Surveyor 3 spacecraft required a high degree of precision. Small errors in navigation or guidance could cause the crew to land beyond walking distance from the Surveyor target. The LM computer was not very bright by today's standards, so shortly after starting the lunar descent the Trench, using radar tracking data, computed an update to the landing site range for the LM computer. The update was voiced to the LM crew for manual entry into the computer. As usual, Bill Tindall was in the middle as the developer of the technique, one that I considered pretty fancy for only our second landing.

Griffin was the launch flight director — for the first time. No matter how much you train, you will never forget a moment of that first launch.

As you approach liftoff, you just pray that all goes well. If you have a problem, you hope it is one you have aced before in training. I was plugged into the assistant flight director's console, sitting behind Griffin on the steps to his left. Every launch, especially a Saturn, was a major, awesome event. They never lost their thrill — or their risk; seven and one half million pounds of thrust is a hell of a lot of energy.

The weather at the Cape was marginal at best, with heavy rain, but Gerry did not show any strain under the pressure he must have been experiencing. A hold was called twenty-two minutes before launch to review the rules and the weather forecast. A weather decision is a classic risk-versus-gain trade-off. Griffin listened closely to his controllers, mindful of his launch window and recycle options. Walt Kapryan, one of the members of the original Mercury team, moved into launch operations and, like Griffin, was in the hot seat for the first time. Kappy had moved up a notch into the launch test conductor chain when the previous launch director moved to headquarters.

As I listened to the weather briefing I could clearly visualize the conditions; low ceilings and intermittent cloud cover were the bad news. The good news was that the winds were light at all altitudes. There were no thunderstorm or lightning reports. During the hold, the public affairs officers commented that President Nixon was in the VIP area, his first and only visit to the Cape to view a launch.

There were few black-and-white decisions in

launch control and I did not envy Kapryan. We had had a near perfect countdown. Every scrub and recycle takes its toll in flight hardware and in people, increasing the chance that we will have a hardware failure, or slip into the next monthly launch window. Kappy gave the Go to continue the count. The test conductors started their polling, the clock started its countdown through the final twenty minutes. I had been there before and thought, "Well done, Kappy." At times it takes more guts to say Go than NoGo.

Griffin's controllers had purged their kidneys during the weather briefing hold; the enormous front screen displays were called up, glowing brilliantly against their black background. After the traditional command to "lock the control room doors" the team settled in, intently scanning the displays during the final seconds as Griffin called, "Recorders to flight speed!" Kapryan had committed Apollo 12 to launch. Within seconds, Pete Conrad's crew and the command module, *Yankee Clipper*, would be in the hands of Mission Control.

At 10:22 A.M. in Houston, Apollo 12 began its journey to the moon's Oceanus Procellarum, the Ocean of Storms, thundering from the launch pad into an overcast sky. Conrad could not restrain his glee at once again leaving the Earth, reporting, "This baby's really moving!" Within seconds, the Saturn disappeared into the overcast. The clouds muted the sound and glowed for a few seconds with a red-orange fire. The tempo of the air-ground communications indicated the *Yankee Clipper* was off to a good start.

Dick Gordon reported, "Looking good. The sky is getting brighter." This message was followed by a brief, "Uh-ohhh!" At that instant, the controllers saw a brief glitch on their TV displays.

In the command module, glaring amber lights in the upper right quadrant of the caution-and-warning panel flashed on. Conrad yelled to Gordon and Bean, "What the hell was that?" (We didn't hear this exclamation on the ground because the crew had an internal intercom that allowed them to talk to one another without Houston listening in. We only heard this later when we reviewed the tapes that recorded those onboard exchanges.)

"We lost a bunch of stuff," Gordon responded on the closed loop of the internal intercom. "We had a whole bunch of the buses drop off." (Electrical power is provided by the CSM batteries and fuel cells to a bus or distribution point. There are two main buses and ten secondary buses for power distribution to the CSM equipment.)

Thirty-six seconds after launch, observers saw a brilliant flash of lightning in the vicinity of the launch complex. Initially, they did not report it to us because we were just too busy.

On the Mission Control systems row, John Aaron was seated at the console in front of Griffin, monitoring the cabin pressure as the Saturn continued its ascent. Rapidly scanning his displays and event lights, John was about to advise Griffin on the cabin pressure status when his displays stopped updating. Data drop-outs were not uncommon during launch, but when the data returned a few seconds later many of his electrical measurements were scrambled. Aaron had seen this unique pattern only once before in his life. During a pad test a year earlier a technician inadvertently switched off a power supply, which scrambled the data. Intrigued by the data funny, John traced it to a power supply operated by a little-used switch. The switch had two positions, primary and auxiliary, and ultimately provided power to fifty-one CSM telemetry measurements.

Griffin needed answers. Nearly everyone was scrambling to nail down the source of the data loss. Sixty seconds after launch, and twenty-four seconds after the data drop-out, Conrad said, with surprising calm, "We got a bunch of alarms. We've lost our platform. I don't know what happened." (A platform is a set of gyroscopes that provide a reference for navigation. The platform is aligned using reference stars and is essential to determining spacecraft orientation and velocity, and in performing maneuvers.) Platform loss during launch phase is a serious problem.

The master alarm and caution and warning reports from the crew indicated big troubles on board the *Yankee Clipper*. With the navigation system unusable, the crew was down to the backup system in case of an abort. The only thing keeping the launch phase going was the Saturn guidance and computer system at the booster's forward end. The CSM gyros were tumbling, useless as a reference for either the crew or the guidance system. The crew was literally flying blind, without instruments they could trust.

Gerry Carr, the CapCom, relayed the reassuring news that the Saturn was still accelerating on the proper trajectory toward orbit. This was the only piece of good luck so far. In the command module, every electrical warning light was glowing. As the seconds clicked by, time was not on Aaron's side. The backup batteries had taken over and Aaron prayed that whatever happened had not shut off the flow of the oxygen and hydrogen to the fuel cells. If the fuel cell valves had closed, the lunar mission was over unless the fuel cell flow could be restored within 120 seconds.

John Aaron likened his role that day to that of a medical intern treating a gunshot victim in an emergency room at midnight, plugging the hole in a man's heart with his finger to stop the hemorrhaging as his emergency room team sprang into action. John's next call made him a legend in Mission Control.

As Conrad's voice reports continued, Aaron suddenly remembered the instrumentation funny from a year earlier. John now translated this single obscure event into a train of actions that would save the Apollo 12 mission.

Griffin was no longer writing in the log. His hand was now clenching a black government ballpoint pen. Like Aaron, he knew he had little time to make a decision. "How is it looking . . . EECOM, what do you see?"

Aaron paused in the middle of an exchange

with his support staff, stared at his displays, then made the decisive call, "Flight, have the crew take the SCE to Aux."* The words tumbling over the loop from Aaron were alien to Griffin, alien to the entire team. Taken aback Griffin stated, "Say again, SCE to Aux?" ending his statement with a question mark. This time more firmly and slowly, Aaron repeated himself.

When the CapCom, Gerry Carr, passed on Aaron's recommendation, it made little sense to Conrad, who blurted out, "What the hell is that?"

Carr repeated the instruction, emphasizing "S-C-E to Auxiliary." During powered flight, the command module switches and controls are allocated to the crewman who can see and reach them. This switch was Al Bean's responsibility. Reaching forward, Al firmly toggled the switch down, and confirmed, "SCE is in Aux."

Moments later, Aaron announced, "I got valid data, Flight. It is looking good." Interspersed with the discussion, the Trench continued to rock through the abort mode calls. For a few seconds, Griffin worried that Aaron might give him an abort call, but when none came, Griffin exhaled a loud sigh of relief. In less than sixty seconds, the fuel cells were back on line, and Griffin had a Go from all his controllers to keep

* The signal-conditioning equipment (SCE) is a small redundant power supply that provided voltage to forty-six critical instrumentation points in the electrical, booster, control, fuel cell, and cryogenic systems. If the normal power supply fails, an auxiliary supply can be switched on.

pressing on to orbit.

Aboard the command module, Conrad reported, "We're pretty well straightened out now. Not sure what happened. I think we got hit by lightning." Conrad's suspicion was soon proved to be correct.

The CSM power was back, but with the status of the spacecraft uncertain, Griffin's team gave a sigh of relief when the third stage of the rocket pushed Apollo 12 into orbit. "That may be one of the better sims," was Conrad's appraisal after achieving orbit. "We were chuckling about it up here. We had so many lights we couldn't read them all."

The Saturn guidance and propulsion had done a fantastic job. Griffin's team settled down and started a meticulous check-out of the spacecraft. The question now was, Could we muster enough confidence in the spacecraft to fire up the engines and shoot for the Moon?

As the spacecraft flew toward Carnarvon, Australia, the Trench got the next shock. Radar tracking was minutes earlier than expected at Carnarvon. This was not good news, since it could occur only if the command module was in a much lower orbit than expected.

With a sinking feeling, Jay Greene conferred briefly with Deiterich, trying to resolve two incompatible pieces of data. Griffin had enough problems, so they decided to keep their concern to themselves while they anxiously waited for more tracking data, finally grunting a sigh of relief when they had confirmation that the *Yankee Clipper* was in the correct orbit. They trusted their gut instincts and they were right. An atmospheric anomaly had bent the tracking data, faking the radar into believing the command module had crossed the horizon early.

During a brief conversation over Carnarvon, Conrad left no doubt that they had been hit by lightning. "At the time of the event, about thirty seconds into flight, and again at about fifty seconds, I saw some illumination out the window. Inside the spaceship, everything went dark." At least one of the two flashes of lightning seen by observers at the Cape had hit the spacecraft. Jim McDivitt, who stepped in as Apollo program manager when George Low moved to headquarters, had seen a ground-lightning strike near the launch pad. The strike's path went from a cloud to the *Yankee Clipper* and then via the rocket's exhaust to the ground around the pad. This was the strike that took place thirty seconds into launch. It was not clear if the illumination that took place at fifty seconds was also a strike. Reports of the strikes had been withheld until after the critical launch phase was over.

Sig Sjoberg, Kraft's deputy and a gifted and intuitive design engineer, was deeply concerned over the report, visualizing the havoc that a lightning strike could cause in the Apollo spacecraft. Concerned over the inability to make a complete check-out of the CSM and booster prior to the translunar injection, he walked down to the systems row of controllers, talking briefly with the GNC, then with EECOM John Aaron.

Moving into the Trench, Sig approached Bostick: "Jerry, if you feel uncomfortable in any way about the TLI [translunar injection], speak out." Fidgeting a bit, he continued, "I will support you if you give a NoGo today." He then left and moved to the booster controllers, giving them the same speech.

Kraft, standing next to Griffin, offered the same advice. "Young man, we don't have to go to the Moon today. It's your call." The input of Kraft and Sjoberg immediately removed all political pressure from the decision. Griffin knew all he had to do was make the right technical call. There could be no other way.

It was impossible to check out the entire spacecraft; that can be done only on the ground. In the short time available Griffin's team ran a pre-maneuver checklist, realigned the CSM platform, and then, after much discussion with the crew, gave Conrad, Bean, and Gordon their translunar injection Go over Carnarvon. Throughout the mission the MCC and North American, the CSM contractor, continued to analyze the lightning strike, assessing any critical circuitry that might have been damaged or would prematurely fail. The pyrotechnic systems were the principal systems that could not be checked out. Since they were needed only for normal entry, their status had no bearing on the decision to go to the Moon.

Conrad and Bean made their pinpoint landing on the Moon next to the Surveyor, establishing a new set of space records, increasing the duration of the lunar surface activity — and surviving two lightning strikes. Aaron's "SCE to Aux" call became legendary and Griffin survived his first launch. All in all, it was a damn good mission.

Christmas 1969

Four tough missions, the first lunar landing, and the Apollo 12 save — yes, 1969 had been good to us. A year of world-class performance under pressure and a perfect track record. We were Super Bowl champs and it was time to party.

After the long and irregular hours, the controllers were at a disadvantage when it came to the splashdown parties after the Apollo missions. After the congratulatory handshake, and puffing the traditional cigar, we secured the MCC consoles and called our wives. Exhausted by the demands of the mission but still pumped up by the adrenaline rush that comes from getting another crew home safely, everyone elected to attend the splashdown parties after the Apollo missions. The controllers' splashdown parties normally started at the Officers Club at Ellington Air Force Base, about a ten-minute drive from the MCC. Occasionally the Air Force Reserve Squadron would fly up to Maine for crew navigation training purposes, stopping long enough to get a bunch of live lobsters and have them cooked and waiting for the splashdown.

Aldrich, a properly raised New Englander, provided instructions on the correct way to eat this delicacy. After finishing at Ellington, we moved to either the Singing Wheel or the Flintlock, both located in the small city of Webster, about a three-minute drive from MCC. I think the Flintlock was in cahoots with the Webster cops. The bartenders would keep shoving us the drinks, then when we left for another party, the Webster police were waiting as we pulled onto the highway. Controllers contributed a bunch of bucks to the Webster municipal treasury.

The debriefing parties were a more private, males-only affair restricted to crews and controllers. (Women engineers inspired by the space program joined the ranks starting in 1971 and now make up about 40 percent of the MCC teams. Four women have become flight directors, three of whom are currently active.) The Hofbraugarten in Dickinson, ten miles south of the control center, was the rallying point. The remote Dickinson location got us away from the crowd, and the Galveston County Sheriff's Department often looked the other way. The restaurant had a large open-air biergarten, a bakery, and a butcher shop. The formal mission debriefings were not for the thin-skinned, so a few liters of beer softened the edge as we cooked sausages, drank, and continued into the informal debrief.

Awarding the "dumb shit medals" (DSM) was the focus of the festivity. Flight directors, controllers, and crew compiled a list of errors, both perceived and actual, during the course of the mission. In an elaborate and highly graphic fashion, we stepped forward to make a speech or accept our honors. The Hofbraugarten oompah band often joined in, playing a dirge as the stories got longer and wilder. The awards took many forms — elaborate certificates, dented and broken equipment, photographs, and multicolored ribbons to be worn around the neck. By the end of Gemini, I had enough awards that the controllers presented me a set formatted like the bars of military campaign ribbons.

One of the highest-order dumb shit medals passed out at the debriefing party was for anyone who missed a pre-sleep checklist item and then had to wake the crew to correct a switch position or pump up the pressures in the tanks. My awards ranged from triggering a fire alarm during a mission when I emptied my ashtray into the wastebasket, to locking the control room doors for launch before all my team members had returned to the room. A common DSM among the flight directors was awarded for leaving the console log behind at the press conference, or for a poor selection of crew wake-up music.

The festivities often included a chug-a-lug contest or some old-fashioned Indian arm wrestling. The controllers and crews put forward their own champions. The parties reminded me of the fighter pilot hijinks back at the Officers Club in Osan, Korea.

Emulating the traditions of a fighter squadron, I decided that Flight Control needed to fashion a beer mug. Maureen Bowen, secretary and den mother to Mel Brooks and the Experiment Systems Branch, was recruited to work with the Balfour Mug Company to design a mug for the flight control team. In a typical engineering fashion we provided some specifications: the mug must hold one and a half liters of beer, be decorated with a copy of the crew's mission patch, and contain the controller's name and MCC console position. By the time we finished, the beer mug had become grand and unique, containing crew signatures and Armstrong's words from the Moon landing.

Maureen started collecting the money, and within weeks she had over $5,000. We had moved beyond the normal coffee pot finances into the big time. She did not want to keep the money around the office so she opened a checking account in the Nassau Bay Bank across from NASA.

Everything was going nicely. The mugs were ordered and we had raised enough money so that we could afford to throw a party at the Hofbraugarten to christen the mugs when they arrived. Then the roof fell in. The NASA inspector general, located in the Manned Spacecraft Center, and two members of the regional inspector's office entered Maureen's office quoting the fines and jail time for violation of NASA directives on the use of the Apollo 11 astronaut badge. It looked as though Maureen, a young secretary, would be terribly old and poor by the time she got out of jail. By the time the bureaucrats were done, Maureen and I were charged with violation of many NASA directives. (The NASA seal, insignia, logo, program, and astronaut and mission operations badges are protected from commercial use or sale.) Maureen was more concerned that the NASA inspectors had confiscated her checkbook, and a lot of bills for the mugs were coming due.

Neil Armstrong asked Mike Collins to refer all of the mug data to headquarters, confirming that the Apollo 11 crew endorsed the design and gave us permission to use their patch. Once he saw the guns aligned on him, the inspector who started the flap backed off. He even purchased a mug for himself.

The lesson, as with any mission, was well learned. Over the years, we in the Flight Control Division managed to build the biggest party fund in NASA, and when it grew too big, we donated a lot of money to charity. We sold mugs, lapel pins, and sweatshirts, and threw good-sized parties at fancy places. Although the NASA legal folks watched us, we never had any further problems with the inspector general.

In 1999, on the thirtieth anniversary of the first lunar landing, we cast the mug for the final time, then broke the mold.

18

THE AGE OF AQUARIUS

April 1970

"Aquarius/Let the Sun Shine In," a song from the rock musical *Hair*, boomed from the stereo speakers of my Cougar daily as I pulled into the parking lot behind Mission Control. The song had temporarily replaced "The Stars and Stripes Forever" as my going-to-work music. The version sung by the group called the 5th Dimension was picked up by the Apollo 13 crew and controllers as symbolic of the energy and momentum of the Apollo lunar program. The song's signature words, "This is the dawning of the Age of Aquarius," symbolized the first mission of the new decade as well as the challenge and excitement of the increasingly difficult and risky lunar missions. When the Apollo 13 crew named their LM *Aquarius*, the song moved to the "top of the pops" for the controllers. The CSM was dubbed *Odyssey*.

Lunar exploration began in earnest after the pinpoint landing of Apollo 12. Mission targeting moved to more difficult and hazardous landing areas. The landing point for Apollo 13 was a target 3,000 feet in diameter, located north of a large crater dubbed Fra Mauro. The crater was located in a geologic formation known as the Imbrium Basin. The basin, one of the largest on the Moon, had been formed by a gigantic cosmic collision. Scientists hoped that samples of the material ejected during the collision would establish the date of the Imbrium event.

Veteran astronaut Jim Lovell commanded the mission. His experience on Gemini 7 and 12, as well as his being one of the first humans to orbit the Moon on Apollo 8, made him a logical candidate to lead a rookie crew. The LM pilot was Freddo (Fred) Haise, a member of the fifth class of astronauts, who had graduated from test pilot school in 1954. Fred knew the LM, especially the software, like the back of his hand. Ken Mattingly, the command module pilot, was a favorite among the controllers for his in-depth knowledge of every aspect of the business. But two days prior to the launch Ken was scrubbed from the mission because he had been exposed to measles. He was replaced by Jack Swigert, a member of the backup crew. During the pre-mission meetings and in training we had spent a lot of time with the backup crews, so Jack was no stranger to the MCC teams. After two days of refresher training he was ready to go.

I was the lead flight director on Apollo 13, a transition mission in many ways. The new flight directors, Griffin, Frank, and Windler, were pulling more weight, preparing to alternate the lead responsibility for the final four missions. Charlesworth had flown his last mission on

Apollo 12 and was forming the Earth Resources Project Office as part of a plan to apply space technology to Earth's problems. Operations was my business and I liked teaching the young controllers, watching them grow during their four-year training period as they progressed from the back room to the MCC main control room. Every new controller was assigned a mentor to test his knowledge, build his confidence, and prepare him for the painful and necessary lessons he would learn from SimSup. When controllers make it to the ranks of the front room and meet the flight director, they fully understand that the price of their admission is *Excellence*, and that a spartan set of standards will govern their conduct. Most of all they understand "that suddenly and unexpectedly we may find ourselves in a role where our performance has ultimate consequences." (*From The Foundations of Mission Control*, a one-page statement summarizing the values essential to a controller attaining excellence. The text was written by flight director Pete Frank. See page 655.)

Failure does not exist in the lexicon of a flight controller. The universal characteristic of a controller is that he will never give up until he has an answer or another option. By the time someone graduated to the front room consoles either he was ready — or he was gone before he got there.

The Apollo 13 flight director chemistry was unique. Windler and I were jet fighter pilots; Griffin flew as a radar operator. For the first time we were working together on a mission. Lunney, the fourth flight director, was the last of the original flight dynamics officers, the master of his craft.

April 11, 1970, Apollo 13

Milt Windler, a veteran of three Apollo missions, drew the launch flight director's assignment. He had earned his spurs as a test director in Kraft's recovery division and not in Mission Control. His transition was smooth. Absolutely unruffled at the console, Milt emulated Charlesworth's low-key and patient demeanor. He was fully in command of his team when the moment came to light the fire on Apollo 13.

The liftoff occurred at 13:13 Central Daylight Time, and proceeded uneventfully through first-stage flight. The five engines of the second stage of the Saturn V ignited and burned smoothly for five and a half minutes. Then the center engine unexpectedly shut down. Milt's team quickly reviewed the status of the remaining four engines, ran the computations for the new engine cutoff times, and passed them to the crew. When the second-stage engines shut down, the S-IVB stage ignited and got the spacecraft to orbit. After the CSM orbital check-out and updating of the trajectory parameters, Windler gave the Go for translunar injection. We heaved sighs of relief, thinking we had gotten through what probably would be the one major glitch in the mission.

The crew and control teams rapidly settled into the routine. During the early shifts, we watched and worked with Jack Swigert, calibrating his performance and finding him a very capable stand-in for Ken Mattingly.

During the translunar coast period both crew and controllers prepared for the events scheduled for lunar orbit when things would get quite busy. As the meticulous check-out of spacecraft and trajectory systems continued, the controllers settled into a state of relaxed alertness. The easy banter among flight director, team, and crew would leave a bystander thinking that none of these guys had a care in the world, when in fact they were maintaining gimlet-eyed focus on the job at hand while gathering their reserves for what lay ahead.

With the exception of the live TV broadcast from Apollo 13, my second shift of the mission was also uneventful. Mattingly had been pestering us for access to the MCC, his medical status still indefinite. I decided that if he was showing symptoms of measles at the time of the EVA, we would put him at the network console on the floor of the control room that was not being used on the mission, directly below us, giving him a chance to listen in but not exposing people to contagion. As the crew concluded its onboard TV broadcast just before 8:00 P.M. Houston time I glanced up to the viewing room, and could see Lovell's and Haise's wives and families leaving. Swigert was a bachelor.

Lunney's Black Team was arriving in the control room and there was a rising hum of conversation as the shift handover process began. After talking to his controllers in the trench, Lunney moved into the seat next to me, reading the flight director log for all events since his last shift. I began preparing my handover summary for him while we were getting the crew and spacecraft configured for the sleep period.

We zipped through the pre-sleep checklist, verifying that each system was set up to enable us to watch over the crew while they slept, monitoring the switch positions and dumping the telemetry records, making sure that once the crew members were asleep we did not have to awaken them. The flight activities officer got a verbal confirmation from the crew for the completion of each checklist page. With little else to do, I was following the checklist closely.

Earlier on the shift we had a worrisome but minor communications glitch. For a brief period, the CSM high-gain antenna did not work in either of two automatic modes and had to be positioned manually. Then when a spacecraft roll maneuver was performed the antenna abruptly locked up. Now, all of a sudden, it was working properly. There was insufficient time to troubleshoot this glitch prior to the crew's going to sleep. I hated to leave this as an open item for Lunney.

The crew continued the close-out, terminating the command module battery charge. During the previous sleep period, an alarm monitoring pressure in a hydrogen tank had awakened the crew. After considerable debate in the MCC, we did not reset the alarm out of concern that it might inadvertently trigger again and wake up the crew. As a result, a cryo pressure warning indication was illuminated in the spacecraft and also on Sy Liebergot's console. During the translunar phase we were in continuous voice and data communications, so Sy intended to stand watch over the pressures from the ground during crew sleep.

Liebergot was my EECOM, having moved up to the front line during Apollo 8. Now, after a year's experience, he was considered a veteran controller. He had a second glitch he was working. The telemetry gauge in oxygen tank 2 had been reading normal at 80 percent through the mission, then during our shift the gauge went through four rapid up-and-down cycles, finally failing and sticking at a constant reading of 100 percent. We no longer had a valid reading from the sensor.

Cryogenic oxygen and hydrogen mixed and reacted in the three fuel cells in the service module to provide electrical power. The reaction also provided pure water used for drinking and for cooling the CSM systems. The only other source of power to the command module was the three reentry batteries, normally used only for the final two hours of the mission. The oxygen and hydrogen, maintained in a liquid state at temperatures below –300 degrees Fahrenheit, were stored as liquids in spherical tanks insulated by a vacuum between the outer and inner walls. As the mission progressed, the oxygen and hydrogen went through a progressive change from a liquid to a gas.

At the time of launch the cryogenics in the service module tanks were a dense super-cold liquid, but now, two days into Apollo 13, the cryos were a thick soupy vapor, part liquid and part gas. Fans were located internal to the cryogenic hydrogen and oxygen tanks. The fans were periodically activated by the crew, at the request of the MCC, to stir up the mixture and allow precise tank quantity measurements. Heaters were located in the tanks to raise the tank pressure. The heaters could be activated either automatically or manually by the crew.

Sy Liebergot, wrestling with the oxygen pressure management problem and hoping to avoid an alarm during the sleep period, decided to request a cryo tank stir prior to sleep. The stir request was passed to the crew, with Swigert acknowledging the request. As Swigert started the stir at 55:53, Liebergot focused his attention on the TV monitors displaying the fuel cell currents to nail down the exact time the stir started.

(What we could not know at this time was that a design flaw existed in the heater circuit of the cryogenic tanks. During pre-launch testing, the heater thermostat switch closed and, due to the design flaw, the Teflon insulation on the wiring internal to the tank was damaged. When the tank was serviced for the mission, the bare copper wires in the tank electrical system were submerged in liquid oxygen. Two days after liftoff two of the three conditions for an explosion existed, gaseous oxygen and damaged tank insulation. All that was required to initiate an explosion was a spark.)

Nothing happened for sixteen seconds after Swigert started the cryo stir. Then, inside tank 2, a spark jumped between the wires of the heater circuit. The pressure in the tank rose rapidly. Preoccupied with monitoring the fuel cell currents, Sy Liebergot did not notice the oxygen flow measurements on all three fuel cells fluctuating slowly for eighteen seconds. Then the pressure in oxygen tank 2 began to rise rapidly, but failed to set off a high-pressure alarm in the command module or at Liebergot's console because the cryo pressure alarm had been disabled for the crew sleep. The tank pressure continued its upward climb, then dropped rapidly. The temperatures in the spherical tanks began to rise rapidly. One minute and fifty-three seconds after the stir began, there was a three-second telemetry data loss. When the data returned, the tank 2 pressure read 19 psi, temperature +84 degrees Fahrenheit. The normal pressure reading is 865–935 psi.

The time was now 55 hours 55 minutes and 04 seconds from launch.

Like rolling thunder, my voice loops came alive: "Flight, we've had a computer restart." Then in the blink of an eye, Swigert said, "We have a problem." Then other controllers chimed in with bad news. At this point Lovell uttered the ominous words: "Okay, Houston, we have a problem!" In both the MCC and on board the spacecraft, voices were normal, but heart rates had picked up. Seconds later Haise reported, "Right now, Houston, the voltage is — is looking good. And we had a pretty large bang associated with the Caution and Warning."

In the MCC, you can't see, smell, or touch a crisis except through the telemetry and the crew's voice reports. But you can feel some instinct kicking in when something very wrong is going on. By the time I heard Lovell's report, three controllers had related problems. I was wondering which problem Lovell was reporting, as he started relaying the long list of warning indications from the spacecraft displays. The reports and our experience indicated an electrical glitch. I believed we would quickly nail the problem and get back on track.

I was wrong.

A crisis had begun. Events followed in rapid succession, escalating and complicating the problems as the crew's situation became increasingly perilous. It was fifteen minutes before we began to comprehend the full scope of the crisis. Once we understood it, we realized that there was not going to be a lunar mission. The mission had become one of survival.

The reports continued, but nothing made sense. Each controller stared incredulously at his display and reported new pieces to add to the puzzle. It took extra seconds sorting out what was real and credible. It appeared we were losing our oxygen and with it the fuel cells, the major source of power. When that happened we would lose control of the main propulsion system. Nothing remotely like this had ever happened in simulation.

As we watched the command module's life-sustaining resources disappearing, like blood draining from a body, the voices of the crew were unbelievably calm and restrained. It was as if they were reporting something that was no big deal. From all sides of the cockpit, Haise, Swigert, and Lovell were continuing the dialogue, giving us the cockpit meter readings and warning light indications.

I had heard about the fog of battle, but I had never experienced it until now. The early minutes were confusing; all reports and data were suspect. Small firefights occurred as individual problems were corrected, but we had no sense of the big picture. With both electrical buses in an undervoltage condition, the crew was working independently of the control team to restore power to the craft. We were seconds behind them, slowly responding.

I remembered the call from INCO (instrumentation and communications), Gary Scott's call, that the antenna had switched beam width at the exact time of the power problem. I became convinced that we had an electrical short caused by another antenna glitch. Again I took the wrong fork in the road, believing we would be back on track shortly.

Five minutes after the event, the significance of the crew's words, "We had a pretty large bang . . ." hit me. GNC Buck Willoughby, unflappable, started speaking to me slowly, evenly, and without a hint of emotion, "Flight, have the crew verify that the Quad D helium valves are open. I suspect that the big bang shocked the valves shut, cutting off fuel to the attitude thrusters." Buck's call started me down a different path. On Apollo 9, I was flight director when the pyrotechnic shock occasioned while separating the CSM from the Saturn S-IVB booster closed the fuel valves. That gave us a few bad moments then. The bang heard by the 13 crew must have been awfully solid to do the same, closing the propellant valves. From this moment on, I proceeded more deliberately and methodically. We were five minutes into the crisis.

CapCom Jack Lousma, frustrated, came up on the loop. "Flight, is there any kind of lead we can give them? Is it instrumentation or have we got real problems or what?" Lousma echoed everyone's feelings. We were making no progress, virtually every controller still had problems, but no one could see a pattern in all this. It was like living a bad dream, with every event taking place in slow motion. The frustration of the crew and controllers was starting to creep into their voices. Everything we knew about our spacecraft, all that we had learned about the design, precluded the kind of massive failures we were seeing. The data told us we were looking at multiple simultaneous failures. Two, possibly three fuel cells were down, both oxygen tanks depleted, and we had an undetermined attitude control problem that was pushing the two spacecraft around. Soon we would lose power. When that happened, we would lose everything.

The teamwork in the MCC under a crisis is spectacular. While Liebergot, Lousma, and I worked the electrical options with the crew, the remaining controllers were making their inputs to the CapCom, correcting their smaller problems. While sensing the urgency of the electrical problems, they tended their own business, protecting their systems and giving crisp, brief reports so as not to disturb or aggravate the resolution of the main problem — whatever it was.

INCO Gary Scott watched the antenna signal strengths like a hawk. He knew that the crew did not have time to point and select antennas. Gary recommended a fallback to the less powerful but adequate Omni antennas. There are four Omni antennas on the spacecraft. Through the critical first hour, until help arrived, he called out each antenna switch, protecting this vital link as the docked CSM and LM drifted out of control and were pushed around by some force we couldn't identify. If he had missed once, we would have lost communications, diverting the attention of crew and control team from critical tasks. Scott, like many others, made hero category by his patient, timely, undistracted management of the data stream while everything else was falling down about my team.

It was now ten minutes into the crisis; all the bosses had gone home after the crew's TV show. I needed to notify top management that we had a hell of a problem on our hands and that we didn't fully understand what it was. Turning to Lunney, I asked him to call Kraft. Glynn handed me the phone as Chris's wife, Betty Ann, answered. In response to my request she explained that he was in the shower. I said, "Betty Ann, get him out, I need to talk to him." When a still-dripping Kraft got on the phone, I told him that we had a major electrical problem and that I believed we had lost one or more fuel cells. I concluded on a somewhat desperate note: "Chris, you better get out here quick; I think we've had it!"

GNC and GUIDO, Willoughby and Will Fenner, had been quietly watching the crew struggle to control the spacecraft attitude and avoid "gimbal lock." This grave problem would come about if the rings that support the whirling wheels of the gyroscope all aligned in the same position. We would then no longer have a usable reading from the gyroscopic platform. In gimbal lock we would be unable to maneuver or point the spacecraft. We would be literally adrift in space until the crew took a fix on certain stars to realign the gyros, much in the way a nineteenth-century sailing ship figured out its position. Every time the crew got close to the danger point, Willoughby, in a hushed but forceful voice, would call, "Flight, they are getting close to gimbal lock." Lousma would advise the crew, who then used the CSM hand controller and attitude jets to maneuver away from disaster.

The team was now functioning well; we were fourteen minutes into the crisis, fighting a delaying action until we figured out what was going on and what to do about it. Most of the problems seemed to rest on Liebergot's shoulders. He was responsible for the systems needed to sustain life, power, water, oxygen, and pressure. But no matter what we tried, we were unable to stanch the hemorrhage of the fuel cell oxygen reactants.

Then, abruptly, all the pieces of the puzzle

came together. Lovell reported, "It looks to me, looking out the window, that we are venting something." Then with emphasis he said, "We are venting something out into the — into space — it looks like a gas!" A shock rippled through the room as we recognized that an explosion somewhere in the service module had taken out our cryogenics and fuel cells. The controllers felt they were toppling into an abyss. Needless to say, the lunar mission was now a NoGo. The only thought on my mind was survival, how to buy the seconds and minutes to give the crew a chance to return to Earth.

Now I was damn angry that I had wasted fifteen precious minutes by not assembling the pieces earlier. I should have seen it. Somewhere, somehow, an oxygen tank exploded and it caused a lot of collateral damage. The feeling of self-reproach passed quickly; I became icy cold, my mind reaching out for options as my training kicked in.

Our objective from here on was survival. The crew's only hope was Mission Control. My team had to start the turnaround. With two flight controller teams in the room, the level of chatter was distracting. My team needed to get back on the voice comm and get focused. I finally took charge. Standing up I yelled across the top of the consoles, "Okay, all flight controllers, cut the chatter. I want every member of the White Team to settle down and get back on the voice loops — the rest of you shut up!

"Now, let's everybody keep cool. The LM is still attached, the spacecraft is good. So if we need to get back home, we have the LM to do a good portion of it with.

"Let's make sure that we don't blow the [remaining] command module electrical power with the batteries, or do anything that would cause us to lose fuel cell 2. We have to keep the oxygen working and would like to save the attitude control propellants. We are in good shape to get home.

"Let's solve the problem, team . . . let's not make it any worse by guessing."

The team focused on keeping the crew alive and finding a way to get them home. Our determination was evident as we calculated the limited resources available in the damaged spacecraft. For the moment the power and the oxygen in the CSM could keep the crew alive but the LM was ultimately the only safe haven, even though it had been designed to accommodate only two men for two days.

I knew I had to move quickly to stabilize the situation and then hand over the remnants of the mission to Lunney's team. I wanted time to review all the data. I had the absolutely chilling fear that I had missed something important. I hoped that some fresh minds might pick up on it. I wanted to get the White Team off-line, get them together in a quiet corner, nail down the cause, and then start on a plan to rescue the crew. We were the lead team. It was our responsibility to take over management of the crisis.

My console was a mess, littered with schematics, procedures, the console log, and cigarette butts. Lunney's team was scurrying around the room preparing for handover. Clint Burton, Liebergot's replacement, nervously awaited his turn in what had become the hot seat. Ed Fendell, who managed communications, joined Gary Scott at the console. Together they would keep up the communications, the key to an orderly transfer to the lunar module. Fendell had been at home and had just happened to have the radio on. He heard the news, jumped in his car, and came in. Racing his Corvette through the back streets of Clear Lake, Fendell arrived in a cloud of dust and parked in the middle of the exit lane. He joined in the battle with Scott, making sure that communication with the crew would be maintained without interruption throughout the crisis. I was glad to see him. I did not know how much longer Scott could continue running solo and pitching a perfect game with the communications.

Kraft arrived as we were starting the second phase of the power-down of the CSM. Liebergot signaled the next phase of the withdrawal with a simple suggestion. "Flight, I think we better get going in powering up the LM. We're running out of time." He then gave Lousma the call to have the crew secure the command module entry oxygen system, a small oxygen bottle used during the final two hours of the mission. We were putting together a lifeboat; what did we need to make it work?

Kraft plugged into my console. I glanced up momentarily and said, "Chris, we're in deep shit." Moments later, Liebergot began to lay out the bad news, the whole nine yards: "Flight, I hate to tell you this, but I think we've lost fuel cells 1 and 3." I nodded, still thinking that maybe fuel cell 2 and one of the oxygen tanks might be salvageable and could be added to our get-home resources.

Lunney had been down in the Trench reviewing the get-home options. At the time of the explosion, Apollo 13 was 200,000 miles from Earth, 45,000 miles from the surface of the Moon. We were entering the phase of the mission where lunar gravity becomes stronger than Earth's gravity; we call it "entering the lunar sphere of influence."

When Lunney came back up to the console, Kraft stepped down from his position behind me. In a hushed tone, Glynn said, "I had the Trench look at maneuvers with ignition about three hours from now. We have two basic options, a direct abort and one going around the Moon. The fastest direct abort gets us home in thirty-four hours. We fly in front of the Moon but we have to jettison the LM and use all the main engine fuel. We have several options that fly around the Moon. The best one takes two days longer, but we don't use the main engine and we can keep the LM." We rapidly went through the mathematics; the lunar module was good for two crewmen for two days. A quick estimate using the LM powered-down checklists and taking the path around the Moon left us at least thirty-six hours short on battery power.

Windler, the leader of the Maroon Team, now joined us at the console. He believed the shortest and fastest path back to Earth was the best. He seemed to favor the direct abort. Lunney and I disagreed. I said, "I don't want to jettison the lunar module. We haven't nailed down the exact cause of the explosion or the extent of the damage. The main engine or its control systems may have been damaged. We need more time to work out the procedures for the return."

Lunney chimed in, "Keeping the LM buys time. We don't have a second chance and if we jettison the LM we cut off a lot of options. Whatever we do, we damned well better do it right."

I wrapped up the discussion: "We should hold on to the lunar module and go around the Moon and take our chances with the LM power. I believe we will come up with a plan that gets us home."

Debates among flight directors are not uncommon. We all arrived at the flight director position along different paths. Given a few minutes, the rapid pooling of experience is often the quickest way to firm up our direction. The discussion was brief, intense, and conclusive. I wanted to get every option and opinion out on the table before we selected the return path. The Trench was nervous about pulling off a direct abort so close to the Moon. I knew Lunney would fight to the death for the long return after talking to the troops in the Trench. Controllers clustered about the console as we talked, recognizing a decision was imminent. Bostick and Deiterich were joined by my FIDO, Bill Stoval, from the Trench. Lousma crowded in, representing the crews.

I vividly remembered the EVA flap from

Gemini 9, when I left instructions on the console about what to do only to have top management intervene, thus putting us on a risky course. With that in mind I was not about to leave the trajectory plan undefined. We had all the players at the console and I did not want to open the subject to further debate. I looked directly at Kraft. "Chris, I don't trust the CSM service propulsion system. It's in the back end, where we had the explosion, and we won't know if it is good until we try it. Then it may be too late. We need to buy some time to think and to build the come-home procedures. I believe we can find the power. Our only real option is to go around the Moon."

Kraft had been listening; he looked at Lunney and then nodded. Lunney said, "I agree. The direct abort closes out our options. We should keep the lunar module."

The Trench had been standing by, faces grim, hoping they would not be told to pull off a direct abort at this late time. When they saw the decision coming down in favor of their preferred option, they smiled for the first time in a long while, nodding in agreement and relief. Through some miracle, a burst of intuition, something we had all seen, heard, or felt now told us, "Don't use the main engine." To this day I still can't explain why I felt so strongly about this option.

We did not have much time to debate, and I was glad that there had been immediate agreement. Many people were unaware of the options, but I believed that the systems controllers thought I had made the wrong decision.

They favored the fastest way home, a direct abort.

Missions run on trust. Trust allows the crew and team to make the minutes and seconds count in a crisis. In the scramble to secure the command module, we didn't have a chance to brief the crew or even get their opinion on the return path. In my mind I knew the crew would fight to hang on to the lunar module. I felt Lousma, as their representative, would speak out if needed. Kraft went up to brief the NASA brass, who had congregated in the viewing room, on our plan to get Apollo 13 home. The Trench returned to their consoles to start developing the return trajectory plan and brief their back room. The systems guys would have to find a way back with what we had.

Fifty-three minutes after the explosion, the plan was becoming clear. The retreat to the LM was proceeding, the trajectory path chosen, and the handover to Lunney was accomplished. I signed off in the log at 57:05 mission elapsed time, one hour and ten minutes after the explosion. It had been the longest hour in my life.

When I left the control room the remaining cryo oxygen tank pressure was down to 100 pounds per square inch. Time was running out. In less than two hours the command module would die. The situation was not yet stable, but the direction was clear. Lunney presided over the retreat to the LM, saving as many of the CSM resources as we could. After transferring the navigation data to the lunar module computer, Glynn, with a resigned shudder, told the crew to turn off the CSM systems and the computer.

Walking down to the data room to meet my team, I thought of the work ahead. We had to put together in a few hours a set of procedures that normally would take weeks to work out. We would operate outside all known design and test boundaries of the space systems.

The White Team (called the Tiger Team by the media) assembled at 10:30 P.M. CDT in the data room on the second floor. The room was large, about thirty by thirty feet and essentially bare. The only furniture was gray government tables, two overhead TV monitors, and a single intercom panel. The controllers used the room occasionally for team meetings, systems troubleshooting, and working sessions with design engineers. When I walked into the data room I was greeted by my augmented flight control team. The controllers' arms were filled with orange-colored recorder paper. They were kneeling on the floor, the paper strewn all over the place. In pairs, controllers marked the time annotation, measuring the squiggly traces and rapidly scanning for anything that could pinpoint the cause of the explosion. The room was noisy and smoky, the tension in the air palpable. Engineers and program office personnel, as well as key managers from Grumman, the LM designer, and North American, the CSM designer, were sitting on the tables, since there weren't many chairs. It was a working room, used principally when there was trouble. Since we had no shortage of trouble, everybody knew it was the place to be.

There were three pieces to the puzzle of the return journey. The command module was the reentry vessel; it had the heat shield but very limited electrical power. The lunar module, which would be used as the lifeboat, was designed to support two crewmen for two days on the lunar surface and was our source for power, life support, and propulsion. The third piece was the damaged service module, the true extent of the damage still unknown. With these pieces we had to fly around the Moon, perform maneuvers, support three crewmen for more than four days, and then, at the very last moment, evacuate the crew into the command module. Then we had to separate the pieces so they would follow different trajectories for reentry.

Electrical power, water, and oxygen were critical. There was no way to stretch the power unless the Trench came up with options to speed up the return after we passed the Moon. The return plan split into two phases: In less than eighteen hours we needed a maneuver plan and procedures to speed up the return journey. Once this was completed we needed the entry plan sixty hours later. Everything else had to fit in between these two critical events. So far all we had done was to buy the time to give the crew a chance; now we had to deliver. I said a prayer for the crew and the teams that had to do the work.

Walking to the front I motioned to three of my controllers saying, "Aldrich, Bill Peters, and Aaron, come on up front where everyone can see you. The rest of you knock it off and find a place to sit." I had worked many missions with Aldrich and Aaron and knew their capabilities. Peters, the TELMU on Griffin's gold team, I knew as one of the best analytic minds in flight control if you gave him a bit of coaching. Some of the controllers rolled up their recorder paper and moved to sit on the tables surrounding the room. Others curled up cross-legged and sat on the floor and still others continued to roll out the records, reading the timing marks and placing notations on the edges.

The room was getting crowded and noisy as even more people tried to force their way inside. Quickly scanning the crowd, and then motioning to my White Team controllers, I said, "We have to cut this team down to size and get some order. There is plenty to do and this room is too full to get anything done. I want the White Team to look around the room and if you see anyone that you don't need, send them back to their consoles." Several of the controllers, program managers, and spectators left voluntarily, gingerly stepping around the recorder paper strewn on the floor.

I turned to Chuck Stough, my trusted flight planner. "Chuck, you're the recorder. I want you to get with the Trench and the flight planners and build me a work plan on every decision we need and when we need to do it." Then I addressed the people remaining in the room.

"Okay, team, we have a hell of a problem. There has been some type of explosion on board the spacecraft. We still don't know what happened. We are on the long return around the

Moon and it is our job to figure out how to get them home.

"From now on the White Team is off-line. Lunney, Griffin, and Windler will sit the console shifts. We will return only for two major events. The first will be a maneuver, if we decide to do one, after we have passed the Moon. The second will be the final reentry. The odds are damned long, but we're damned good.

"My three leads will be Aldrich, Peters, and Aaron. Make sure everyone, and I mean everyone, knows the mandate I'm giving them. Aldrich will be the master of the integrated checklist for the reentry phase. He will build the checklist for the CSM from the time we start power-up until the crew is on the water.

"John Aaron will develop the checklist strategy and has the spacecraft resources. He will build and control the budgets for the electrical, water, life support, and any other resources to get us home. Whatever he says goes. He has absolute veto authority over any use of our consumables.

"Bill Peters will focus on the lunar module lifeboat. There are probably a lot of things we have not considered and he will lead the effort on how to turn a two-man, two-day spacecraft into one that will last for four days with three men.

"Whatever any of these three ask of you, you will do.

"Now, I'm addressing myself to the program office and design engineers in the room. Aldrich, Aaron, and Peters need numbers, answers to questions, and unlimited access to your resources. They will ask for things you never thought you would be called on to do and to answer questions you never expected to be asked. I want nothing held back, no margins, no reserves. If you don't have an answer, they need your best judgment and they need it now. Whatever happens we will not second-guess you. Everything goes in the pot.

"My message to everyone is: rely on your own judgment, update your data as you go along. If you are not the right person, step aside and send me someone who is. When you leave this room you will pass no uncertainty to our people. They must become believers if we are to succeed."

I moved to the blackboard. "Okay, now let's get down to work and come up with our initial shopping list. Then I want Aldrich, Aaron, and Peters to select a work area and pull their pieces of the job together."

In real time I used the same brainstorming techniques used in mission rules or training debriefings, thinking out loud so that everyone understood the options, alternatives, risks, and uncertainties of every path. The controllers, engineers, and support team chipped in, correcting me, bringing up new alternatives, and challenging my intended direction. This approach had been perfected over years, but it had to be disciplined, not a free-for-all. The lead controllers and I acted as moderators, sometimes brusquely terminating discussions with "Close it," or "It'll take too much," or "We've tried it before, it didn't work!" Often the response was simply, "Save it for our last-ditch try."

With a team working in this fashion, not concerned with voicing their opinions freely and without worrying about hurting anyone's feelings, we saved time. Everyone became a part of the solution. For the next fifteen minutes, we brainstormed out loud. Astronauts and engineers were assigned to teams, rooms selected, and working schedules arranged. The session concluded with a brief update from the Trench on the trajectory options. Then I took a deep breath and concluded the meeting.

"Okay, listen up. When you leave this room, you must leave believing that *this crew is coming home.* I don't give a damn about the odds and I don't give a damn that we've never done anything like this before. Flight control will *never* lose an American in space. You've got to believe, your people have got to believe, that this crew is coming home. Now let's get going!"

In the control room upstairs, Lunney had completed the evacuation of the crew to the LM and was preparing for a small maneuver that would place the spacecraft on a path to return to Earth. After completing the Tiger Team meeting, I went back to the control room to get a status update from Lunney. Glynn was now concerned about powering down the navigation system. He had been advised by astronauts Tom Stafford and Gene Cernan, in the simulators, about the difficulties of performing an alignment while docked, using the LM optics. The navigation platform (gyros) is aligned using stars and a sextantlike device. Sunlight reflecting off the CSM made it difficult to recognize the navigation stars.

The crew's report that they could not recognize stars due to the cloud of debris surrounding the spacecraft further convinced Lunney that we needed to keep the LM computer and display system powered up until we completed the get-home maneuver. Then we could power down and coast back to Earth. The risk was high, trading electrical power to keep the computer and navigation system on line against the possibility that we might be unable to realign the navigation platform. Without a navigation platform we could not perform the maneuver to accelerate the return to Earth. If we ran out of power, we could not get into position for reentry. We were playing showdown poker; we needed to get a better hand.

I tracked Aaron down and gave him this new complication, telling him that he should make it his number one priority. Aaron said, "I don't have to run the numbers all out, but I can tell you that if you take this approach, you will have to power down to a survival-level, limited telemetry, a very late power-up for entry, and the possibility that some of the systems may fail due to the cold. It is going to be tough on the crew. Some of the systems may freeze up."

"John, the best judgment we have," I replied, "says that we cannot realign the navigation system. I think this is our only option. Get on it!" Aaron's problem just got a lot tougher. His group set up their camp in the support rooms adjacent to my home base in the data room. John Aaron would not return to the console until the final shift four days later.

The next set of estimates were grim. We were twenty hours short in electrical power and thirty-six hours short in water for the return trip. I kicked myself for overlooking water as the most critical resource. Water is used for cooling LM equipment, for drinking, and for food preparation. I dug rapidly into a "how goes it" sheet Peters had made for me. We had water for about sixty hours available in the LM descent stage and fourteen hours in the ascent. We weren't even close. I was now grateful for the time we had spent before the mission in the mission rules, flight planning, and procedures meetings developing options and work-arounds for all conceivable spacecraft failures. We knew when the chips were down we could use the command module survival water, condensed sweat, and even the crew's urine in place of the LM water to cool the systems.

There is no such thing as a first team in mission control. All teams must be balanced, equally competent, equally capable of sustaining the effort. Over the next four days the flight control teams would routinely shift every eight hours, with Lunney, Griffin, and Windler steering the return course. But at this moment the battle for the return of Apollo 13 shifted to the back rooms and factories where the components were assembled and tested. We needed their data and we needed it fast. We needed tests in the laboratories and crews in the simulators to prove the procedures we were writing. Engineers hastily recalled from sleep and still rubbing their eyes were given the challenge to get the tests running, dig out the data, bring up the simulators.

My immediate job now centered on developing the procedures for the get-home maneuver using the engine designed for the lunar landing. The maneuver would have to take place two hours after the CSM and LM passed the lowest point of the orbit around the Moon (pericynthion). My next set of decisions involved determining how aggressively to pursue a rapid return to Earth. The most aggressive option would cut twenty-four hours off the return journey and require jettisoning the damaged service module and using all of the LM descent propellants. Several other options were available and there were advocates of each option. Glynn and I kept our powder dry, abstaining from the debate, knowing Kraft would ultimately turn to us for the final decision. Griffin strongly pushed for any option that would get the splashdown point moved from the Indian Ocean to the Pacific, where we had full recovery capability. We vetoed two faster return options as too time-critical, leaving no downstream options. Some folks lobbied to jettison the service module, but since we had not worked out jettison techniques, that was a moot point.

Almost since the instant of the explosion

Glynn and I were pretty much in agreement on not using the main engine. We seemed to be running with some intuitive link that surprised our team members. When we needed to synchronize our thinking we would pass messages via runners or a brief phone call or meeting. Kraft was running interference for the flight directors and we agreed to let the NASA managers play with the options and alternatives. During a crisis every boss wants to get in on the act. Letting them think like flight directors for a few hours kept them out of our hair, and prepared them for when we laid the real plan on the table. The flight directors would get together privately, generally in the second-floor viewing room over a cup of coffee, and discuss our positions before we went to management meetings. The last thing we wanted to do was let the brass think there was any real disagreement in our group or uncertainty about our recommendations.

We were ten hours into the crisis when Lunney handed over to Griffin. I joined them at the console, reviewing the maneuver options in preparation for Kraft's meeting in the viewing room on the second floor, where the shift change briefings for management were held. By now Glynn and I had settled on the maneuver option that got us back to a landing in the Pacific at 142 hours MET. This was a middle-of-the-road option, cutting only twelve hours off the return journey. Because we had doubts about our ability to check the LM navigation with crew fixes on stars, we chose this approach, which gave us a greater margin for error in maneuvers and reserved some propellant for correction maneuvers on the return. The road to safety would prove to be long and cold and dark.

19

COMING HOME

April 14, 1970

As I arrived with the White Team at 3:00 P.M. Houston time for the get-home maneuver I glanced up at the viewing room and chuckled. Two members of the press (one print and one TV) and their public affairs escort were now firmly compressed in a small glass booth about the size of a large desk at the far corner of a viewing room. The reporters, their noses pressed to the glass, were listening to our communications. Headsets, reference data, pencils and paper, and all kinds of the tools of their trade were visible on the desktop. Jack Riley was the PAO chaperoning this duo in the viewing room. I had no doubt he would have preferred to be on the floor. I just hoped they were pumping air into the room or we might have another emergency to deal with. From now on we were living in a fishbowl. Everything we said and did was going directly into the homes of America and the world.

Griffin had set us up well for the maneuver and, after a brief handover, the White Team was back on console. As the spacecraft passed the

Moon, the lunar gravity pulled it in an arc toward a rendezvous point with Earth. Our job was to hasten the rendezvous. We briefed the crew on the maneuver procedures, mission rules, consumables, and the return strategy on the remote chance we would lose communications during the return. Throughout the briefings I continually stressed to the controllers and to the crew that the burn start time was not critical. We were already on a return path and if anything did not look right we could NoGo the burn until everyone was confident about proceeding with it. I had a high degree of confidence in this maneuver, since it was a variation of an LM engine burn we had executed on Apollo 9.

There was an air of expectancy in the room as the maneuver time approached. The viewing and control rooms were filled to the brim. The maneuver was a turning point in the struggle to get our guys home, anchoring the return time and placing many tough decisions behind us. Two hours after Apollo 13 passed behind the Moon, the crew ignited the small descent engine designed for the Moon landing, burning for four and a half minutes and increasing the return velocity by almost 1,000 feet per second. The execution by the crew was perfect, fixing the landing time for 142 hours and moving the landing point from the Indian Ocean to the Pacific near Samoa. The aircraft carrier *Iwo Jima* was dispatched to the landing point to recover the crew and spacecraft.

Getting back on the console with my team felt good. By the process of elimination, we were whittling down the work yet to be done and improving the crew's chance of survival. I had one more thing to do before finishing the shift. The systems controllers had been watching the temperatures at various locations on the LM (and by inference the CSM) since we had started using the LM as a lifeboat. We initially believed that a random drift would keep the temperatures in limits, and conserve power, water, and propellant. The designers disagreed. Prior to the return maneuver I spent some time in the spacecraft analysis room reviewing their data. I left the meeting convinced that we had to execute a passive thermal control (PTC) maneuver before we powered down and sent the crew to sleep.

The PTC is a kind of rotisserie maneuver that slowly spins the spacecraft on its long axis so the sun can heat all sides. The maneuver is used routinely to ensure equal heating of all surfaces and the systems under the skin of the spacecraft. Setting up this roll maneuver is not easy. After getting properly oriented the spacecraft must become perfectly motionless, then the small control jets are briefly fired, setting up the spin. If the procedure is not done perfectly, the spin will rapidly turn into a wobble and diverge from the sun line. This procedure had never been done in a docked configuration using the LM jets, which were not favorably located for this procedure.

As I was discussing the procedure with the White Team, Slayton and Kraft approached the console. When Slayton growled a hoarse "Gene," it was obvious they had something on their minds, so I stood and turned to talk to them. Slayton didn't waste a second. Jabbing me with his finger, he said, "I want you to get my crew to sleep. They are too damn tired, they are going to make a mistake."

I, too, was tired as I looked into Slayton's grizzled face; his heavy black beard stubble was starting to show. We glowered for a few seconds, but before I could respond, Kraft started in. "I want you to get the spacecraft powered down. You're calling it too damn close on the batteries!" My team was approaching our thirty-fourth hour; we also were running out of gas. I turned to Slayton and snapped, "Crew sleep and power-down are gonna have to wait. We won't get them home if we let everything freeze up. I'm gonna do the PTC." Slayton and Kraft were not used to being shouted at. Before they could respond further, I motioned them away and returned to my seat at the console. "CapCom, read the crew the PTC procedures," I ordered.

Slowly, Kraft and Slayton retreated to a position at the console above me. Max Faget, the chief of engineering, sat down next to Kraft. I think spacecraft analysis had sent him out to lobby on behalf of our plan if the argument got too heated. Faget's earnest Cajun voice carried. "Chris, that's the right thing to do," he said. "There is no use getting the spacecraft back if the systems don't work when we reenter." I envisioned Max patting Kraft on the hand, telling him to settle down.

Every second of delay stole time, power, and badly needed sleep for the crew. The spin-up maneuver had to be perfectly set up, very deliberate. Doing it using the LM jets was like trying to thread a needle with bad eyesight. After about forty minutes the crew fired the thruster to start the roll. We watched the resultant motion, and within minutes it started wobbling. To head off a repeat visit from Kraft and Slayton, I roared, "Okay, CapCom, tell the crew we're gonna have to do it again!" I could hear Kraft and Slayton grumbling on the console above me, probably muttering in my direction, "We told you the crew was too damn tired!" This was a good test of the flight director's mandate. I respected both Kraft and Slayton for not second-guessing me and letting me get on with my job.

The second spin-up attempt worked, and we initiated the complete power-down of the spacecraft. It was now twenty-six hours after the explosion. The crew and my team could finally get to sleep. I tried not to think about how cold it was going to get for the crew, but I knew we had decided on the right option, staying powered up until we went around the Moon.

With the shift over, I called a brief meeting of the White Team in Room 210. Aaron's battle plan had the crew powering down to a twelve-amp load (about a quarter of the power consumed by a household vacuum cleaner) for the return journey. It would be an ordeal for the crew. It would get cold, damned cold, but there was nothing we could do. Crew comfort was our last priority; they would have to tough it out. The power numbers had improved as a result of the work of Peters and Aaron, but we were still too tight on water. When we ran out of LM water we planned to use waste water and urine for cooling, if needed.

Engineering gave a status report indicating that they were close to a solution on another problem. The crew's breathing was slowly poisoning the cabin atmosphere with carbon dioxide. We had run out of the cylindrical air scrubbers used in the lunar module, and engineering was testing an adapter for the square command module canister that was being fabricated from cardboard, a plastic bag, a sock, and a hose from one of the crew's pressure suits. You have to picture a plasticized flight plan cover, to funnel air flow, curved over the top of a lithium hydroxide air scrubber (for removing CO_2) and a hose attached to the scrubber's bottom, which in turn ran down to a small fan, which pulled air through the scrubber and sent it through the sock, which served as a filter. The device was all held together by duct tape, a commodity which, fortunately, was always carried in the spacecraft.

By the time we arrived at this rather bizarre but functional contraption, we had been awake for a day and a half, so I told the White Team to get six hours' sleep. Then we would start working out the final set of procedures for the reentry phase. I had developed a habit on previous missions of resting in the viewing room when we had problems. The room was as cold as a meat locker, quiet except for the crew and flight director voice loops, and with few occupants except during major events. I staked out the upper corner of the viewing room as my home base when I wanted to rest, and after a thirty- to forty-five-minute catnap generally got back on track quickly. Since there were a lot of people in the third-floor room, I went down to the second-floor viewing room. It was also close to the action if someone needed me.

The final phase of the struggle to return the crew now began. Flight Control had fought a delaying action. We had stabilized the situation and protected the options. We had a pretty good idea of the resources available in both spacecraft. The show now belonged to Aldrich, Peters, and Aaron. Their job was to manage the resources, trade off the options, build margins wherever possible, and finalize the detailed procedures for the final entry phase of the mission. If ever there was a trio prepared for battle it was this one.

The ebb and flow among the design, test, and operations communities provided answers to the questions we had yet to ask, problems we had yet to identify. Aaron Cohen and Owen Morris, the NASA spacecraft program chiefs, rolled up their sleeves and joined with their counterparts Dale Myers from North American and Tom Kelly from Grumman. Together they directed a superb effort to solve a complex technical problem in a very tight time frame. These four engineers were the highly respected generals who commanded the engineers in the factories, laboratories, and test facilities. I believed that this team could move mountains. The flight directors had worked with all of them during the spacecraft redesign after the Apollo fire and subsequently in preparation for the missions. The trust among program manager, designer, and mission control was absolute.

Added to this respected group were two other great engineers, Don Arabian and Scott Simpkinson, whose pedigrees traced back to the early days in Mercury Control. They were well versed in real-time troubleshooting and were fully aware of the high-stakes poker we were playing. Above all, they knew that you had to have answers before the clock ran out.

This task force worked in the SPAN (spacecraft analysis) room, focusing on only one thing — how to get the crew home. They provided the missing pieces we needed. The handovers between engineering and operations were smoother than on an Olympic relay team and we did it repeatedly for almost four days. There were a lot of heroes but the SPAN team never got the gold medal and the recognition it deserved.

Aldrich was the scribe, watching the clock, assembling the pieces, listening to the debates, then deciding when enough was enough and it was time to put the plans on paper. Aaron was the accountant, keeping a meticulous balance sheet on the precious resources. Aaron became critical for power, Peters critical for water. Aaron checked every procedure entry, exercising his sole and absolute power of veto, often sending the controllers back to square one, telling them, "Your input was not good enough. Give it another shot and be back to me in an hour with your bottom line." Aaron, with his veto authority, soon became the dominant player in the return planning. The LM water available for cooling dictated an extremely low power level during the return journey. As a result it soon became clear that we could make it home with the LM battery power.

When Aaron recognized we would now have some power to spare he wanted to recharge the command module batteries. The three CM batteries would be the sole power source for the final hours of reentry. Since the batteries had provided the CSM power in the minutes after the explosion, they were no longer fully charged. Aaron wanted to find a way to charge them to maximum capacity.

When the two spacecraft were designed, it was never envisioned that we would need to charge the command module batteries from the LM. But now the controllers started looking at ways to use the LM heater cable in the reverse direction to charge the batteries for the final entry phase. Aaron and Aldrich now started bartering with Peters for the excess power in the LM batteries. The controllers were intensely debating the risks to both systems, trading off options but keeping an eye on the clock. Aaron finally resolved the issue: "We're going to charge the CSM batteries. I can't see leaving any power in the LM when we jettison it. I want a test rig set up to verify the procedures and to measure the power loss during charging." With the decision made, he turned to the SPAN team to set up a test rig to prove the procedure.

While we labored in Mission Control, SPAN

continued to dig out the answers and give us their best judgments about tough, critical questions that would lead to irrevocable decisions. "How cold can the thrusters get and still fire?" "How many amp hours are really in the battery beyond the spec values?" "We don't want to chance skipping off the Earth's atmosphere because our trajectory is too shallow — how critical is the reentry angle?" These questions triggered other questions; discussions of alternatives abounded; engineers wanted priorities. The engineers needed to know how their piece of data fit in to that of other engineers working on related problems.

A 100 percent correct answer, too late to be of use, was worthless. The White Team needed answers quickly to develop the procedures, integrate them in the simulators, and voice them up to the crew. Personally, I wanted a few hours to sit and think before the White Team and crew started the final eight hours of entry.

I don't think Aaron got any sleep in the last forty-eight hours. He had delegated well, but he knew where the buck stopped. His intuition was incredible. He kept turning up at the place where the logjams were building. With a few words he cleared the jam, then moved to another room, another debate. His prescience was almost mystical.

Throughout this period, astronauts in simulators tested the entry procedures, looking for traps that could endanger a near-freezing, deadly tired, and dehydrated crew. We all knew that cold and dehydration impair cognitive and motor responses — and it was now damn cold in the spacecraft.

In the final thirty-six hours the White Team came together at four-hour intervals with Aldrich, Aaron, and Peters to review the progress. Buck Willoughby, my GNC officer, was concerned about his thrusters. The LM would provide attitude control until jettisoned, then control would switch to the CM. Without heat since the explosion, the CM thrusters were dangerously cold, the propellant valves sluggish. Willoughby wanted a "hot fire" test to make sure they were all working before separating from the LM. The Trench joined in supporting the request. Slayton lobbied to power up early, using excess power to warm the spacecraft and his crew.

I vetoed most departures from the agreed procedures, stating that we had to keep them simple, and I wanted to be able to function in case of an LM battery failure. I intended keeping everything possible in reserve until I knew we had it made. Then and only then would I consider other options. The flight planners started a shopping list to be used when power became available. The hot-fire and early power-up were put at the top.

Since the White Team would handle the final shift for reentry, my deadline to Aaron and Aldrich to complete the procedures was landing minus twenty-four hours. As the deadline approached, the crews in the simulators wanted more time to check out the "final-final" set of procedures, which were in the tenth revision. Thirty-nine pages in length and containing more than 400 entries, they were the ticket home for our crew. The astronauts in the simulator were bothered by the continual changes and the frequent updates. They wanted a run-through with the final set of procedures. I froze further changes to the procedures and agreed to give the simulator crews six more hours to give me their okay.

With this delay, Lovell finally showed his exasperation with the entire process. The crew had been living in an icebox that was hurtling toward the Earth. Other than a brief overview of the intended sequence of the final eight hours, Mission Control had given them nothing but the reassurances that "the procedures are coming along." Lovell wanted specifics, not vague reassurances.

Aldrich kept the master copy of the procedures in his personal possession, identifying each update by a revision number. He guarded them as if they were the tablets Moses brought down from Mount Sinai. Others may have had mark-ups, but his procedures were the ones that would be executed. His great fear was that he would misplace them as he moved between the meeting areas. At the time of Lovell's prodding, Aldrich was working on the third revision for Thursday, April 16. We were less than twenty-four hours from entry.

In the final hours the flight planners, John O'Neill and Tommy Holloway, became the last link in the chain to get the crew back to Earth.

They established a loop between the crews in the simulators, the controllers, and the work being done by Aldrich and Aaron. They tracked the instructions voiced by the CapCom to the crew. Their checks and balances virtually guaranteed that in the rush to brief the crews nothing would be overlooked. They were the guardian angels, always hovering near and making sure that we gave the crew the right information at the right time.

April 16, 1970

Shortly after 5:00 P.M. CDT, the White Team took their positions next to Griffin's Gold Team members. By now my guys had been working almost continuously for about eighty hours. We had had a brief rest for about four to six hours after we passed the Moon and then snatched rest when we could. I remember their eyes, dull with fatigue and shadowed by anxiety. But their confidence and focus never wavered. As controllers plugged in their headsets, they shifted the papers and notebooks on the consoles. It was tough to find a place to work. As soon as CapCom Vance Brand started the entry checklist read-up, I was bombarded by calls from the controllers. Then I realized that in the rush to start the read-up to the crew, we had not made copies of the procedures for every team member. I told Brand to stop while we went out for copies. This was a vexing time for the crew. Time was becoming the most critical element, and with exasperation, frustration, and exhaustion gnawing at all of us, we had to wait for another half hour while copies were made for the controllers. Aldrich took this brief opportunity to incorporate two minor revisions into the final procedures.

Slayton, standing by in the MCC, had sensed the pressure and came on line to the crew. With just the right tone, his reassuring presence calmed our deadly tired crew. Deke was a pilot's pilot, an operator's operator, a straight shooter. Deke reassured Lovell, Swigert, and Haise that all was well with the procedures, and he kept up the chitchat as the minutes passed with agonizing slowness. Coffee was the substance that kept us going. Our surgeons had offered us something stronger, but we were all concerned about our performance deteriorating when the stimulants wore off. Most of us decided to make it on caffeine and cigarettes.

Brand began the final read-ups eighteen hours prior to entry, continuing into Windler's Maroon Team shift. Although I was concerned that something might get lost with three teams vying for the console, we had no option but to continue. A single slip anywhere could be fatal. We were out of time and out of options. This was our last shot. Ken Mattingly and Joe Kerwin, aces among aces as astronauts and my CapComs for the final shift, stood behind Vance Brand and Charlie Duke at their console during the read-up. They listened to the astronauts' questions, their voices and inflections, making sure that they fully comprehended every step and the rationale behind it. The read-up to the crew was concluded six and a half hours before the final entry procedures were to begin, not enough time for any of us to get any rest, just time to back off a bit before the final charge. I went to the viewing room downstairs for a brief nap.

Now, with a surplus of power, Windler gave the order to start power-up two hours early to try to get some heat into the spacecraft to give the crew a brief respite from the cold. Throughout the entire mission I had believed in my heart that we would get the crew home; now it was becoming a reality. Generating options was our business, and options remained as long as there was power, water, oxygen, and propellant. My controllers kept finding options.

April 17, 1970, Apollo 13 – Reentry

Three hours before dawn, the White Team took its place next to Windler's Maroon Team controllers. The eighty hours of uncertainty were now past and we were down to the final shift. During the return we had twice fought a shallowing trajectory, a glitch in a lifeboat battery, and a brush with typhoon Helen at the landing site. A last-minute maneuver returned us to the reentry trajectory. But we maintained our course as hour by hour we closed in on our objective, Earth.

The most chilling discussion came a few hours before entry, as the crew jettisoned the damaged service module and then maneuvered to observe the damage.

Lovell: "Okay, I've got her, Houston . . . there is one whole side of the spacecraft missing. Right by the . . . Look out there, will you? Right by the high-gain antenna, the whole panel is blown off, almost from the base of the engine."

CapCom (Joe Kerwin): "Copy that!"

Haise: "Yes, it looks like it got the SPS [main engine] bell, too. That's the way it looks, unless that's just a dark brown streak. It's really a mess."

Lovell: "And, Joe, looks like a lot of debris is just hanging out the side near the antenna."

Since the time of the explosion, I had deliberately avoided any discussion of damage to the command module, the reentry spacecraft. Briefly my thoughts focused on our decision not to trust the engine. Was it just a lucky guess or was there some gut instinct that Kraft, Lunney, and I shared? The heat shield scare with John Glenn's Mercury mission was never far from my mind, but I gave this no further consideration. If there had been any damage to the command module heat shield, there was nothing we could do about it now. At a certain point the human factor has accomplished all it can. Then things rest in the hands of a higher power.

As I went around the horn for the final Go NoGo check for entry, I felt a sense of loneliness in the room. We were getting ready to turn the crew loose. Once in blackout they were on their own; no more help from the team, no one watching over their shoulder. During the last twenty-four hours we could vividly imagine how desperate the atmosphere must have been in the spacecraft, how cold and how close to the edge the crew must have been.

Joe Kerwin, my entry CapCom, was an astronaut and medical doctor. His bedside manner with the crew during the final hours was spectacular. He was coach, mentor, doctor, friend, and partner to the crew. At times I felt he was virtually on board the spacecraft, nudging the crew through its checklist. With my final status check and the "Go for entry," a feeling of melancholy filled the air in the control room. This crew was special. We just could not lose them. Once again, failure was not an option.

It was tough to express feelings on the air-ground loop with the whole world listening. On this loop both the ground and the crew try to maintain a professional, almost unemotional tone and demeanor. Because we must be, we're conditioned to be hard on the outside, show no emotions in our response, and never betray any uncertainty. But at times the emotional charge passes among team members like a ray of sun breaking through the clouds, then it is gone again. When a mission is over and the crew is safe, my feelings of relief and pride make me choke back tears. This was not yet the time; we had a way to go — but we were close. In less than thirty minutes, the saga of Apollo 13 would be concluded.

On board the spacecraft, Jack Swigert, a rookie, finally broke the silence. You could feel the emotion in his voice as he said, "I know all of us here want to thank all you guys down there for the very fine job you did!"

Lovell chimed in, "That's affirm, Joe."

Kerwin's response indicated how close it was:

"I'll tell you we had a good time doing it" . . . pause . . . "Just for your information, battery C will fail about the time your parachutes come out. You have enough in the other two for landing." Moments later, after a brief burst of static, we were in blackout.

As I've indicated, the blackout is the toughest time in a mission for the teams. Every member does his soul searching, reviewing the decisions and the data, knowing we had to be damn near perfect and knowing how tough perfection is. Every member of our team on the ground, whether at the consoles, in the back rooms, or seated with SimSup, shared this common agony. Lovell's description of the damage to the service module made this agony particularly acute. Controllers were trained not to worry about things over which we have no control. We were now in the hands of God and a deadly tired crew, executing a set of procedures written on scraps of paper in the command module, procedures that had not existed eighteen hours ago. The teams knew the fragile hold we had on the many variables, the many decisions we had made in the four days since the explosion. But this is the nature of our business — to live with risk.

Everything now was irreversible. As the spacecraft and crew went through the final braking in the lowest part of the atmosphere, the heat was intense, preventing communications. The aerodynamic braking slowed the command module from five miles a second to less than 100 miles per hour when the chutes opened. The glow of the ionized atmosphere surrounded the crew in brilliant fire-orange as the temperatures soared outside the spacecraft.

The control room was absolutely silent. The only noises were the hum of the electronics, the buzz of the air conditioning, and the occasional click of a Zippo lighter snapping open, followed by the rasp of the lighter wheel against flint. No one moved, as if everyone were chained to his console. Cigarette smoke filled the room, creating a blue haze as we watched the track on the big world map tracing the path of the spacecraft to Earth. All eyes were on the clocks counting down to the end of blackout. Blackout was an eternity. I always said a prayer for the crew at this time.

We were pretty good at computing the blackout times, nailing the start and stop to within seconds. I worked it out in my mind; the beginning of blackout occurred over Australia, as RETRO had predicted, so the end of the blackout time should be on the nose. As the minutes passed, all eyes turned with a thousand-yard stare to the wall clocks as they counted down the final few seconds.

When it hit all zeroes, I told Kerwin, "Joe, give them a call." Kerwin responded immediately: "*Odyssey*, Houston standing by." There was no response, only static. More seconds passed and we called again. There was only static. Controllers pressed their earpieces farther into their ears, listening for the faintest signal. Kerwin called again. We were now almost a minute past the expected signal acquisition time. Still no response. Seconds turned into minutes and minutes into infinity. A sinking feeling, almost a dread, filled the room. When the wall clock rolled past one minute, we wondered what the hell had gone wrong. I wanted to smash something, hold on to something. Was there some screwup in the communications setup or relay? I told myself: they are there; we just are not hearing them.

There was one irrevocable piece of data yet to come. There would be a sonic boom as the command module reentered the atmosphere. When we received the report we would know the crew was coming back to Earth. Quietly, in hushed tones I called Deiterich, my RETRO: "Chuck, were the clocks good?" In a whisper he responded, "They're good, Flight." We waited. The world waited. We were 1:28 past the expected acquisition time when a crackly report from a downrange aircraft broke the tension: "ARIA 4 has acquisition." I pounded the edge of the console; the room erupted, then quieted down quickly.

In the movies, the controllers always stand up and cheer each mission event, but if a controller ever did that before the mission was over and the crew was on the carrier, that would be the last time he sat at a console. There was only one thought now on our minds; all we need now are the parachutes, just the parachutes. The crew was almost home.

Kerwin called again and a few seconds later we heard, "Okay, Joe!" Just two words, but the intensity of the relief was overwhelming. The viewing room, the back rooms, and our instructors erupted again as they saw the chutes blossom on the TV. In the control room each controller has his moment of emotional climax. I find myself crying unabashedly, then I try to suck it in, realizing this is inappropriate. But it doesn't work; it only gets worse. I was standing at the console crying. When the crew hits the water we once again sit at our consoles. Our job is over only when the crew is on the carrier and we have handed our responsibility to the aircraft carrier task force commander.

When this happened on Apollo 13 we finally realized that Flight Control and the people in the back rooms, factories, and laboratories had won the day. Our crew was home. We — crew, contractors, controllers — had done the impossible. The human factor had carried the day.

I was totally unprepared for the events of the next two weeks. The day following landing the flight directors and I stood on a platform with the wives and families of our crew as we received the Presidential Medal of Freedom on behalf of the mission operations teams. The cheers of our teammates, NASA engineers, and our families rose in crescendo as President Nixon concluded reading the award citation, "Three brave astronauts are alive and on Earth because of their [the mission operations teams'] dedication, and because at the critical moments the people of that team were wise enough and self-possessed enough to make the right decisions. Their extraordinary feat is a tribute to man's ingenuity, to his resourcefulness and to his courage." When our glittering technology failed us, our resourcefulness and courage, as well as every bit of the experience gained since the abortive four-inch Mercury launch, had carried the day.

Two weeks later on May 1, 1970, the flight directors and our wives flew to Chicago on the NASA Administrator's private aircraft for a luncheon and ticker-tape parade. As the aircraft pulled to a stop, the throbbing pulsing tempo of "Aquarius" played by city high school bands filled the air on the tarmac. The tempo matched my heartbeat as we waved to the airport crowd and were greeted by the City Council. During a luncheon at Chicago's Palmer House, Lovell was awarded the city's Medal of Merit and we were each given the key to Chicago by Mayor Richard Daley. After lunch Marta and the wives took their place in the reviewing stand as we were escorted to the parade automobiles.

Jim Lovell and Jack Swigert took the lead in a Lincoln convertible. Sig Sjoberg took the second while the flight directors were paired in Cadillac convertibles. Milt Windler and I rode in the third car and were followed by Lunney and Griffin. (Fred Haise did not attend, as he was still recovering from the flight.) The convertibles were flanked by three layers of police, plainclothes men next to the convertible, a second row in crisp uniforms with white gloves, and a third row of police motorcycle escort. Screaming sirens and flashing red lights of the police outriders added to the cacophony. As we passed around the Chicago Loop, fireboats in the Chicago River sent streams of water into the air. The cheers of the crowd made it impossible to talk.

Chicago will always remain in my memory as a class city, and I thanked them for the moments they gave us to bask in the eye of the public. When we returned to the airplane after the exhausting day, a box containing a silver punch bowl was on each of the wives' seats. (Upon return we each had to submit paperwork to the NASA legal folks to determine whether we could keep the gift from Mayor Daley and the City of Chicago. Since the punch bowls were engraved, the lawyers decided we could keep them.)

I think everyone, once in his life, should be given a ticker-tape parade.

The Apollo 13 debriefing had few surprises. We learned that the tank failure was due to a combination of a design flaw, mishandling during change-out, a draining procedure after a test that damaged the heater circuit, and a poor selection of the telemetry measurement range for the heater temperatures.

The debriefing party at the Hofbraugarten was merciless, beginning with a parody of the mission. The tape prepared by the Apollo 13 backup crew and the CapComs was not for the thin-skinned. The parody began and ended with the "immortal words" Liebergot and I exchanged early in the crisis.

Kranz: "I don't understand that, Sy."

Liebergot: "I think we may have had an instrumentation problem, Flight."

The clips of the voice tapes from the mission and the press conferences were interwoven with a Spike Jones record, gospel music, and various sound effects. No one was spared. As the tape continued, the crews and controllers roared and poured more beer. The tape took a shot at every flight director and crew member, as well as Slayton, Kraft, and even President Nixon. By the time the evening was over the words "I don't understand that, Sy" were forever embedded in memory.

There were no more missions in 1970. After we snatched victory from the jaws of defeat on Apollo 13, the rest of the year was a time of change, hard work, and frustration as further cuts were ordered in the flight schedule. Winners, however, persevere. We had a job to do and we sure as hell were going to do it. We had four more lunar missions; we had to get the crews to the Moon, attain our objectives, and get those crews back.

20

SHEPARD'S RETURN

The downtime needed for redesign of the service module gave us an opportunity to take a breath and look around. The world was a mess and so was our country. White adults were attacking black children being bused to school. Black Panthers were shooting it out with police in our cities. Four students were killed by the National Guard on the campus at Kent State University. Egypt and Israel were at war, airliners were being bombed or hijacked, and civil wars were erupting around the world.

I was frustrated by the lack of national leadership, the absence of individuals capable of rallying the many voices, putting the pieces back together. I had my own doubts about the war in Vietnam and the course set by the President and political leaders, but I refused to dump the blame for the way the war was going on the military.

The space program was also suffering. The lunar program was coming to an end. With the cancellations of the last Apollo missions — 18, 19, and 20 — I felt betrayed. It was as if Congress was ripping our heart out, gutting the program we had fought so hard to build. Leadership is fragile. It is more a matter of mind and heart than resources, and it seemed that we no longer had the heart for those things that demanded discipline, commitment, and risk.

The future of our space program after Apollo was a small Earth-orbiting space station dubbed Skylab. Its mission included astronomy, life sciences, Earth studies, and a grab bag of other experiments. The Skylab space station would use the leftover hardware from the canceled Apollo missions.

During the period after the Apollo 13 mission, a small team of controllers continued to follow the redesign of the oxygen system, while others were reassigned to the developing Skylab program. John Llewellyn was one of the controllers reassigned. I believed his trajectory skills could be put to good use in the Skylab Earth studies. John initially was not happy with the reassignment, but I was convinced that he would eventually come around.

With two programs, the computing services (not to mention our budget) were tight, and all computer runs were prioritized based on need and schedule. One afternoon I got a call from a computer operator asking if I had authorized some runs by Llewellyn. My response was short. "Not that I know of, but it is possible they are for his Earth resources project." The operator said, "Gene, you better look at these. They are for a lunar trajectory that lands on the back side of the Moon."

"There aren't any sites on the back side," I

said, "and I don't know what in the hell John is doing. Send the computer run requests up to the office."

In short order, John was standing at attention in front of my desk. He was never one who stammered or tried to mince words. He came right at you, and you better be ready for every emotion except regret. John never apologized. He believed that offense was the best defense. I found it hard to keep a straight face as my judo partner proceeded to explain why he was studying landings on the back side of the Moon. As he talked he paced the room, gesturing wildly in patriotic fervor. "We think the program is pretty well fucked up. This cancellation of the rest of the Apollo missions is a bunch of shit, and we're trying to do something about it."

John both challenged me and piqued my curiosity. We stood nose-to-nose. "John, just who the hell is 'we'?"

He ignored my question and continued, "Gene, can't you see what the hell is going on? The pogues" — John's favorite word for bureaucrats — "are taking over and pretty damned soon there won't be anything left of the space program. I know you had to put someone in this crappy job you gave me, but you better be aware that I am A RETRO first, and the section chief for Earth resources second!" John then stormed out. My office echoed from his shouting, and I still did not know what had set him off and what he was thinking.

An hour later, I received a visit from our geologist-astronaut, Jack Schmitt. He knocked on the door, politely walking into the office. "I understand you just had a talk with Llewellyn," he said. Now I was really confused, but I was starting to suspect that I had uncovered something that Schmitt probably had started. My suspicion was confirmed when he said, "I've got a small study group going on alternate lunar missions. We meet after work in my apartment. I provide the refreshments."

Jack Schmitt was the astronaut most like a member of the flight control team. He was a geologist, and the son of a geologist, who had explored Indian reservations in his native New Mexico as a boy. Jack had assembled some of the early composite lunar photographs while he was working in Flagstaff, Arizona. Accepted as a scientist-astronaut in 1965, he finished second in his class of fifty in Air Force flight school. At NASA, Jack helped develop the scoops, shovels, and other tools that were used to dig samples from the lunar crust.

Jack was unique, an intellectual with degrees from the California Institute of Technology and Harvard but with the soul of an adventurer. He was at every flight control party, celebrating each victory, big or small. A favorite of the controllers, he was one of the few astronauts who really put a few away at our parties — the others nursed their drinks. He was loud, effervescent, brash, not quite my image of the typical scientist. Schmitt seemed to have no limit to his interests, no end to his enthusiasms. He was an instigator who dropped a few well-chosen words in receptive ears and then let events roll on to what he knew would be a stormy, noisy, and wild conclusion. Currently without a mission assignment, Jack wanted to make sure that he got to the Moon, and the more missions on the schedule, the better his chances.

I went to Jack's next study session and was not surprised at finding Llewellyn and a handful of my flight controllers and flight designers. As I watched them work, I had the impression they were a bunch of Boy Scouts setting up tents and starting campfires. It was the same impression I had had of a similar bunch when I joined the Space Task Group. It was crowded in the apartment and the cross-talk was lively. One moment they were busily sketching out mission options, and then debating the pros and cons of missions to the back side of the Moon for the final Apollo flights. The team believed that if we could pull off something spectacular, something that had never been done before, we might recapture the interest of the American public and get the canceled missions back in the program. After all, the space hardware was already bought and paid for, and the team did not want to let the Saturn boosters and capsules end up as displays in museums.

The risks involved in a back-side landing might well create compelling drama. The risks would again put the lunar program on the front pages of the newspapers, and for a few days we would capture the public's interest. During a back-side landing, Mission Control could not give the crew any help. The crew would be on their own in a virtually uncharted world and, like the early explorers, living by courage and ingenuity alone.

We would not even know whether they landed or crashed until the CSM relayed the status a half hour later. These would be explorers like Byrd, Scott, Peary, and Cook. Schmitt's team continued its work; Llewellyn got his computer time, and when I had a chance, I joined the discussions. I wondered if meetings like this had happened before the master mariner Christopher Columbus decided to find the Indies by sailing west.

The plan never had a chance, never got to the attention of NASA management, but Llewellyn, Schmitt, and their team members believed it was better to go down fighting than to meekly accept defeat. Schmitt wasn't going to let the Apollo program come to an end without making sure that a real geologist set foot on the Moon. Mission Control, and the small group that worked in his apartment, cheered Schmitt the day he got his assignment to Apollo 17, the mission that would close the era of lunar exploration.

January 31, 1971, Apollo 14

Mission Control was a world bounded by math and physics, a world of statistics and probabilities. We were not superstitious, but we knew that every time we flew we were rolling the dice. We had beaten the odds on the last three missions. Probability said that someday we would run out of luck — as we almost did with Apollo 13. So we treated every mission as if it were our very first one.

I was out of the flight director rotation again. Frank, Griffin, Lunney, and Windler led the Apollo 14 teams. With a nod to the Wright brothers, the crew of Apollo 14 had named the spaceship the *Kitty Hawk*. Apollo 14 lifted off on the last day in January, headed for the Apollo 13 landing site, Fra Mauro. Alan Shepard, the first American in space, was the mission commander. Shepard had been scheduled to lead one of the first Gemini missions, but had been grounded by a rare inner ear disorder that caused severe vertigo. Like Slayton, he had been looking out for the interests of the astronaut corps. Then, in the summer of 1968, when he was partially deaf, he took a chance on experimental ear surgery to correct the vertigo and, against the odds, it worked.

There wasn't enough time to get sentimental, but all of us shared the pleasure of Shepard's comeback. Ten years earlier, while the nation watched and prayed and wondered, he soared into space in a capsule called *Freedom 7*. In just fifteen minutes the ride was over, but we had opened the door to space just that much wider.

Now forty-seven, Alan's appearance had changed little. He still resembled the actor Steve McQueen, and had a direct, no-nonsense, I-am-what-I-am kind of air about him. He always looked you right in the eye, and you felt he was looking right through you. (He would go on to become an admiral, one of the few who attained that rank without ever commanding a major ship. The list was small but distinguished, and two of the other names on it were the explorer Richard Byrd and Hyman Rickover, the pioneer of the nuclear submarine program.)

Shepard's compatriot in Apollo 14's lunar module, the *Antares*, was Ed Mitchell, a naval aviator flying his first mission and a virtual unknown to most of the controllers. Mitchell dabbled in psychic phenomena and for once I was sorry that I had moved Llewellyn out of the RETRO job. I would have liked to see if they could pull off a retrofire data exchange via mental telepathy. Stu Roosa, the command module pilot, had been my CapCom on Apollo 9 and had won the "most valuable player" award. At mission completion, when we light up the cigars, an eighteen-inch replica of the crew's mission patch is hung on the wall of the control room. The flight directors, by consensus, select the single controller considered most valuable to hang the patch. Like many of my controllers, Roosa came from Oklahoma, bringing with him the cheerful exuberance of a farm boy. You might say we had a mixed bag of characters playing in this one.

Roosa faced the first challenge when the docking system malfunctioned. In spite of Roosa's precise maneuvering, three docking attempts in two hours failed to dock the CSM with the LM. Even the tiniest debris lodged in the mechanism would account for the latches failing to engage. Running out of time, fuel, and options, MCC decided to advise him to come in fast and dock with a bang. Roosa would ram the lunar module, in effect, doing a ring-to-ring docking. Shepard turned to Roosa and said, "Stu, just forget about trying to conserve fuel. This time, juice it." While Roosa thrusted against the docking ring, Shepard manually fired the latches. The technique worked, and after some hair-raising moments, the mission was back on track.

Spaceflight rarely gives you a second chance, but Apollo 14 was the exception. Crisis management textbooks use the term "prodrome" to define a warning or intimation of an impending crisis. I define it as the signal that causes the hair on the back of your neck to rise. With luck, someone picks up the signal, recognizes something is wrong, and starts an action that short-circuits the crisis.

The mission had gone well, and Shepard and Mitchell were in the LM, setting up the switches and computer for the descent to Fra Mauro. The timeline had been followed almost perfectly, the crew and ground were in sync. It was the way we like it when we are getting ready to land on the Moon.

Bordered with black and yellow tape and centered in the LM control panel was a round, red push-button switch. The white letters ABORT on the in button distinguished this switch as the one that started an irreversible process to terminate a mission. It was used only when there was no other alternative. The switch had electrical contacts to issue signals to the LM engines, computer, and abort electronics. When the abort switch for Apollo 14's LM had been manufactured, a small piece of metal had been left in the switch. Now, in zero gravity, and with both crew and ground oblivious, this piece of metal was floating among the contacts of the switch, randomly making intermittent connections.

Dick Thorson was the LM control engineer for the descent and landing. Using his console television display, he was tracking the crew's progress through the checklist. The CSM had undocked and separated from the LM. Shepard and Mitchell in the LM were passing on the front side of the Moon on the final orbit, one hour before starting the lunar descent. All was looking good in both spacecraft and in the MCC.

After Apollo 13, the controllers' console warning light logic had been reversed to aid the controller in rapidly recognizing a changing status in critical systems. Of the hundreds of event lights on Thorson's panel, only a few were now illuminated. The earlier problems with the docking system were long forgotten; the focus now was keeping to the precise timing for the landing trajectory.

Out of the corner of his eye, Dick noted a change in his status panel. He glanced up quickly to see a red light at his console that indicated the crew had pushed the LM abort switch. This did not make sense. Thinking there might be a telemetry patching error to the light panel at his console, he selected a TV display that let him look into the guts of the computer. Rapidly scanning down the list he saw that Computer Channel 30, Bit 1 (Abort) had been set on. He kept his cool and called the back room to get a reading directly from the data stream. The technician confirmed, "Channel 30, Bit 1 is set." There was no doubt now, this was a valid indication. If the LM engine had been running, the abort bit would have signaled the computer to change operating modes from the landing mode to the ascent/rendezvous mode. When this occurred, the descent engine would automatically throttle up. If the LM was close to the lunar surface and near fuel exhaustion, the computer would command a fire-in-the-hole staging. This staging sequence would shut down the descent engine, fire the explosive bolts to separate the ascent and descent stages, and ignite the ascent engine, while simultaneously changing computer programs and switching electrical power and control to the ascent stage. When you started the ascent engine, which was buried in a cavity inside the descent stage, under these conditions, you had a fire-in-the-hole situation. All events occurred in fractions of a second. The sequence had to work perfectly.

If the crew was near the surface and the abort bit set, it would, at best, eliminate the possibility of landing. Coupled with other malfunctions, it could lead to a lunar crash. The abort bit now illuminated on Thorson's console quickly got his full attention.

In the SPAN room, two of Thorson's counterparts, Hal Loden and Bob Carlton, also noticed that the abort bit had been set. Without hesitation Carlton, the Silver Fox, leaned over Loden's shoulder. In a laconic drawl he said, "We should get the crew to knock on the panel."

Great controllers seem to have a prodigious

memory and the gift of excellent recall. Carlton remembered a NASA ALERT bulletin, that cited problems with internal switch contamination. Loden concurred with Carlton's rather unusual troubleshooting and told Thorson on the intercom, "Dick, have the crew knock on the panel by the switch and let's see if the abort indication goes away. We may have a contaminated switch." Thorson momentarily wondered if this wasn't a crazy idea, but he didn't have any alternative to offer. He stood up, stretching his headset cord, and walked to the side of flight director Gerry Griffin's console. In the high-tech age of Apollo, he was embarrassed to resort to shadetree mechanic fixes in front of the whole world.

Leaning toward Griffin, Dick puffed on his cigarette and hoarsely muttered, "Gerry, I'm seeing an abort indication in the lunar module. Have the crew verify that the button is not depressed." Thorson now had Griffin's attention as he continued, "If they say negative, have them knock on the panel while we watch it."

The CapCom, Fred Haise, passed the instruction to the crew. The crew, unaware of the potential gravity of the situation, acknowledged the call. Mitchell reached over and tapped on the switch with a flashlight.

At Thorson's console the light and TV indication disappeared. Thorson called out, "Gerry, I'm NoGo. I've got some problems here and it is going to take some time to work them out." Griffin waved off the landing attempt, and Apollo 14 had dodged a bullet. The crew and control team received another chance. The crisis was real, but with advance warning the team could develop options to save the mission. The critical element once again was time. We had precious few hours to work on the problem. Thorson's dilemma was a thorny one: to land, we needed to bypass the switch, but if we had problems during landing, we needed the switch to abort. It was a hell of a risk-gain trade.

While developing the mission rules, the LM controllers studied every switch and circuit breaker in the spacecraft, assessing the failure potential of each and options to work around a switch failure. This included the case where a failed switch would set the abort bit in the computer. Thorson's team had developed a set of instructions to tell the computer to ignore the abort switch. Thorson now reached into his bag of tricks, addressing Griffin: "Flight, I've got a software patch for the LM computer that will disable Channel 30, Bit 1, in the computer. It will lock out the abort switch for the landing."

Griffin listened intently, then frowned as Thorson concluded by saying, "There is only one problem. If the crew has to abort, they will have to use the backup system or the computer keyboard to manually enter the abort program for the primary system." Griffin was willing to buy the extra risk, and he knew without asking that Shepard would, too. "Dick, dig out the patch," he instructed, "and run it through SPAN and get with the simulator people. If it works, I'm going to give it a Go." Astronauts were soon clambering into the simulators, with the results to be relayed instantly to CapCom Haise.

Satisfied with the ongoing action, Griffin called his team to attention, resetting the timeline for the landing. The answers were needed in less than two hours.

In the Flight Dynamics staff room, Jack Garman listened to the conversation. Garman was the computer expert who had helped Steve Bales out of his hole on Apollo 11. Now he had to come up with the answer to a different problem. Jack was normally excitable and when there was an option to use "his" software to save a mission, he leaped into action. Garman talked with every part of his body, eyes and hands constantly in motion. Within seconds, he was on the voice loop to the software team at MIT. Like Griffin, he was worried about the time it would take the crew to make the computer entries if they had to abort. Garman felt that the crew needed a software patch to protect against the switch failure that would still give them the ability to use the abort button. He didn't know whether such a patch could be developed, but he sure as hell was going to give it a try.

MIT was listening to the astronaut voice loops and had heard the discussion about the software patch to bypass the abort switch. They had already cranked into action. (MIT's Draper Laboratory developed the guidance and navigation systems for Apollo. To assist in rapid troubleshooting during a mission, the MCC was in direct communications with the lab.) Within an hour the Draper Lab came up with a procedure that gave us an option to bypass the abort switch at engine start and then reenable the switch. The procedure was complex and time-critical. Sight unseen, Griffin elected to give the new procedure a shot.

Communications were a mess during the final front-side pass. Static crackled and punctuated Haise's instructions to Mitchell. The procedure required the crew to start the engine at a low power, using the acceleration to move the contaminating piece of metal away from the switch contacts. Once the engine was started, Mitchell would insert a string of sixteen computer commands to enable guidance and provide steering.

When this was completed, another string of sixteen commands would disable the abort program, and another fourteen commands would lock into the landing radar and the descent software. This entire sequence would occur as the crew was descending to the Moon. The mission now rested on an emergency patch to the flight software that was less than two hours old, had been simulated only once, and was being performed by a crew that had never practiced it. Every step had to be executed precisely on time and in sequence.

Sitting on the step behind Griffin, I looked into the viewing room. I could see the spectators buzzing. Our words and procedures were gibberish to them. In the mission operations control room, it was just another day, another final exam, as the controllers calmly chipped away at the final procedures and counted the seconds until engine start. The controllers' ability to focus at times like this was nothing short of a miracle, a miracle of ingenuity, discipline, and training.

Shepard sounded just as he had all those years ago when he first went into space. Marvelously calm, his voice was flat and emotionless as Mitchell read the checklist, verified the switches, and entered data to the computer. With split-second teamwork, we started down to the Moon. Mitchell announced, "Engine start." Thorson, staring at the abort switch display, added, "Ten percent thrust . . . throttle up. I see no abort indications."

Mitchell quickly entered the first string of commands to bypass the abort bit. With the successful entry of the data, you could hear the relief in Shepard's voice: "Thank you, Houston, nice job down there."

Just another day at Mission Control.

Thorson's eyeballs, which had been locked on to the abort bit, now swung over to his engine systems. "Gerry, I'm Go," he confirmed. "All of the data is in correctly. The abort program is bypassed." Fred Haise, the Apollo 13 LM pilot, then talked Mitchell through the rest of the work-around procedure.

With a lyrical comment from Alan Shepard — "It's a beautiful day to land at Fra Mauro" — the LM *Antares* was on its way to the Moon. We had skated across thin ice and reached the other side.

But the battle was not over. The team would be challenged once more.

Prior to starting the descent the guidance officer provided the navigation and target data to the lunar module computer for landing. The computer then developed the guidance commands to reach the target during descent. In the final phase of the landing, the LM computer needed more accurate data than that provided by Mission Control. There might be an error of several thousand feet between the altitude data provided by the MCC and the true altitude provided by the LM radar. Mission rules required an abort if the radar data was not obtained before descending to 10,000 feet. Landing without a good hack at the altitude would be worse than landing on a carrier on the ocean on a dark night; there was no good way to judge height. LM landing fuel was tight, so that a "grope and feel" approach would deplete fuel prior to touchdown, leaving the crew in a low-down fire-in-the-hole abort. Not a good situation.

Passing through 32,000 feet, Mitchell started looking for the landing radar data. In Mission Control, guidance officer Will Presley cued up his displays. He had about ninety seconds to make a judgment to accept the data or, if it was not in limits, abort the mission. The radar data did not show up when expected, and Presley in the MCC and Shepard in the LM both had to be thinking, "Where the hell is the radar?"

In a voice lacking conviction, Presley gave Griffin a Go to continue descent at five minutes, knowing that in the next sixty seconds he would have to call an abort if he didn't get radar. I knew his stomach had to be churning. The LM radar was Thorson's responsibility. He and Griffin were also watching for the indications that the radar was tracking the surface.

Griffin was the first to move. "Dick, you got

anything you want to try?"

Thorson reached into his bag for the only thing that could be done in a few seconds: "Flight, have them cycle the circuit breaker."

Quickly, Shepard acknowledged, "Cycled."

Seconds later, a jubilant Will Presley shouted, "Flight, we got radar lockup!" Every controller had been holding his breath. For a few seconds the voice comm was noisy as all exhaled, some even whistled. Presley's next words were virtually shouted, tripping over each other as he blurted, "Altitude data is Go, accept the radar." From there on, the landing was a piece of cake.

Of course, Alan Shepard walked on the Moon and left behind two souvenirs for some future explorers to find. Somewhere in the craters of the Moon are two golf balls, the first ever hit in outer space. He attached the head of a Spaulding 6-iron to a tool used to scoop up lunar soil. This was done for the highly scientific purpose of seeing how far a golf ball would travel in gravity that was one sixth the Earth's. Actually, he duffed his first shot. "It got more dirt than ball," he confessed to our controllers.

"That looked like a slice to me, Al," came the reply.

His second shot traveled, by his estimate, "miles and miles." The experiment did not appear on anyone's manifest, but Shepard had cleared it with Deke Slayton. The agreement was that he would do it only if the landing had gone well, and it had. Years later, Al would reflect, "I'm probably a lot more famous for being the guy who hit the golf ball on the Moon, than I am for being the first American in space."

After the post-mission debriefing, Mitchell invited Thorson and several members of the Trench to dinner. Mitchell, the astronaut who believed in psychic phenomena, said that he knew moments before the call that Thorson was going to have the crew cycle the landing radar circuit breaker. Griffin and Thorson were never sure whether Ed was kidding them or not. They were just happy it all worked out.

After the debriefing, Shepard took Griffin aside and confided, "I had come too far to abandon the Moon. I would have continued the approach even without the radar." On Apollo 14 the error in the LM computers' knowledge of the actual altitude was almost 4,000 feet before the landing radar data update. With an error this great in the computer, Griffin and the Trench were convinced Shepard would have run out of fuel before landing. But everyone who knew Al never doubted he would have given it a shot. We also never doubted he would have had to abort. The fuel budget was just too tight.

In the three Moon landings, the crews and controllers had become masters of improvisation. With three lunar missions to go, we were pretty cocky, feeling that there was no emergency we could not handle and nothing that would defeat us.

21

WHAT DO YOU DO AFTER THE MOON?

Spring 1971

One morning in the spring of 1971 Gerry Griffin walked into my office. Briskly, he said, "The 15 crew thinks it would be good for you to take a break and get out of this stuffy cell. How would you like to go on the next field geology trip with the crew and myself?"

I felt my pulse speed up a bit. This was something new, something different, something pure fun. I knew nothing of field geology, but since we had won the race to the Moon, the shift to lunar science dominated much of our effort. This trip might provide an opportunity to get smart on a new aspect of the business. Lately, I had spent much of my time with the lunar scientists and, although they were much older than my controllers, they had a similar kind of enthusiasm, energy, and commitment to the future of manned spaceflight. Knowing how important the data collected in the Apollo missions would be to our understanding of unknowns like the formation of the solar system, I made sure that my controllers served them well in the planning and execution of their experiments. So when they asked me to work with and learn from them, I jumped at the chance.

The field geology trips provided the practical training ground to complement the astronauts' classroom training. By their nature, astronauts were curious, and many became dedicated to studying the space sciences. Scott, Griffin, and I made for a colorful trio, to put it mildly. All three of us had flown jet fighters. Scott could have served as a poster for astronaut recruitment. He and Griffin shared a cheery exuberance, a perpetual optimism, and a zest for their work.

The field sites were the laboratories for the astronauts preparing to explore the lunar surface. No terrestrial site could replicate the lunar surface, but there were locations where the rugged terrain could provide conditions similar to those found on the Moon. While training the astronauts, the geology instructors used sites that covered the globe, from the volcanic areas of Iceland to the Grand Canyon, from the mountains of New Mexico to the craters of Hawaii. This training would be vital in selecting the materials to be returned to Earth from the Moon and in answering such questions as the Moon's age and composition.

I had first met the lunar surface science teams in the conferences to set up their operating structure. I got to know more of them when my Flight Control Division inherited the operation of the Experiment Packages placed on the Moon during each lunar mission.

Our scientists didn't pay much attention to bureaucratic structure. They wanted to work directly with the crews and controllers to establish a mutual understanding and supportiveness that would make their work on the Moon much more productive.

No one among them impressed me more than Lee Silver. He stood out as we were setting up the surface science rooms, and again during the skull sessions at the apartment of Jack Schmitt. Silver had taught Schmitt as a student at CalTech, and it was obvious they shared the same passions. Silver's academic credentials were formidable, but the man was even more impressive. You can tell great teachers by their demeanor, how they talk, how they always seem effortlessly in control. Silver appeared born to roam the deserts and mountains, reading the land.

William Muehlberger, from the University of Texas, had the same characteristics. Each geologist approached his work passionately; you felt this passion when you were around them. Schmitt, Silver, and Muehlberger inspired and motivated others much in the same way as my other great teachers like Harry Carroll at McDonnell in St. Louis, Jack Coleman in flight training, Ralph Saylor at Holloman, and Chris Kraft in the MCC, all supremely confident and capable leaders and teachers.

Since I had trained in fighters at Nellis Air Force Base in Nevada, I was familiar with the general terrain as our helicopter lifted off at dawn for the flight to the north end of Frenchman's Flat, ninety miles northwest of Las Vegas. I had flown over this area en route to the gunnery and bombing ranges but I had never really looked closely at the surface. From the helicopter, I was amazed at the results of the nuclear tests that had pockmarked the desert as far as the eye could see, the craters a myriad of desert colors, the rocks and boulders arrayed from the blast point. Flying low over the area, I imagined this was what the surface of the Moon might have looked like, except for the colors, as Armstrong and Aldrin described it during and after the landing.

The bond between teacher and student astronaut was clearly evident on the way to the site as Silver and his students engaged in a lively Q&A session. There was not a wasted moment from the time we crossed the snowy mountains and descended to the test site.

The Apollo 15 crew, Scott and Irwin, attacked their role as surface geologists with the intensity and enthusiasm they demonstrated in learning to fly a new spacecraft. To them, the surface experiments were just another form of flight test. They were exploring a new world full of riches for the scientists, and were in competition with their predecessors and themselves.

Griffin and I were really just along for the ride, but we found out quickly that Silver had other ideas. After a brief summary of the training objectives, Lee gave me a quick course in field geology 101, then sent me off to find as many different materials as possible within walking distance of our landing site. I had no clue where to start, but I felt obligated to give it my best shot. Griffin tagged along with the crew, listening as Silver laid out their project for the morning. I learned to use my eyes to detect subtle changes in hue, composition, and texture of the land . . . to see the parts and assemble the whole, then to work it in reverse, striving always for the big picture, looking for cause and effect. After Silver's brief training, I could visualize the great collisions as the meteors showered the Moon, instantly forming the craters and hurtling the rocks and boulders enormous distances in the low gravity of that airless sphere.

On the final Apollo missions, a Jeep-like vehicle called the Lunar Rover extended the range of operations on the surface. The Rover had a television camera mounted on it, one that Ed Fendell and his team would control from 240,000 miles away to bring history into the homes and offices of the world. The first priority was to use that camera for studying the lunar surface. We could now see what the astronauts saw, close up and in real time. It was like being able to look over the shoulders of Lewis and Clark as they trekked into the great unknown of their era. Scott and Irwin would be the first humans to walk the vast and incredibly lonely Hadley-Apennine region. At the atomic test site, I could visualize what they would see, touch, and do.

Back at MCC, the challenge of the final three missions was emotional as much as technical.

My controllers, average age now twenty-seven, were asking themselves, "What do you do after you have been to the Moon?" They had come to us at the beginning of Apollo, in their early twenties. Now, with NASA limiting the program to only three more missions, they were taking it the hardest. Mission Control was their portal to the stars; they believed we had taken only that first "giant step for mankind" and could not understand why we were not taking the next leap forward. I knew how they felt. When I won my wings, I believed I would fly fighters forever. When my dream ended, my world folded. So I had to pick myself up and get on with life, and find a new vision. In the process, I took a lucky fork in the road that got me first into flight testing and then into the space program. It was that one-in-a-million chance you take in life that pays off.

In 1971 a big part of my job was convincing my young controllers that there was a damned interesting and challenging world after Apollo. Together we would take the next fork in the road and blaze a path into the next era of space. I am a dreamer, believing that the mark of a champion is the ability to thrive in tough times. I was convinced that Mission Control would evolve, adapt, and exploit every opportunity. We can make the future ours if we believe and fight to make it happen.

Change was again in the wind. The high visibility of my controllers provided them with opportunities for top-level leadership roles. In 1970 Glynn Lunney had traveled with Robert

Gilruth to Moscow to determine the degree of Russia's interest in a joint space mission and the resources the Russians would make available for it. Now after Apollo 15, Lunney joined Hodge and Charlesworth in top management. While I lost extremely capable leaders, I had the opportunity to bring along another new generation of young leaders in the Mission Control "leadership lab." Every young man or woman coming in at the bottom could take a shot at a flight director, division, or branch slot in the first ten years of their career.

In the final days of Apollo I was fortunate to be knee-deep in mission-ready leaders, most veterans of twenty or more space missions, including those of Gemini and Apollo. My only fear, with the constant juggling of priorities and people, was the loss of focus on the final three missions. I also knew that any major mission glitch would give those who were nervous about the risk an opportunity to argue that since the Moon had been reached there was no need for the remaining missions. I was glad the final flights would be led in Mission Control by three former aviators, who understood how to live with and manage risk. Maybe it was our fighter aircraft mentality, or maybe it was our confidence in the human factor, but flight directors Gerry Griffin, Pete Frank, and I believed the nation had sacrificed too much to surrender to the increasingly conservative national leadership. As we opened the era of extended lunar operations, we felt fully capable of meeting every challenge that we and our crews would face during the final missions.

To extend the range of the lunar expeditions, modifications had been made in the LM to provide stowage for the battery-powered, Jeep-like Rover. We shaved our mission rule margins and, with extra oxygen and batteries, we extended the surface duration to almost three days. Extravehicular activity was planned for each day the crew spent on the lunar surface.

The CSM also received a facelift. A full bay had been filled with instruments to map the lunar surface, study its physical environment, and investigate its gravitational and magnetic properties. The controllers and the new lunar orbital scientists learned from each other about the science and operations of space exploration, developing the rapport needed for successful missions.

Approaching the end of Apollo, my frustration often surfaced. No one in America seemed to care that we were giving up, surrendering the future of the next generation of young people with stars in their eyes. Often I sat silently, somewhat moodily in my office, rereading President Kennedy's words, "The United States was not built by those who waited and rested and wished to look behind them. This country was conquered by those who moved forward, and so will space." But what if we had not pushed ahead, exploring and opening the American frontier? What sort of nation would we have been? What were the implications for our decision not to push ahead now? How I wished John F. Kennedy were still alive, challenging us to dare and to dream. I feel the same way today; the boldness and scope of his vision is not to be found today in our space program and in our nation.

July 1971

As the budgets tightened and public support waned, NASA was resigned to the cancellation of the Apollo 18 and 19 missions. The tough choices prompted "the grand old man" of lunar science, Harold Urey, to write in the *Washington Post*, "We wish to finish a job which has beautifully begun and now we get stingy. Because of the additional cost of 25 cents per year for each of us, we are dropping the final two flights to the Moon. How foolish and shortsighted from the view of history can we be?"

The final three missions were no easier than the earlier ones. The complexity of the spacecraft systems as well as the objectives for each mission made it unlikely that any mission would be trouble-free. In retrospect I am still amazed that we risked so much, so often, and came through unscathed again and again. At times, I believed Providence watched over us all.

Kraft was now preoccupied with the future and his inevitable promotion to MSC director. Seated in Kraft's chair since Apollo 13 was Sig Sjoberg. In many ways, I felt vaguely uncomfortable without Kraft in charge, almost as if he were our bearer of good luck, our talisman in Mission Control.

Sig, invariably accompanied by Bill Tindall,

was radically different from Kraft. He was accustomed to being the deputy, not the sheriff. I could get mad at Kraft, standing face-to-face, pounding out my position. Chris had a short fuse. His expression would change to the incredulous, then his face would turn red and he would invariably start with, "I don't give a damn what you think!" Then he would lay out his position. But then Kraft would always listen. It was great having a boss who felt so emotional about the job, and was willing to engage in a brawl if needed to get to the best answer.

It was tough, almost impossible, to get mad at Sjoberg. He was just too nice. He reminded me of a grandfather with his grandchildren, always giving a kind word of encouragement. Only infrequently did he ever admonish his charges, and then quietly. When I first met Sig, at Langley, my family was growing. One day, he arrived at work and asked me out to his car. When he opened the trunk, I saw a large old tricycle, almost the size of a rickshaw. I had never seen one so large or sturdy. Sig said, "We don't need it anymore. I would like your kids to have it." This remarkable man had not only this sweetness of character but real depth and earnestness.

Sig Sjoberg took over the four divisions of Flight Operations in 1969 after Kraft became Dr. Gilruth's deputy director. Sjoberg and Kraft were born to be together. We were sure that Sig's position as our boss was only temporary and that he would become Kraft's deputy for the MSC when Gilruth retired and Chris took over.

July 26, 1971, Apollo 15

After a glitchless countdown, the Cape launch team handed Griffin a virtually perfect command module, named *Endeavor*. Conscious of the importance of science for their mission, Dave Scott, Jim Irwin, and Al Worden named their craft for the ship commanded by Captain James Cook. That *Endeavor* sailed in 1768 from England to Tahiti to observe the passage of the planet Venus between the Earth and the Sun. The LM was dubbed *Falcon*, in honor of the Air Force mascot for an all–Air Force crew.

The launch and orbital checkout of the CSM and booster clocked off in the normally intense timeline. After TLI booster engine cutoff a series of critical events takes place for a half hour, starting with the separation of the CSM from the booster, followed by the turnaround and docking with the lunar spacecraft nestled atop the booster. While taking a breather before extracting the LM, Al Worden looked around the cockpit, casually noting, "The main engine thrust light on the entry monitoring system panel is on. I'm not sure when it came on."

Instantly, GNC Gary Coen snapped, "Flight, Panel 8, have the crew pull both pilot valve circuit breakers."

Worden replied, "Okay, they're pulled."

Gary continued, "Flight, the engine is now safe. The thrust light indicated the CSM main rocket engine was armed and ready to fire." An engine start signal triggering actual engine ignition during the critical turnaround, docking, and extraction sequence could have crashed the CSM into the LM or the booster. Pausing briefly, Coen continued, "I think we have an electrical short in the engine start circuit." With MCC's preliminary diagnosis and the rocket engine now safe, the astronauts continued the timeline, firing the pyros to release the LM and firing thrusters to maneuver away from the booster stage.

The initial hours on the outward journey of a mission are always busy. There are many housekeeping items, and when they are completed the crew and control teams settle into a groove for the three-day transit to the Moon. Every glitch must be closed out so the work is distributed among the teams. Much of the systems analysis work, like determining the cause and cure of this engine electronics glitch, is assigned to the shift that is on duty when the crew is sleeping.

Griffin handed the thrust problem to Windler. After a brief period of troubleshooting, Windler passed the problem to Lunney's team to develop the workaround procedures. The workaround had to have three parts: protecting against an unplanned engine start, keeping the engine running during the maneuver, and cutting the engine off at the correct time.

Every system on a spacecraft is critical, but when you had to make up your mind whether or not to go into lunar orbit, the service propulsion system (SPS) in the CSM moved to the top of the list. It must work to get into lunar orbit, and once in orbit it was the crew's ticket home. The SPS design provided redundant electronics, electrical power, and propellant feed systems, but there was only one engine nozzle and a single set of propellant tanks, so any leaks were cause for rapid mission termination.

Mission rules require full redundancy of the engine control electronics systems to enter lunar orbit and to allow LM separation once in lunar orbit. The job fell to Lunney's team to make sure that the SPS was fully operable. By the end of Lunney's first shift, troubleshooting limited the problem to the A engine control circuit. The B control circuit was fully operable. Time is one of the most precious resources of flight directors and, for a change, time was on our side during the three-day translunar coast. Lunney's team GNC, Joe DeAtkine, was short, quiet, young, and unassuming. Flying his first mission as GNC, Joe felt the weight of the world on his shoulders.

A flight control team develops a keen awareness of situations in which someone needs help. Gary Coen and Jack Kamman were on each side of DeAtkine's shift. Each extended his shift duties so that they overlapped, giving Joe the coaching and mentoring he needed to survive his first crisis. After developing the workaround Joe emerged with the confidence that he could do it on his own the next time.

Griffin executed the maneuver with the amended procedures and, most importantly, met the criteria for continuing the mission. This uncanny ability at assimilating data, making judgments, and balancing risk versus gain reached maturity in Apollo.

The Hadley Rille landing site was in a mountainous region of the Moon well north of the equator and on the edge of a mile-wide canyon. Scott and Irwin guided the *Falcon* through the descent, surprised to find that the landmarks were less sharply defined than expected. As they continued the descent, they searched for their specific landmarks, redesignating the landing location several times, steering to remain short and north of the one feature they recognized, Hadley Rille, a mile-wide canyon seven miles from 18,000-foot-high Mount Hadley. In the final fifty feet, they flew blind through the lunar dust. The round blue lunar contact light triggered Scott to cut off the landing rocket. The LM fell the final feet to the surface and then lurched in an unsettling motion, tilting back and to its right. Two of the landing pads were settled into the edge of a small crater.

My White Team started its shift work during the lunar orbit phase of the mission. A bay of the service module had been outfitted with a pallet of scientific instruments to obtain photographs of the Moon's surface and map its chemical composition. I worked the same shift schedule as Gerry Griffin. Griffin's team supported the lunar EVAs, while my team supported astronaut Al Worden's operation of the service module experiments.

The mission continued smoothly through the first EVA period. Scott and Irwin were out to prove that an astronaut was capable of performing in a scientist's arena. They were open to every change, driving to set a standard higher than that of any previous crew, and determined to prove that they were up to the demands of Lee Silver and his team.

Griffin, Mission Control, and the lunar scientists were not aware, however, of the price Scott and Irwin were paying to maintain a very heavy workload. By the end of the first EVA, the crew's hands ached as if they were arthritic. As they continued, the skin under their nails hemorrhaged and turned black, the fingertips tearing from the constant rubbing against the gloves. Every task on the Moon demanded dexterity; this crew was not about to let physical discomfort get in the way of achieving the mission's objectives.

After their second EVA Scott and Irwin began an intense timeline in their final twenty-four hours. After a six-and-a-half-hour sleep period, the crew was awakened to begin preparation for the third and final activity. The timeline was already short. As the crew in the LM slept, Lunney had worked with Bill Muehlberger and the science team to replan the EVA. To get a reasonable duration, the controllers started cutting bits and pieces of margin from the timeline. Fifteen minutes were taken from sleep, another fifteen from the period after eating, and twenty-five minutes more were snatched from the stowage and ascent preparation to allow a four-hour and thirty-minute EVA and still meet the scheduled lunar liftoff time.

Nearing the end of the final EVA Scott and Irwin had one last, sad duty to perform before they lifted off and returned to *Endeavor*. Four weeks before their flight, three Russian cosmonauts had died during the reentry of Soyuz 11. They had been in orbit for a record 233U4 days (570 hours), and the spaceship made an apparently flawless landing. The three were found unmarked, reclining in their seats as if asleep, killed almost instantly and silently by an oxygen leak. It was an eerie reminder, as if one were needed, of the unpredictable nature of space voyages.

Scott and Irwin left on the Moon a plaque with the names of the three Russian astronauts, adding theirs to the honor roll of others — the Apollo 1 astronauts, the Soyuz 1 astronaut, and all the rest — who had lost their lives in the quest to explore the universe.

Windler launched the crew off the surface, rendezvoused, and then docked before handing over to Lunney on the fifty-first revolution. The crew was on the timeline and there were few apparent problems for Lunney, the kind of shift you pray for. This was a time to catch a breath, clean up the spacecraft problems, and get ready for the trip home. The crew had been awake for sixteen hours, performing physically demanding work, which was followed by the intensity of a lunar liftoff, rendezvous, and docking. The principal activity of Lunney's shift was to jettison the lunar module and then get the crew to sleep.

Lunney's problems started during the crew's seventeenth hour awake. As a precaution for a loss of pressure during the LM jettison, the crew donned helmets and gloves and performed a suit pressure check. Then, to make sure there were no leaks in the hatch seal, the crew gradually depressurized the tunnel connecting the command module (CM) to the lunar module (LM) while the CM pressure was monitored for any decrease.

The initial suit integrity check failed due to a pressure suit leak at the fitting where water is fed to the liquid-cooled garment. After both suits were plugged, the suit integrity test was passed satisfactorily. After a brief verbal update on the depress sequence, the crew continued the preparation for the jettison.

During communications with Mission Control, Scott commented on the difference between the pressures in the command module and in the tunnel. "The tunnel pressure was at 2.7 a while ago," he said, "and now it is down to 2.0." Scott was reading the pressure from a small gauge in the tunnel, one normally used by the crew prior to opening the hatch, or when separating the spacecraft. Lunney's hairs stood on end; many things must be right in space, and cabin pressure is at the top of the list. Lunney, now concerned about the decrease in pressure, ascribed it to a possible hatch seal leak. Given the earlier problems in the suit integrity check, he scrubbed the LM jettison. The crew backed out of the configuration, removed the hatch, and visually inspected the seals. They were then given precise instructions for another command module pressure check.

While the astronauts were performing the pressure checks I was in the orbital science back room talking to the controllers and scientists prior to coming on shift. Dick Koos, my Apollo 11 SimSup, now operating in a new role as an experiments engineer, motioned me to his console. As I leaned over he said, "Lunney's having a hell of a problem getting the crew through the separation checklist. Something is out of whack." I thanked Dick for his heads-up and quickly moved to the control room.

At acquisition of the CSM telemetry and voice communications on orbit fifty-three, Scott had unexpectedly vented the tunnel. Lunney, absolutely unruffled, told the crew to pump the tunnel back up. Glynn was getting frustrated; he knew something was wrong and forcefully reminded his controllers to call out every step of the crew's procedures as they were performed. He wanted his team's eyeballs in the cockpit with the crew.

Lunney continued with his usual superb, unbridled confidence, his voice never exposing any emotion, so his people never sensed his frustration. Now satisfied with the suit and cabin-pressure check, Glynn gave the "Go for jettison." I was spooked just listening. Even in the most bloodcurdling simulation I had never seen the crew and ground so out of phase.

After LM jettison, Scott delayed the separation maneuver, believing he would collide with the lunar module if he went with the MCC plan. To reassure Dave, Glynn replanned the maneuver, executing it by a standard rule-of-thumb cooked up by Bill Stoval, his young FIDO. "Glynn, have the crew stay in front of the LM, point at it, and thrust away for two-feet-per-second velocity. This will give us enough clearance."

When Scott executed the maneuver, Lunney showed his only emotion, inscribing in the log, "Hurrah — I felt I was in one of those bad dreams where you can't wake up and you can't get anything to go right!" Even though I had come in only for the last two hours, I had the same creepy feeling. At no time on the console had I ever felt so apprehensive.

Throughout the pressure check and maneuver fiasco, Dr. Chuck Berry had been standing and talking to Dr. Gilruth and Kraft at the console behind us. With the maneuver completed, Berry approached the flight director's console. I was sitting next to Glynn reading his log and preparing for shift handover when Dr. Berry pulled up a chair. When the surgeon visited the flight director for a powwow, you knew he was unhappy about something. In a hushed voice he said, "Glynn, we saw a bunch of heart irregularities on Irwin. We also saw some on the moon during the third EVA."

Those sitting next to the flight director cocked an ear, edging over to hear what was going on. Berry continued his discussion, now using words that were new and strange to us. He talked of a "bigeminal" rhythm, where both chambers of the heart try to contract at the same time. Berry said they had also seen PVCs, premature ventricular contractions, probably caused by the crew's working to near-exhaustion levels. Glynn and I were doing a slow burn; we should have been told about this much sooner.

To control the risks of spaceflight, the flight director must have all the facts from his team members, and he must get them in a timely fashion. In this case he did not. If Lunney had been aware of the medical problems, he would have given the crew a rest period, delayed the jettison, or simply had the crew go to sleep. We were going to remain in lunar orbit for two more days; we could have given the crew some slack — if we were given the information in a more timely manner. The surgeons' concerns about medical privacy, and their consequent reluctance to give the flight directors the full story, almost got us into a heap of trouble.

Slayton had been previously alerted to the medical problem. After instructing Irwin to downlink the biomed data, Deke got on the air-to-ground loop: "I want the commander and the lunar module pilot to each take a Seconal and get a good night's sleep." This was the typical Slayton imperative. Irwin said, "Thanks, Deke."

Two hours later, they finally signed off. The crew had been awake for twenty-two hours. Lunney's handoff to me was brief. "Good luck," he said, wearily.

My team spent most of the night reviewing the data on the suit and cabin pressure to make sure that the systems were fully operational. We remained in lunar orbit for almost two more days, mapping the surface, assessing the radiation environment, deploying a small satellite. Finally, after six days at the Moon, I gave the Go for trans-Earth injection.

For the first time in spaceflight, I had been truly rattled. Working with a chronically fatigued crew was bad enough, but when you added disorientation and memory loss the crew could have been experiencing because of dehydration and changes in blood chemistry (especially potassium deficiency) due to exertion, you were skating on very thin ice. I thought of the sign in my office:

> Aviation in itself is not inherently dangerous. But to an even greater degree than the sea, it is terribly unforgiving of any carelessness, incapacity, or neglect.

I now mentally added a word to the text: "ignorance."

Fall 1971

The work on the final two missions continued unabated, each one more difficult than the last, with landings at more rugged and desolate sites. It seemed to be a "can you top this" contest among the lunar geologists, with the astronauts cheering them on. Given the focus on science in the Apollo program it was certain that a scientist would soon fly. The controllers — Llewellyn and I in particular — were ecstatic when Jack Schmitt was selected for the final mission, Apollo 17. In my mind, no one deserved a flight more.

The cheering didn't stop there. Ron Evans, the command module pilot, was one of the

CapComs most familiar to the Mission Control team. Along with Ken Mattingly, Charlie Duke, and Fred Haise, Evans had been the CapCom for four Apollo missions.

The commander on 17 would be "Captain America" — Gene Cernan, the Navy's red-white-and-blue answer to Dave Scott. I thought this was fitting. I had launched Cernan on his first mission on Gemini 9 and now we would fly our last mission together.

Lunney, now in the Apollo program office, "borrowed" Aldrich for a trip to Russia to set up the joint U.S. and Russian working groups for a planned 1975 rendezvous mission. Aldrich, a pioneer in operations and developer of many great MCC systems engineers, was about to move on.

January 1972

The new year got off to a gloomy start. We were told there would be no hiring and no promotions for the entire year. The misery continued as Apollo 16 was delayed for a variety of technical problems related to the LM batteries, pyrotechnics, experiments, and space suits. Another reason for the delay was that Charlie Duke, the Apollo 16 lunar module pilot, caught a flu bug and was unable to train for the mission.

John Young, Duke, and Ken Mattingly were assigned to Apollo 16, aboard the command ship *Casper*, with a landing target at the edge of the Descartes Mountains. Technically and scientifically, this would rate among our most successful missions — and one of the least remembered.

April 16, 1972, Apollo 16

My third Saturn launch was routine, if launching the world's most powerful machine is ever routine. But, after achieving Earth orbit, we had one of those failures that the designer claims will never happen. Both regulators on the attitude control system were dumping gas overboard. With one eye on the Saturn IVB's attitude control fuel and the other on the clock we raced through the Earth orbital check-out and briefed the crew on assuming manual attitude control. It was a tight race to get the spacecraft injected to the Moon and to extract the lunar module. Once again we lucked out.

The mission continued normally through the lunar orbit and the preparation for landing. The White Team had the shift preceding Griffin's landing shift and had worked the usual "nits." Phil Shaffer, an ex-FIDO now training as flight director, had been working with my team throughout the mission, assessing the impact of various anomalies and making sure that there were no modifications to the landing plan.

After the mission debriefing with Kraft and Dr. Berry on Apollo 15, the flight directors were once again in the loop on crew status, and crew potassium intake was now the main concern of the surgeons. We hoped to prevent problems of the type experienced on Apollo 15, so the astronauts were provided orange juice spiked with potassium that added electrolytes to the fluid.

The concoction did not taste quite like nectar, and John Young was quick to inform us that it made his crew gassy and nauseated, not a good state for a confined cockpit in zero-G. The crew began a semicomic rebellion, with sharply reduced fluids and a reluctance to increase the orange juice intake.

The crew's orange juice protest was becoming the "ditty" for the mission press conferences. To better understand the crew's problems, and answer the questions at the press conference, I asked my ever-patient White Team flight surgeon to get me some of the juice. Within minutes a courier arrived from his back room with the infamous OJ. One dose was enough for me to get the crew's point. It tasted thick, heavy, almost metallic. I offered the remainder to Shaffer, who wisely demurred. I leaned over the console and yelled to my surgeon, Dr. Z (John Zieglschmid), "John, it tastes like crap. How about taking some to the press conference and let them take a shot?" Fortunately, there was no press conference scheduled for the shift and Dr. Z was not about to add to the controversy.

Shaffer and I concluded the handover to Griffin noting, "Looks like both spacecraft are clean going into activation and descent. Good luck."

The lunar module power-up, undocking, and visual inspection went off without a hiccup. At undocking, we swung into a dual team operation. Don Puddy, formerly a TELMU, responsible for life support, electrical, mechanical, and EVA systems on the LM, picked up flight director duties for the command module while Griffin followed the LM. The aerial ballet continued, with Puddy giving Mattingly the "Go for the circular maneuver" on the Moon's back side during the twelfth orbit.

On revolution thirteen, Jay Greene and Chuck Deiterich in the Trench were the first to see that something was wrong. Tracking data indicated that Mattingly had not performed the scheduled maneuver. Their suspicion was confirmed when Mattingly's voice broke through the static. "I scrubbed the burn," he said. "TVC number 2 was unstable."

Thrust vector control, TVC, was the steering system used to keep the spacecraft oriented during the course of a burn. If TVC was lost during a maneuver, the spacecraft would use precious attitude control fuel and in a long burn could start tumbling.

Mattingly continued to describe the problem symptoms, his troubleshooting, and the results. Ken was an expert on the CSM spacecraft systems, recognized by both designer and controller as the most knowledgeable of the CSM pilots. In Mission Control Ken worked well with the teams. As a CapCom he was a natural and intuitive pilot-engineer who asked the kind of questions that I wish I had asked.

From the tone of Mattingly's voice, you could tell he was feeling embattled. After being scrubbed from Apollo 13, he had finally landed his trip to the Moon. Now a malfunction deep in the guts of the CSM engine control system was on the brink of denying his team their lunar landing. He knew the rules and grudgingly accepted the need for redundancy.

Ken also knew that the control team would press to find a way out of the current problem. If there was no fix, the controllers would reassess the rules, but they would likely arrive at the same position as they had before the mission. (For loss of redundancy in the CSM propulsion systems, the CSM would re-dock with the LM. The lunar landing would be scrubbed and the lunar phase of the mission would be terminated. The TEI maneuver would be performed while docked to the LM, like on Apollo 13, to provide a backup engine to return to earth.) When the burn for the maneuver was scrubbed, Griffin temporarily waved off the LM descent preparation.

The LM was already in the correct orbit to begin the descent to the Moon, so Greene planned a maneuver to return the CSM to the vicinity of the LM. If they solved the engine control problem this would give the crew the option to immediately swing into another landing attempt without additional LM maneuvers.

The problem hanging up the mission fell into GNC Larry Canin's lap. It was like giving a piece of raw meat to a hungry tiger. The TVC system was Canin's flight control specialty. Without waiting for the recorded telemetry from Mattingly's testing to be processed, Canin informed Griffin: "Flight, tell Mattingly I want to run another TVC test. Give me a few seconds to get my team on line and get paper in my recorders and then let's get going."

Within minutes of Mattingly's initial report

the test was in progress. As it went forward, Larry's gut feeling told him that the problem was an open circuit, a broken wire somewhere in the control system. Working with his backroom staff, Canin set out to develop a test to further isolate the problem. Canin had worked with me during the Apollo 9 mission when we had run a series of in-flight tests to determine what would happen under similar malfunction conditions.

Two hours after the wave-off, Griffin had powered down the lunar module. Both spacecraft were now flying in close formation. After reviewing the Apollo 9 test data, Canin moved to Griffin's console for a private one-on-one. "Gerry, if the problem is in the control circuit," he said, "I think we can give the Go for the separation and landing. We can electrically drive the engine nozzle into position for the maneuver and then lock it in place with the drive clutch." Listening in, Kraft interrupted, "That's not what North American engineers say." Canin responded, "Let's get them on the voice loop and go over the test data from Apollo 9." Griffin, feeling better after Canin's input, called his team to order, and reset the landing for revolution sixteen. Shortly thereafter, lab testing re-created the problem by cutting the control circuit wires. The designers agreed with Larry Canin's proposed solution to use the drive clutch to hold the nozzle stable during the burn.

In less than four hours after the alarm, Griffin got everyone marching again. "Okay, Gold Team, settle down," he said, "the mission is back on. We're going for the landing at Descartes."

Ken Mattingly was ecstatic when he got the word along with the revised procedures. Coming from behind the Moon on revolution sixteen, after the maneuver, Mattingly reported, "*Casper* [CSM] did it this time!" Then he continued in his more customary, casual voice with his standard post-burn report. Happiness reigned in both spacecraft and Mission Control.

Pete Frank took a short shift after landing to get the crew into their sleep period and then handed over to my White Team. With the delay in the landing, my job was to lay out the surface plan, establish team schedules, and anchor the lunar liftoff time.

Because of the delayed landing, my Team's flight planners quickly reassessed the mission. We had at least two and a half days on the surface, cutting only slightly into our water reserve. With their input I advised the control team to plan for a three-EVA mission, and fixed the lunar liftoff for revolution fifty-two. The final step in the process was to post the schedules for the four control teams. My White Team would do a whifferdill, performing two shifts on the second EVA day to get my team into the cycle for lunar liftoff. Midway in the shift I woke up the LM crew, started the EVA preparation, and then handed over to Pete Frank's Orange Team for the EVA.

The Lunar Rover, used on the final three lunar missions, added a new dimension to the surface operations. It was a miracle of engineering, a battery-powered version of an off-road sport utility vehicle. The Rover deployed from Bay I of the descent stage by a series of cams, pulleys, and cables, unfolding like a collapsible baby carriage. When the Moon buggy landed, it carried everything the crew needed. Like the gold prospector's burro, the Rover carried the crew and its equipment to the exploration sites. With wire-mesh wheels, four-wheel steering, television and equipment stowage, it was Young and Duke's magic carpet.

Surface operations are hard work and the first EVA is the toughest. The high-priority science objectives are taken care of first in case something happens to cut the mission short. There are always glitches; Young and Duke had their share as they plowed through the initial surface operations. Then, to highlight their first EVA, they climbed aboard the Rover for the ride of their lives. Their initial targets were named "Spook," "Buster," and "Plum" (small craters used as landmarks by the crew). Young and Duke had learned a great deal about pacing themselves and avoiding exhaustion thanks to the Scott and Irwin debriefing. This imperative to conserve energy had been emphasized by the flight surgeons at every opportunity.

My ascent team, like the pilots in Korea, remained on "strip alert" close to the MCC. Throughout the entire lunar surface stay we were at ease, but cocked and ready to get a team in place to plan and execute a lunar liftoff in less than two hours. If a controller could not get from his home to the MCC in less than thirty minutes he was required to stay in the MCC sleeping quarters. If any problems occurred with either spacecraft it was our job to get into Mission Control, assess the options, and get the crew into lunar orbit.

The first EVA went well. Griffin debriefed the crew, and then I picked up the next shift. I was surprised that Griffin's handover notes indicated that instructions had come down from the program office's daily management meeting to shorten the final liftoff by one revolution, and end the mission a day early. My flight planners and the LM backroom team were already in the process of marking up the flight plan and checklists for the early liftoff when I arrived for my shift. I was damn angry and told them to stop doing the mark-ups. I intended to stay with the EVA and liftoff plan we had established on our prior shift.

There had been a rumor circulating after landing that the mission would be limited to two EVAs as a result of the landing delay but I had paid it little attention. Now someone in the management chain was making decisions normally reserved for the flight directors, and the decisions were contrary to the mission rules.

If the mission was cut short, I knew that Young and Duke on the surface and Mattingly in the command module would go for broke to complete every objective. The crew's drive to get as much done as possible would put us into the kind of exhaustion and resulting physiological problems that we had experienced with Scott and Irwin. I also did not want to compress the preparation for the lunar liftoff. Departing the Moon is one thing that you don't want to do in a hurry, one of the times when you proceed with deliberate caution.

For the second time in my MCC career, I lost my temper. I turned to Bill Tindall seated behind me at the Flight Operations director console, and asked, "Bill, do you know where this bullshit plan to lift off early came from?" He raised his hands. "I think it came down from today's program management meeting." My reply was curt. "Does Kraft know this?" Bill nodded affirmatively. Glancing into the viewing room I saw Chris, Sjoberg, and McDivitt. "Let's take a walk," I motioned to Tindall.

This time I was a lot smarter. I sketched out for Kraft the surface plan that led to my selection of the liftoff time. Then citing the crew workload and the Apollo 15 experience, I said that I thought it "unsafe" to press for an earlier liftoff. I used "unsafe" about every fifth word, since I had watched Kraft play the safety card at many meetings and now I decided to steal his trump. I finally put it pretty bluntly: "Chris, why don't you leave the mission planning to the team? What the hell is the hurry to get off the Moon?"

I studied his face, got the feedback I wanted, and then continued. Turning down the emotion: "A lunar ascent is the time when everything has to go well. I want two hours of data for the controllers and time for the crew to wind down from the EVA and transition to ascent thinking. We've had a good mission so far, why push it? This plan increases the risk with no payback."

I had argued often with Kraft before, but seldom had I ever caught him so far off base, agreeing to a change of plan without hearing the other side. After a brief exchange on keeping a reasonable crew workload, Chris nodded, "So be it."

Kraft's heart was still at the console. He was still "Flight" no matter what his position in management, and while the crew medical status was foremost in the minds of every top-level NASA leader, he gave his team the benefit of the doubt. The experience on Apollo 15, especially the lack of specific knowledge of the hour-by-hour crew status, colored our judgment. Neither crew nor controllers had the experience or the data to know how close we were to the edge.

After three days on the surface, the White Team gave the Go for the lunar liftoff. To the delight of the science team, Young and Duke had performed three EVAs at Descartes. We had waved off the first Apollo 16 landing attempt, worked out the problem, and ultimately achieved our objectives.

Two of the young controllers, Don Puddy and Phil Shaffer, members of the new generation, had earned the title "Flight" and would lead the teams into the next era of space.

Apollo had one more mission to go. The end was no longer beyond the horizon. For this one, hotel rooms at the Cape and in Houston would be reserved for months ahead. The airlines had a waiting list on flights into every airport in the area. The topless go-go bars on the beach would be so crowded that the cops would not be able to get into them to close them down.

22

THE LAST LIFTOFF

Fall 1972

Apollo 17 was going to be a tough, dramatic, and melancholy mission, the last lunar strike. It would mark the end of an all-too-quick decade in our American history, where we grabbed for the brass ring and got it on our first try. The lunar missions thoroughly absorbed us, and, in our haste, we never took the opportunity to savor the moment. Lunney often said that we were drinking wine before its time.

When Apollo ended, so would my life as a flight director. A new generation of flight directors was trained — the "top guns" of Mission Control, smarter, quicker, more responsive than we were. I was the last of the initial group of flight directors, an anachronism. I had the experience and the mission judgment, but I could see that I was compromised by events and near misses. There is a saying in aviation: "There are old pilots, and there are bold pilots, but there are no old, bold pilots." It is doubly tough when you have to make decisions about another person's life. As flight directors, we had made calls that only by the grace of God turned out right. But that was our life. The new flight directors were in a better position to manage this risk more objectively, balancing the odds and pressing to the objective. Watching Phil Shaffer, Skinny Lewis, Don Puddy, and Neil Hutchinson during their training, I felt old at age thirty-nine.

The preparation for Apollo 17 kicked in during the first week of September. We began our bittersweet planning for the final mission to the mountains of Taurus, near the giant crater Littrow. (These sites were named for the constellation Taurus, the bull, and nineteenth-century astronomer Johann von Littrow.) The landing site was a flat-floored valley four miles wide, bordered on three sides by high mountains. The valley contained numerous craters that might be volcanic in origin and of great interest to the scientists. It was going to be one hell of a mission, the grand finale to Apollo. I was proud that we would plant a sixth American flag on the Moon. But I could only wonder when and where we would plant the next flag.

Gene Cernan, the Apollo 17 commander, and Dave Scott, on Apollo 15, were probably the most outspoken in their feelings about our country, patriotism, and commitment. Maybe it was just the times we lived in that we needed reminders of what we stood for. The world around us seemed to be going haywire. With the gradual withdrawal of U.S. ground troops from Vietnam, it was left to the Navy and Air Force pilots to pound North Vietnam, while the South tried to stabilize its defenses. Any military man knew it was a lost cause, and I was ashamed of the way our nation hung South Vietnam out to dry. The Munich massacre of Israeli athletes during the Summer Olympics, protests at the national political conventions, and the attempted assassination of former Alabama governor George Wallace were the background for the final Apollo mission.

December 6, 1972, Apollo 17

There were the usual notes in the flight director's log when I saddled up on the console at 5:53 P.M. Central Standard Time to launch Apollo 17. This was an Apollo first, a Saturn V launch that would be seen rising on a pillar of fire in the night sky, before pitching over and starting its thrusting to the east, toward Africa, and carrying Cernan, Jack Schmitt, and Ron Evans on their way, our last ambassadors to the Moon.

Griffin had left several notes on the countdown funnies. He had reviewed a master alarm with Evans on the phone and closed out a battery problem as acceptable for launch. Trust is key in our business. If someone says a problem is closed, it is closed. In flight control, if you didn't have trust in your fellow controllers you could not get the job done.

In the last hour the launch countdown became a nightmare. Data faults began occurring around the world, then the Mission Control Center experienced a series of power glitches and the display system failed while I was conducting my launch status check. The maintenance team swung into action while the controllers dug for the procedures in case they had to move to adjacent working consoles. It was almost as if our ground system were reluctant to send the final Apollo on its last journey to the Moon. The problem was fixed by launch minus seven minutes and the count continued.

Sixteen seconds to launch, a Saturn auto sequence cutoff was issued. The liquid oxygen tank pressurization did not start automatically. Again the team swung into action to provide the launch updates to the Cape team.

During a launch hold, the trajectory ground track changes on a minute-by-minute basis throughout the launch window, due to the Earth's rotation and the changing geometry of the lunar trajectory. As soon as the launch hold is confirmed, the recovery forces and communications relay aircraft start moving southward, perpendicular to the ground track at maximum speed.

The countdown went through two recycles before the problems were corrected, finally lifting off almost at the end of the four-hour launch window. We were hard pressed to keep every aspect of the system in sync, but finally Mission Control was "Go for launch" of Apollo 17.

At 11:33 P.M. CST, the Saturn literally glowed as it left the launch pad. Night became day in a brilliant flash, a beacon for all America, all the world to see, as a symbol of the power of a free and open society.

The terse, calm controllers' reports wiped away my last vestige of sentiment and emotion on the world's final launch of a Saturn V rocket. Only during my final shift as a flight director, during the lunar phase, did emotion creep back in.

Launch and translunar injection were flawless. From then on it was a series of farewells for the "elite team" in Mission Control. The Saturn booster team, assigned by the Marshall Space Flight Center to my division, signed off for the final time after maneuvering the booster for the lunar impact. This crackerjack team, led by Scott Hamner, Chuck Casey, and Frank Van Renssalaer, had broken through the political and intercenter rivalries that stretched from Alabama to Texas to Florida, welding a partnership with my division and especially the Trench. "They was us" was Llewellyn's simple tribute to the booster team as it left the console for the final time.

Cernan's crew was probably the most relaxed of the Apollo crews. The day before the launch, the three of them had left the rules in shambles, violating their isolation to go duck hunting on a nearby farm. A pack of reporters was on their heels, but agreed not to write anything until after the flight, a reflection of the team spirit that sometimes infected even the press. Besides, this was the last roundup. Everyone understood that the usual protocol did not apply. The mellow attitudes of the Apollo 17 crew were also due in part to the fact they had no crises in the early going that threatened the mission. Another factor may have been the temperament of Cernan, a Chicago native, Navy captain, and space walker (Gemini 9). Gene had a sense of adventure reminiscent of that of the Mercury astronauts.

The teamwork between space and ground had peaked at a perfect time. The long interval between missions had given the crew of Apollo 17 full access to the controllers and training resources. Now the crew treated the control teams to the most vivid descriptions we had ever received of any flight. Even the ultra-quiet Ron Evans joined in broadcasting the account of the night launch. "During each staging the fireball overtook us, then when the engine kicked in we once again flew out of the orange-red cloud into darkness."

The translunar injection started in darkness, the booster propelling them through sunrise. The description was lyrical. There was no doubt the crew was enjoying the ride. During the coast phase, Jack Schmitt waxed philosophical on the origin of life in the universe and man's efforts to extend his realm. Listening to the crew's narrative, I again felt the magic of the Genesis readings of Apollo 8, and of Armstrong's call the day the *Eagle* landed.

If there had been a way to stretch the next few days into a lifetime I believe that Pete, Gerry, and I would have done so. We were in the final hours of our careers as flight directors, and for a few final moments we savored the wine. Pete Frank conducted the three extravehicular activities, the most productive of the lunar program, benefiting from the lessons learned in every previous hour of spaceflight. The crew, controller, and science teams breezed through the EVAs. The crew set records for the longest lunar mission, mass of lunar materials returned, longest lunar EVA time, and greatest lunar surface distance traveled.

In this final mission, the crews and controllers all had time to sense the history in our work. As the final hours approached, I found myself mentally reviewing the early years of space, trying to fathom why we succeeded when by all rights we should have failed. Chris Kraft had pioneered Mission Control and fought the battles in Mercury and Gemini, serving as the role model of the flight director. He proved the need for real-time leadership. In the seconds-critical world of Mission Control, a single individual must assume responsibility to take any actions needed for crew safety and mission success. Kraft's legacy had defined the leadership role.

As the mission went forward, I felt increasingly frustrated and melancholy. I would often sit in the corner of the viewing room, silently watching the teams at work and realizing that I had started my transition to an entirely new role. But I also thought about the legacy of my generation: trust, values, teamwork. I wanted to be a living connection between the new generation of mission controllers, reminding them of how and where it all started with my generation and where theirs might take us in the future.

Bob McCall, in my belief, the premier artist of space, had been sitting on the step to the right of the flight director console, sketching during the final Apollo EVAs. He had designed the Apollo 17 crew patch. When Bob took a break for a cup of coffee, I joined him in the cafeteria. Like Sjoberg, McCall's talent shone because of his sincerity and humility. As we talked, I don't think Bob was surprised when I asked him if he would design an emblem for the Mission Control team. I spoke emotionally, from my heart and gut, about the control teams and crews, and our life in Mission Control. "We fought and won the race in space and listened to the cries of the Apollo 1 crew. With great resolve and personal anger, we picked up the pieces, pounded them together, and went on the attack again. We were the ones in the trenches of space and with only the tools of leadership, trust, and teamwork, we contained the risks and made the conquest of space possible."

Over the next six months, McCall developed the emblem worn proudly by every subsequent generation of mission controller. He inscribed his final rendering of the emblem: "To Mission Control, with great respect and admiration, Bob McCall 1973."

During the final EVA Cernan and Schmitt unveiled the plaque on the LM landing gear that commemorated the conclusion of the first period of exploration of the Moon, voicing the hopes of the astronauts, controllers, designers, factory workers, secretaries, and clerks. Speaking for the Apollo generation, Cernan concluded, "This is our commemoration that will remain here until someone like us, until some of you who are out there, who are the promise of the future, will come back to read it again and to further the exploration and the meaning of Apollo."

There was not a dry eye in Mission Control.

As I accepted the helm of Mission Control from Griffin for my final time, I put on my traditional white vest. I felt somewhat as I had the last time I strapped myself into an F-86 Sabre, relishing the final moments, touching the canopy and instrument panel, hesitating briefly before putting the helmet on. I knew that one life was about to end and another one about to begin.

I finally shrugged and plugged into the console. The time for recalling old memories was over. It was time to get the crew off the lunar surface. After the meeting with McCall, I had the satisfaction that no matter what direction I would take in the future, I, too, had helped to define the legacy of Mission Control.

The White Team picked up console duties at 183:00 MET for lunar liftoff. My thoughts now were on the business at hand, getting Cernan and Schmitt off the Moon and docked to the CSM. There are two times in the mission where the options of the flight directors and crews converge to zero. They are the lunar liftoff, and the subsequent trans-Earth injection. Engine failure in either case is catastrophic. We have options for everything else. A lunar liftoff is unlike any rocket launch from Earth. There are no abort alternatives.

The LM checkout prior to liftoff is exquisite in its detail. The liftoff is a single shot and must work perfectly. Liftoff time is critical, since most of the power, oxygen, and water have been used during the surface period. Decisions and actions must be perfect and instantaneous.

The Trench for the lunar ascent was a curious mixture. My flight dynamics officer was Bill Stoval, a youngster from Casper, Wyoming. Blond, blue-eyed, and cockier than hell, he was perfect for the job. Bill was typical of the new generation scribbling their names on the Captain Refsmmat poster in the hallways. He was matched with Jim I'Anson, an older, lanky Texan, who flew the B-17 Flying Fortresses in the Pacific during World War II.

The ascent countdown was a series of escalating events. Stoval called up the large-screen displays as I laid out various contingency procedures, mentally reviewing the mission rules. The countdown hit ascent minus ten minutes, and the team tightened up as the crew blew the pyrotechnic valves to pressurize the propulsion system. If the tank used to pressurize the ascent propulsion system started to leak, my lunar module team would cry out, "Emergency liftoff!" For a few seconds the suspense held, then I heard the call: "Flight, ascent helium is Go. The system is pressurized, there are no leaks."

The countdown continued. I started my final status of the room at liftoff minus 8:30. We passed the "White Team is Go" to the crew and I opened my launch timeline, mentally running my personal mission rules, going through the final set of options for the final lunar mission. *There are no reasons to delay liftoff,* I thought. *I will switch over for critical program alarms and navigation errors. My ascent status roll calls are at plus one, plus three, and plus five minutes.* Communications checks were Go.

I conducted the final poll at minus two minutes. There were no open items, all mission rules were complied with. The trajectory and data sources were Go.

The White Team was now "negative reporting," that is, in a listening mode as the crew called out, "400 plus . . . master arm . . . abort stage, engine arm ascent."

By the clock at five seconds, I heard Schmitt call out, "Proceed," then from several sources in unison in Mission Control I heard, "Liftoff." Cernan sang out, "We're on our way, Houston . . ."

I noted in the log, "At 188:01:35, the last men left the Moon."

The ascent in the LM *Challenger* was a hell of a ride. Stoval's boyish glee rivaled I'Anson's drawl as they reported events, times, data quality. The loops were crisp, fresh, professional, few wasted words. For ten seconds, the *Challenger* rose vertically on the plume of the 3,500-pound-thrust engine. In the lesser gravity of the Moon, this was the equivalent of 21,000 pounds of thrust on Earth. The small ascent stage moved out smartly from the valley of Taurus-Littrow with Cernan and Schmitt and their precious payload of rocks, as well as our dreams.

The lunar module now pitched forward, gaining velocity as well as altitude in its dash to capture the rendezvous orbit. Ed Fendell had done well again, capturing the liftoff in a blaze of sparks, debris, and motion with the Rover TV camera.

The images of the lunar liftoff, the faces of my control team and their voices, are forever captured in my memory. The finality of our mission was expressed in a simple plaque left on the surface by Cernan: "Here man completed his first explorations of the Moon, December 1972 a.d. May the spirit of peace in which we came be reflected in the lives of all mankind."

The rendezvous ended in a good, tight Navy formation during the CSM visual inspection. And then at docking there was a brief and joyous exchange with Evans. We were not home free yet, but a lot of critical milestones on Apollo 17 were now behind us.

It was traditional since John Glenn's first U.S. manned orbital mission for a message from the President. I had been angry when Kennedy's planned message caught us right at the end of Glenn's first orbit. Kraft's reminder that day, "The President is the boss," still rang in my ears as I reviewed the message from President Nixon to be read to the crew. The presidential messages always seemed to come after a critical event, and when the odds radically improved that the crew would come home safely.

The crew was busier than ever as they prepared to open the hatch and enter the command module. Immediately after the docking, I passed the President's message to the CapCom. The message began with the customary "attaboys," followed by glowing words about the Apollo program's impact on humanity. The message was designed for a spot on the evening news.

The concluding words were the bitter wine: "This may be the last time in this century that men will walk on the Moon!" We had started out with John Kennedy's vision and command: get a man on the Moon by the end of the decade. Nixon's message was, effectively, Apollo's obituary.

Two hours later, I took off my white vest and stowed my headset. My career as a flight director was at an end. Flight directors do not retire to the blast of trumpets or to a roll of drums. There had been no formal change-of-command ceremony for Kraft, Hodge, Lunney, or Charlesworth. One day they just packed up their headsets and left the console. They were not there on the next mission.

Griffin and I decided to do it a bit differently, handing over to the new generation of flight directors in lunar orbit. We felt it was important to pass the torch so that a new generation, born in Apollo, would lead the teams into the future. Griffin's Gold Team became Neil Hutchinson's Silver Team for the return journey to Earth.

Chuck (Skinny) Lewis, one of five college students in the first class of flight controllers, a graduate of the Zanzibar site and my wingman for many missions, assumed command of my White Team, now dubbed the Bronze. Just as

Kraft had passed me the baton in real time, I now passed it to Lewis.

On occasion, flight directors take over the CapCom's job in selecting the crew wake-up music. For Lewis's first prime shift, bringing the crew home from lunar orbit, Griffin and I selected "Light My Fire," by the Doors (listed as the Lettermen in the flight director's log), to welcome a new generation of flight directors. It was time for Lewis to fly solo, so I moved to the viewing room as he gave his Go to bring home the command module, *America*.

For the splashdown, I continued a tradition established in Gemini 9. If my team had done especially well I would wear a celebration vest. Splashy, gaudy, it was my way of saying, "Thanks, well done." Marta knew how I felt about leaving the console. We both shared the pride in my work, in the Mission Control teams, and in America.

Marta made a surprise vest, my final vest as a flight director, for the Apollo 17 landing. It was a spectacular creation, and the favorite of all my vests, made of a metallic thread with broad red, white, and blue stripes, the colors of our flag and also the colors of the first three flight directors. For me the vest stood for America, President Kennedy, outer space, the many firsts, and the Brotherhood of Flight Control.

Proudly displaying my resplendent vest, I said thanks to my bosses and my teams, and, "Thanks, America, for the privilege of serving you." When the crew's feet hit the deck of the carrier, I lit the traditional cigar and cried like a baby. I cried for hours.

Flight directors' colors are retired just like numbers on football jerseys. At retirement, a proclamation is read declaring that the color will never be used again. The proclamation is hung on the wall of the control room in which the flight director last served.

The words of the proclamations are written by one's peers, the only people who matter in our business. Mine read,

> Whereas his leadership and inspiration molded the flight control team, which was vital to the first rendezvous, manned lunar exploration, and the study of man, Earth, stars, and technology.
>
> Be it resolved that on behalf of the personnel of the Flight Control Division, the color "White" be retired from the list of active flight control teams to forever stand in tribute to "White Flight," Eugene F. Kranz.

My proclamation now joined with those of the pioneer flight directors on the wall of the third-floor Mission Control room at the Johnson Space Center. Over the years other proclamations would be added, including one recognizing the honorary Gray flight director, Bill Tindall. We were all members of the Brotherhood who opened the door to space.

EPILOGUE

The success of the early American space program was a tribute to the leadership of a politically adept NASA Administrator and a relatively small number of engineers, scientists, and project managers who formed and led NASA in the early years. This team, with the technologies it created, reached for and attained a goal that many of its peers thought impossible. A clear goal, a powerful mandate, and a unified team allowed the United States to move from a distant second in space into a preeminent position during my tenure at Mission Control.

Entering the twenty-first century, we have an unimaginable array of technology and a generation of young Americans schooled in these technologies. With our powerful economy, we can do anything we set our mind to do. Yet we stand with our feet firmly planted on the ground when we could be exploring the universe.

Three decades ago, in a top story of the century, Americans placed six flags on the Moon. Today we no longer try for new and bold space achievements; instead we celebrate the anniversaries of the past.

In the 1960s just beyond the midpoint of the twentieth century, we were a restless nation when a young President, John F. Kennedy, awakened us to our responsibilities and the opportunities we had to make our nation and our world better. Overnight it seemed we became a nation committed to causes. Young and old marched for civil rights, or journeyed to foreign lands in the newly formed Peace Corps. Pictures of Earth from space gave new emphasis to the environmental movement, and again people marched. While we often moved to different cadences, our nation was alive with ideals. We were in motion. Violence was everywhere but so was a conviction that we must somehow make this a better world.

Thirty years later I feel a sense of frustration that the causes that advanced us so rapidly in the 1950s and 1960s seem to have vanished from the national consciousness. We have become a nation of spectators, unwilling to take risks or act on strong beliefs. Since I grew up in the world of manned space exploration, I am particularly frustrated that we have abandoned the frontier that was opened in the 1960s. The American space industries and the NASA team that built and operated the spacecraft no longer exist. The proud spacecraft design and manufacturing teams at Grumman, North American, and McDonnell are only a memory.

Since my retirement from the space program in 1994 I have spoken to over a hundred thousand Americans in hundreds of business, professional, and civic forums. The story of our early years in space, of tragedy and ultimate triumph, has awakened Americans to the power of a dream and of clear goals. I believe there is a widespread interest in space that can be focused to support a public mandate for space exploration. The four steps needed to return to a visionary space program are:

First, put space on the national agenda. Space is not currently deemed a priority. There is no established constituency to lobby on behalf of space exploration. No current candidate for President has taken a strong position on space.

NASA, its contractors, and the technical space societies represent a sizable asset that with proper leadership and a single voice could bring space back onto the national agenda. As a federal agency, NASA is prohibited from lobbying on its own behalf. It can, however, provide appropriate information and educational materials that its employees and contractors need in order to make the case for a new and long-term effort in space.

A large corps of NASA and contractor alumni exists in every state in the union. The space technical societies and the NASA Alumni League must assume the leadership, and they must be supported by the rank and file of NASA and its contractors. We must all move into the public arena, speak at business and civic forums, and go into the schools in order to reach every sector of American society to carry the message of space. *Unless every person who has ever worked on or dreamed of probing far into outer space is willing to make this commitment we cannot succeed.*

Second, revitalize NASA. Lacking a clear goal the team that placed an American on the Moon, NASA, has become just another federal bureaucracy beset by competing agendas and unable to establish discipline within its structure. Although NASA has an amazing array of technology and the most talented workforce in history, it lacks top-level vision. It began its retreat from the inherent risks of space exploration after the *Challenger* accident. During the last decade its retreat has turned into a rout. The NASA Administrator is appointed by the President and to a great degree represents the current President's views on space. If space is put on the national agenda for the coming national election, a newly elected President will have the opportunity to select new top-level NASA leadership that is committed and willing to take the steps to rebuild the space agency and get America's space program moving again.

Third, develop a long-range plan for space. The last set of clear goals for space was produced in 1986 by a national commission that included Neil Armstrong, Chuck Yeager, Gerard O'Neill (CEO, Geostar Corporation), Jeane Kirkpatrick (former U.S. ambassador to the U.N.), Thomas Paine (former NASA Administrator), and many others. The report, entitled *Pioneering the Space Frontier*, was developed through a series of fifteen public forums held across the country and represented the opinions of a substantial portion of the public on the future of the civilian space program. The goals defined in 1986 are as good today as they were a decade and a half ago. We do not need to engage in another round of studies. We must establish a plan to meet the goals of the National Commission on Space.

The report, which was written shortly after the *Challenger* accident, projected the next fifty years of the space age and deliberated on NASA's goals for the next twenty years. It articulated *A Pioneering Mission for the Twenty-first-Century America*:

> To lead the exploration and development of the space frontier, advancing science, technology, and enterprise, and building institutions and systems that make accessible vast new resources and support human settlements beyond Earth orbit, from the highlands of the Moon to the plains of Mars.

Fourth, engage Congress in the space program. NASA Administrator Daniel Goldin, responding to a Clinton White House foreign policy initiative, brought Russia into the space station program as a major program element without the support of the U.S. Congress. The subsequent redesign of the space station, abrogation of existing contracts, and program delays cost NASA valuable support within Congress. Reinvigorating the space program entails significant costs and cannot happen without strong congressional leadership and support. NASA needs a new Administrator, someone who knows how to represent the space program in the political arena, someone like its second administrator,

James Webb, who was a master of the bureaucratic process and a skilled builder of support alliances. A new Administrator with a clear set of goals, supported by an energized and vocal space alumni, can build a mandate for space.

A long-term national commitment to explore the universe is an essential investment in the future of our nation — and in our beautiful but environmentally challenged planet. An American-led program of multinational space exploration is a critical test of our intention to continue as a world leader in the twenty-first century. Only through such commitments will we inspire the youth of the coming century to step forward to preserve and protect the future of our nation and the rest of mankind. Only in this way will we develop new and difficult technologies, and make the scientific discoveries required to sustain our way of life and to make our world better.

This book began with the dream given to my generation, but I believe that President John F. Kennedy was addressing all generations to come when he said:

> The United States was not built by those who waited and rested and wished to look behind them. This country was conquered by those who moved forward, and so will space. . . .
>
> Well, space is there and we are going to climb it, and the Moon and the planets are there. And as we set sail we ask God's bless-

ing on the most hazardous and dangerous and greatest adventure on which man has ever embarked.

Our work is unfinished.

My wish as I close this book is that one day soon, a new generation of Americans will find the national leadership, the spirit, and the courage to go boldly forward and complete what we started.

Eugene F. Kranz
Dickinson, Texas, December 1999

WHERE THEY ARE

The original astronaut Class of '59 went their separate ways. Four survived to celebrate the thirtieth anniversary of the first lunar landing in July 1999. Astronauts present at the Cape for two days of commemoration and celebration on July 16 and 17 included Buzz Aldrin, Neil Armstrong, Gene Cernan, Wally Schirra, John Young, Charlie Duke, Al Worden, and Walt Cunningham, as well as our leader, George Mueller. Duke, Young, and Armstrong joined us for the celebrations in Houston.

Others were at a funeral that took place the day before the thirtieth anniversary. Pete Conrad, one of the astronauts from the second group selected, was killed in a motorcycle accident. He was deeply missed, another "missing man" in the formation, a formation that included gallant men like Alan Shepard and Deke Slayton (who died of brain cancer in 1993) as well as Ted Freeman, Charlie Bassett, C. C. Williams, and Elliott See, who died in aircraft accidents before they could get their chance to reach for space — and the three others who would live in our collective memories forever,

Gus Grissom, Ed White, and Roger Chaffee.

In 1988, after the *Challenger* accident, we made a change in the Mission Control emblem, inserting a comet, a symbol of risk and sacrifice that served as a reminder of the individuals who gave their lives for space exploration. We lost astronauts and we lost controllers, among them Cliff Charlesworth, Dick Thorson, Tec Roberts, Carl Huss, Ted White, Scott Hamner, and many others. We also lost Bill Tindall and many others who, though not flight controllers, made our work possible. Like comets, they all swept through the sky casting brightness in their wake — and then they were gone.

For those that remained the years were good to us.

• In 1974, I became deputy of NASA Mission Operations and in 1983, director. I continued my work in Mission Control with the controllers, flight designers, planners, and instructors and was assigned additional responsibilities for all aspects of Shuttle flight operations including design and development of MCC and simulation facilities and preparation of the Shuttle flight software.

• My final hours in Mission Control came during the December 1993 Shuttle mission to repair the Hubble Space Telescope. I retired from NASA in March 1994 and have never returned to the Mission Control room.

• In retirement I returned to aviation. I constructed an acrobatic biplane and flew as engineer on a B-17 Flying Fortress. I speak on the space program to at least sixty to seventy professional, civic, and youth groups each year.

• Chris Kraft retired from NASA in 1982. He served in many corporate and civic roles including as director-at-large of the Houston Chamber of Commerce, on the Board of Visitors of Virginia Polytechnic Institute, and on the board of the Manned Space Flight Education Foundation. He now serves as a consultant and corporate board member and is currently writing his memoirs of the space program.

• Glynn Lunney was program manager for the Apollo-Soyuz rendezvous mission, Space Shuttle program manager, and, after NASA retirement, president of Rockwell Space Operations–Houston.

• John Hodge left NASA in 1970 to study advanced transportation systems while working for the Department of Transportation. Twelve years later he reentered NASA as director of the Space Station Task Force. He retired in 1987 and founded a high-tech management consulting firm.

• Gerry Griffin became director of the Johnson Space Center, president of the Houston Chamber of Commerce, consulted on the *Apollo 13* movie, and as an actor has appeared in several space-related movies.

• Jerry Bostick became Space Shuttle program deputy manager, and after NASA retirement, vice president of Grumman Space Operations. He consulted on the *Apollo 13* movie and the TV miniseries *From the Earth to the Moon*.

• Arnie Aldrich was named the NASA headquarters director of the Space Shuttle program after the *Challenger* accident. He led the efforts to return the Shuttle to flight status. He then served as the associate administrator for Aeronautics, Exploration, and Technology. After retirement in 1994, he joined Lockheed's Missile and Space Company as vice president for commercial space programs.

• John Llewellyn opened and operates a cattle and sugar cane ranch and a riverboat touring company in Belize. For a while he owned two satellites recovered during a NASA shuttle mission and today works in the telecommunications industry.

ACKNOWLEDGMENTS

Writing This Book

During my thirty-four years with NASA I kept notes of meetings, mission logs, voice tapes, and post-mission reports. The materials filled seven file cabinets and numerous boxes and bookshelves. When my agent, George Greenfield, approached me to write a book in the summer of 1995 I was well into the construction of an acrobatic biplane and was reluctant to divert my attention from the effort. George persisted and during a trip to Houston convinced me to commit to writing a book. George kept me moving through some difficult times and introduced me to Jim Wade and Bob Bender, who made this book a reality.

Andy Chaikin, author of *A Man on the Moon*, and Al Reinart, one of the scriptwriters for the *Apollo 13* movie, got me started. They coached me on writing a book proposal, developing an outline, and using a storyboard.

The completed outline showed me where I needed more information to write a book on the highly complex events that occurred four decades ago. To fill the gaps I began corresponding with the controllers of the Mercury, Gemini, and Apollo programs.

- Ted White and Arnie Aldrich provided Mercury remote site team reports, technical manuals, photos, and copies of mission messages.
- Dutch Von Ehrenfried contributed manning lists and Gemini EVA mission rules.
- Chuck (Skinny) Lewis contributed Langley and MSC phone books that were invaluable in establishing the organization structure and personnel locations at critical times.
- Jay Greene provided the complete set of Tindallgrams.
- Gerry Griffin provided voice tapes that, when combined with mine, covered every Apollo mission.
- Doug Ward provided press conference transcripts and converted the reel-to-reel voice tapes to cassette.
- Glen Swanson, the NASA-JSC historian, provided biographical materials and researched events in the press transcripts.
- Many other controllers volunteered personal notes, mission rules, console logs, photos, or other memorabilia. Some just answered my questions to help me make complex problems understandable.
- The photos were acquired from many sources, including Mike Gentry in the NASA-JSC Media Resource Center. The controllers' personal photos were processed by

the One Great Photo Lab in Webster, Texas.

The technical content of the story was contained in the voice records, console logs, and mission reports; individual and group interviews were used principally to develop anecdotal data and to capture the gut feelings of the controllers.

The interviews were the most enjoyable part of writing this book. I interviewed controllers in groups to generate the emotional intensity that existed decades ago. I found that the controllers most vividly remembered the best moments, and that time had softened the edges on the bad moments. During the group sessions we sat around a cooler filled with beer, ate pizza, and reminisced. The interview sessions generally lasted about three hours and each involved seven to twelve controllers. I conducted ten group sessions with remote site teams, spacecraft systems engineers, simulation teams, the Trench, and mission designers and flight directors. I conducted individual interviews with Harold Miller (SimSup), John Hodge, Arnie Aldrich, and Ed Pavelka. Each interview was recorded on videotape.

Writing a book is a team effort and as in Mission Control, I needed a lot of help to get the job done.

Jan Pacek Weede, my NASA secretary for two decades, transcribed numerous voice tapes, and typed and updated the manuscript.

Jack Riley, my public affairs officer for many

missions, helped me shape the original story and reviewed every manuscript draft.

My original draft of the book covered over 200,000 words, so Mickey Herskowitz, a *Houston Chronicle* sportswriter, helped me to condense the story and better focus my role in the story.

Throughout every stage of the book there were dedicated readers and reviewers, among them Ed Fendell, Gerry Griffin, Jerry Bostick, Jim Hannigan, Pete Frank, Jack Riley, Chuck Lewis, Glen Swanson (NASA Johnson Space Center History Office), Rebecca Wright and Carol Butler (NASA Johnson Space Center Oral History Project), Ralph Royce from the Lone Star Flight Museum, and my wife, Marta.

Jim Wade, formerly vice president and executive editor of Crown Publishers, accomplished the final structuring and shaping of the book. He is presently a member of the Independent Editors Group. Jim tuned, polished, and structured the manuscript. In the final months, he joined my list of the great mentors.

Bob Bender, my editor at Simon & Schuster, believed in the book from the very start and his editorial work helped to make it what it is today.

There were, remarkably, few disagreements among interviewees concerning events, actions, and principals described in this book. On a few occasions while I was using the MCC voice tapes it was difficult for me to identify the controller or crewman involved in the action. When this happened, I used a combination of the MCC access lists, control room photos, and video tapes, if they were available, to determine the individuals involved. The portrayal of events on the final shift between John Hodge and Chris Kraft preceding the Apollo 1 fire (Chapter 10) and the specific crewmen involved in the final Apollo 11 mission simulation (Chapter 15) represents my best judgment about these events and individuals.

The NASA history series, particularly the books *This New Ocean* (Project Mercury), *On the Shoulders of Titans* (Project Gemini), *Chariots of Apollo* (Project Apollo), and *Stages of Saturn* (Saturn rocket development), were invaluable references in developing the chronology in the book. I recommend them to my readers.

There is no doubt that although four decades have softened many of the emotions, we are still a brotherhood.

Any errors in telling this story are solely mine.

Doing the Job

The one constant in the thirty-four years of my career in NASA was the consistent quality of our people, their dedication, and their willingness to do everything ever asked of them.

Three great leaders directed Flight Operations at the Manned Spacecraft Center in Houston. Chris Kraft, Sig Sjoberg, and Bill Tindall gave us our assignments, and when the chips were down trusted us to get it done. The four divisions under their direction were the Mission Planning and Analysis Division, led by John Mayer and Carl Huss; the Landing and

Recovery Division, led by Jerry Hammack and Pete Armitage; the Flight Support Division, led by Lynwood Dunseith; and my Flight Control Division.

Rod Rose filled many roles in Flight Operations. He was the principal engineer integrating the myriad elements of the telemetry, voice, and trajectory data flow and processing.

Each of these divisions and key individuals, along with many others in the Manned Spacecraft Center, played a major role in the success of our early space ventures. My story is about only one of these great manned spaceflight teams.

The 400 members of the Flight Control Division staffed many of the Mission Control Center's real-time decision positions. I was able to perform my duties as flight director because of a superb division staff who stepped in and ran the division when I was in training or working a mission. I would like to acknowledge the following individuals on the Flight Control Division staff: assistant for operations, Joe Roach; assistant for systems, Mel Brooks; chief of flight directors, Glynn Lunney; MSFC booster engineer office chief, Scott Hamner; technical assistant, Chuck Beers; business manager, Harold Miller; administration officer, Cecil Dorsey, and his assistant, Joyce Gaddy.

Lois Ransdell, "Pink Flight," was my boss secretary and was ably supported by the division office secretaries: Suzanne Miller, flight directors; Carole Helms, booster engineers; Betty

Defferari, business office.

The division had seven branches that corresponded to the technical specialties in Mission Control. The branch chiefs and their deputies worked as controllers while also leading the branch-level organizations. They selected and trained the new controllers in their basic skills, integrated the mission plans and documentation, and supported the spacecraft design. During simulations, they certified their controllers as ready for mission support. The branch chiefs and deputies were Charlie Harlan and Chuck Lewis — Flight Control Operations Branch; Jerry Bostick and Phil Shaffer — Flight Dynamics Branch; Carl Shelley and Gordon Ferguson — Simulation Branch; Richard Hoover and Lou DeLuca — Requirements Branch; Arnie Aldrich, Neil Hutchinson, and Rod Loe — CSM Systems Branch; Jim Hannigan, Don Puddy, and Bob Carlton — LM Systems Branch; Jim Saultz and Gerry Griffith — Experiment Systems Branch.

The branch secretaries also worked in Mission Control, and their work hours and life were as harried as that of the controllers. They were Sue Erwin, Ada Moon, Lucille Booth, Geraldine Taylor, Pat Garza, Maureen Bowen, Dorothy Hamilton, Kathy Spencer, and Elizabeth Pieberhofer.

The section chiefs led groups of five to seven controllers developing the spacecraft handbooks, procedures, and mission rules. They called the cadence, roused their controllers during the long and frustrating hours, and listened to their gripes. The section chiefs were Bill Platt, Ed Fendell, Perry Ealick, Bill Molnar, John Llewellyn, Ed Pavelka, Charlie Parker, Dick Koos, Jay Honeycutt, Lyle White, George Pettit, Charlie Dumis, John Aaron, Buck Willoughby, Gary Coen, John Wegener, Merlin Merritt, Bruce Walton, Harold Loden, Bill Peters, Ted White, Burt Sharpe, and Merrill Lowe.

Several of the section secretaries directly supported us in Mission Control. Among them were Jo Corey, Connie Turner, Sandra Lewis, and Donna Daughrity.

Great contractor teams supported the Flight Control Division. They kept the pipeline of information flowing so the controllers worked with correct and timely design and test data. "Learning by doing" worked because of the network of design engineers cultivated by these lead engineers: Bill "Blaster" Blair — North American; Charles Whitmore — Grumman; Stuart (Stu) Davis — Philco; Bill Harris and Ron Tunnicliff — McDonnell; Richard Freund — AC Electronics; Myron Hayes — IBM; Ron Bradford — Bendix; Jim Elrod — Lockheed; Fred Kuene and Lee Wible — Hamilton Standard.

To the rest of you in the ranks, you are in my heart.

APPENDIX
FOUNDATIONS OF MISSION CONTROL

To instill within ourselves these qualities essential for professional excellence:

Discipline Being able to follow as well as lead, knowing that we must master ourselves before we can master our task.

Competence There being no substitute for total preparation and complete dedication, for space will not tolerate the careless or indifferent.

Confidence Believing in ourselves as well as others, knowing that we must master fear and hesitation before we can succeed.

Responsibility Realizing that it cannot be shifted to others, for it belongs to each of us; we must answer for what we do, or fail to do.

Toughness Taking a stand when we must; to try again, and again, even if it means following a more difficult path.

Teamwork Respecting and utilizing the ability of others, realizing that we work toward a common goal, for success depends on the efforts of all.

To always be aware that suddenly and unexpectedly we may find ourselves in a role where our performance has ultimate consequences.

To recognize that the greatest error is not to have tried and failed, but that in trying, we did not give it our best effort.

GLOSSARY OF TERMS

Abort	A time-critical termination of an event
AFB	Air Force base
AFD	assistant flight director — MCC
ALSEP	Apollo lunar surface experiment package
AOS	acquisition of signal
ATDA	augmented target docking adapter (used as a Gemini rendezvous target)
CapCom	capsule communicator
Cape	Cape Canaveral
CDR	commander — senior astronaut on Apollo mission
CM	command module (reentry portion of the Apollo spacecraft that contains the crew)
CMP	command module pilot
CONTROL	Lunar Module engineer in the

	MCC responsible for propulsion, attitude control, and primary and abort guidance and navigation systems, including computer hardware
Cryo	cryogenic (oxygen and hydrogen fuels stored at very cold temperatures)
CSM	Command and Service Module — Apollo
EECOM	Gemini or CSM engineer in MCC responsible for electrical, environmental, communications, cryogenic, fuel cell, pyrotechnic, and structural systems
EVA	extravehicular activity
FCD	Flight Control Division — provides majority of flight controllers to the MCC
FCOB	Flight Control Operations Branch — provides assistant flight director and procedures, develops mission rules
FIDO	flight dynamics officer (the MCC specialist in launch and orbit trajectories)
FOD	Flight Operations Directorate organization; also, flight operations director in the MCC
FTE	flight test engineer
G	acceleration due to gravity forces
GMT	Greenwich Mean Time, also referred to as Zulu (Z) time

GNC	Gemini/CSM engineer in MCC responsible for propulsion, attitude control, guidance, and navigation systems, including computer hardware
Go NoGo	decision process to continue or abort a mission activity
GRR	guidance reference release
GSFC	Goddard Space Flight Center (Greenbelt, Maryland)
GT-	Gemini-Titan- (followed by mission number)
GUIDO	(pronounced GIDO) MCC specialist in navigation and computer software. During Gemini, this included the Titan II guidance system.
INCO	MCC engineer responsible for combined CSM, LM, EVA, and Rover instrumentation, communications, command, and television systems
KSC	John F. Kennedy Space Center, Florida
LM	lunar module, previously called lunar excursion module (LEM)
LMP	lunar module pilot
LOI	lunar orbit injection (maneuver to enter into lunar orbit)
LOS	loss of signal
LRC	Langley Research Center (Langley Field, Virginia)
MA-	Mercury-Atlas- (followed by mission number)

Mach	ratio of airspeed to the speed of sound at a given altitude
MCC	Mercury Control Center at Cape 1960–65 or Mission Control Center at Houston 1965–1972
MET	mission elapsed time (time since liftoff)
MPAD	Mission Planning and Analysis Division — responsible for analytic trajectory design
MR-	Mercury-Redstone- (followed by mission number)
MSC	Manned Spacecraft Center, Houston, Texas, used through 1973
MSFC	Marshall Space Flight Center (Huntsville, Alabama)
NACA	National Advisory Committee for Aeronautics
NASA	National Aeronautics and Space Administration
NoGo	the decision to cancel a planned event
OD	operations director position in Mercury Control, changed to FOD for Gemini and Apollo
PAO	public affairs officer — position in the MCC to release mission information
POGO	rapid up-and-down maneuver that if continued would destroy launch vehicle
psi	pounds per square inch

PTC	passive thermal control
RCS	Reaction Control System — small propulsion jets for attitude control and small maneuvers
Refsmmat	reference to stable member matrix — technique for conversion between coordinate systems
RETRO	retrofire officer — the MCC specialist in reentry trajectories
RSO	range safety officer — responsible for protecting landmass across the world from errant rockets
SCE	signal conditioning electronics
SimSup	simulation supervisor — the leader of the training team in the MCC
SM	service module — portion of Apollo CSM that contained main engines and power systems
SPAN	spacecraft analysis team — small group in MCC to access design and manufacturing
SPC	stored program command, a command stored in a computer or program that will activate a function at a specific clock time
Stay NoStay	Time-critical decision to remain on the Moon or lift off at the next opportunity
SYSTEMS	Mercury control center engineer responsible for electrical,

	attitude control, display, and structural systems
TEC	trans-Earth coast
TEI	trans-Earth injection (maneuver to return the spacecraft to Earth from the Moon)
TELMU	Lunar Module engineer responsible for electrical, environmental, communications, pyrotechnic, structural, and EVA systems
TLC	translunar coast
TLI	translunar injection (the maneuver to take the spacecraft to the Moon)
TM	telemetry
Trench	the MCC trajectory team consisting of the RETRO, FIDO, and GUIDO
TVC	thrust vector control (rocket steering mechanism)
Z	Zulu — shortened term for Greenwich Mean Time used in logs and messages

Manned Mercury Remote Sites

BDA	Bermuda
ATS/RKV	Atlantic Tracking Ship/*Rose Knot* Victor — designation changed in middle of the program
CYI	Canary Islands
KNO	Kano, Nigeria

ZZB	Zanzibar
IOS/CSQ	Indian Ocean Ship/*Coastal Sentry* Quebec — designation changed in middle of program
MUC	Muchea, Australia
WOM	Woomera, Australia — tracking station in a remote desert location on a military test range
CTN	Canton Island — FAA ground beacon for aircraft navigation in South Pacific
HAW	located at edge of Waimea Canyon on Hawaiian island of Kauai
CAL	California, site located at Point Arguello
GYM	Guaymas, Mexico
TEX	Corpus Christi, Texas

Manned Gemini Remote Sites

CYI	Canary Islands
CRO	Carnarvon, Australia
HAW	located at edge of Waimea Canyon on Hawaiian island of Kauai
GYM	Guaymas, Mexico
CSQ	*Coastal Sentry* Quebec ship
RKV	*Rose Knot* Victor ship
TEX	Used as a training site only

The employees of G.K. Hall hope you have enjoyed this Large Print book. All our Large Print titles are designed for easy reading, and all our books are made to last. Other G.K. Hall books are available at your library, through selected bookstores, or directly from us.

For information about titles, please call:

(800) 223-1244
(800) 223-6121
To share your comments, please write:

Publisher
G.K. Hall & Co.
P.O. Box 159
Thorndike, ME 04986